Understanding Lightwave Transmission
Applications of Fiber Optics

Understanding Lightwave Transmission
Applications of Fiber Optics

William O. Grant

Harcourt Brace Jovanovich, Publishers
Technology Publications

San Diego   New York   Chicago   Austin   Washington, D.C.
London   Sydney   Tokyo   Toronto

ISBN#: 0-15-592874-0

Library of Congress Catalog Card Number: 87-81095

Printed in the United States of America

Production supervision and interior design: WordCrafters Editorial Services, Inc.

en·gi·neer·ing [en je-nir ing]   n.

the application of scientific
principles to practical ends as
the design, construction, and
operation of efficient and
economical structures, equipment,
and systems.

The American Heritage Dictionary
of the English Language

Contents

Preface

Understanding Lightwave Transmission is an introduction to lightwave transmission systems and the equipment and optical fibers utilized in such systems, along with explanations of the basic characteristics of lightwaves and optical fibers themselves.

Intended as introductory reading for those with some technical background in other transmission disciplines but little previous exposure to lightwave transmission, the book establishes a basis from which to pursue further studies.

The text covers the basic elements of lightwave transmission, with particular emphasis on applications in telecommunication systems. Rather than focus on the current state of the art as an end in itself, the intent is to establish a base of knowledge on a level that equips readers to evaluate new technical developments in the field and apply them effectively in a variety of applications.

Understanding Lightwave Transmission
Applications of Fiber Optics

CHAPTER 1

Introduction to Light

In this day and age of acronyms, technical "buzz words," and general public enchantment with every new technology, lightwave transmission fits in beautifully. It has an impressive mystique all its own with connotations of scientific significance and an intriguing vocabulary of terms like "laser," "Megabits," "modal dispersion" and the like. A person speaks, a light blinks, and in an instant, perhaps hundreds of miles away, another person hears the speaker's voice. Surely by any standard of the general public's measure, lightwave transmission is "high tech."

But people have been communicating through "lightwave transmission" since time immemorial. Before the blinking semaphore of ships, before Wadsworth's immortal "One if by land, two if by sea," human beings utilized fire for signaling, if only to guide them home to the caves from the hunt. Even the ubiquitous traffic light, which regulates so much of our everyday movement, depends upon the transmission of light.

But is lightwave transmission definitive?

We can measure a gallon of water with great precision. We can measure the flow of water, the pressure, the temperature, etc. We can measure electricity and its flow, and the pressure producing that flow, and we feel comfortable and confident with the terms and results of such activities.

Can we measure and define light nearly as well? Are we as confident with the results?

THEORIES OF LIGHT

Mankind has long been interested in light—exactly what it is, how it is produced, and how it travels through space. Some of these questions remain completely unanswered, because science does not yet have a complete theory of light. Nevertheless, a great deal has been learned about light in the last three centuries.

Philosophers and scientists have held six major theories concerning the nature of light over the last 3,000 years. These are known as the tactile theory, the emission theory, the corpuscular theory, the wave theory, the electromagnetic theory, and the quantum theory.

The tactile theory is so called because it explained vision by comparing it to the simpler sense of touch. According to this theory, the eye sent out invisible probes to feel objects. The emission theory, on the other hand, suggested that bright objects sent out rays or particles that could bounce off other objects and enter the eye to excite vision. By the eleventh century, the tactile theory had been disproved and the emission theory became generally accepted.

In the seventeenth century, two important but contradictory theories were put forth regarding the nature of light. They were the corpuscular theory of Sir Isaac Newton and the wave theory of the Dutch physicist, Christian Huygens.

Newton observed that light appeared to travel in straight lines and did not seem to bend around corners. He used this observation to arrive at his theory, which stated that light could be described as a system of corpuscles, or particles, sprayed out in all directions from the source and traveling in straight lines. The reflection of light was explained as the particles bouncing off the reflecting surfaces in accordance with Newton's laws of motion.

Huygens disagreed, and pointed out that if light consisted of a stream of particles, two crossed beams of light would cancel each other, just as two crossed jets of water interrupt each other's flow. With regard to light, such cancellation did not seem to occur.

The Wave Theory

Huygens compared light to a wave disturbance, like sound, and explained the laws of reflection and refraction completely on the basis of a wave theory. The main objection to this theory was precisely Newton's original objection: if light were a wave, it would bend around corners. It is now known that light actually does bend around corners, but the wavelength of light is so short that ordinary observations fail to detect the bending.

By 1800 the wave theory was held to be generally the more acceptable, and experiments by the French physicist, Augustin Fresnel, resolved the major objection to light's bending in 1827. The wave theory became so firmly established that the particle theory was largely ignored until about 1900.

At the time of Fresnel, scientists were discovering the nature of electricity and magnetism. It was known that electric currents could produce magnetic fields and also that magnetic fields could induce electric currents. Thus, it was clear that electricity and magnetism were closely related and perhaps might be combined into one unified theory.

Scottish physicist James Clerk Maxwell addressed this task in 1864. In one of the greatest achievements in the history of science, Maxwell combined not only the studies of electricity and magnetism, but also that of light, into one all-embracing theory.

Electromagnetic Theory

According to Maxwell's electromagnetic theory, light was simply one part of the entire system of electromagnetic waves, and the same mathematical techniques could be applied to problems in both optics and electromagnetism. For example, using purely electrical and magnetic constants, Maxwell predicted that the speed of light would be equal to 193,088 miles per second, a figure remarkably close to the experimental value then believed to be correct.

The electromagnetic wave theory, however, could not explain certain effects that were discovered near the end of the nineteenth century. German physicist Heinrich Hertz discovered the photoelectric effect in 1887. He noticed that when light fell on a metal surface, electricity was produced. The application of Maxwell's equations could not explain the number and the energy level of the electrons produced by the photoelectric effect.

The Quantum Theory

In 1900, German physicist Max Planck suggested that light was emitted and absorbed as a succession of bundles of energy. He called the bundles "quanta" and stated that the amount of energy contained in each quantum was proportional to the frequency of the light. Albert Einstein extended Planck's idea and suggested that light not only was emitted and absorbed as individual units, but that it actually traveled in the form of these units, later termed photons. In 1905, applying this concept, Einstein explained and verified the photoelectric effect.

More recently, contributions by many scientists have made it possible to explain almost all behavior of light by means of the quantum theory, which avoids a purely mechanical description of light. Light cannot be thought of as either a system of waves or a system of particles. Sometimes it acts as waves and sometimes as particles, but neither explanation can be considered complete in itself. When light interacts with matter, it exhibits wavelike or particle-like properties, depending upon the nature of the interaction itself.

Light is kinetic energy (energy associated with motion) and thus may be measured in units of energy, such as joules (J) or electron volts (eV). The unit of light energy is the photon, which is a massless bundle of electromagnetic energy. The word massless is used to distinguish photons from objects that have measurable rest mass, such as electrons or atoms.

A photon cannot be captured and put on a scale and weighed. When a photon is "caught," such as when light is absorbed by your skin, the photon ceases to exist as such. Its energy is converted to some other form of energy, for example, thermal energy (heat). When sunlight falls on your skin, it feels warm. It actually is warmer, because the light energy has been converted to heat, and you feel the rise in temperature.

Photons exist as such only when they are moving. They move at the rate of 3.0×10^8 meters per second, the speed of light, in a vacuum, and usually somewhat slower when passing through other materials.

ALTERNATING WAVES

Like radio waves and sound waves, light is electromagnetic energy which travels in a wavefront. Thus, light can be identified within the general electromagnetic spectrum of alternating waves. One complete sequence of values of any electrical alternating wave starting at zero, increasing to a maximum value, decreasing to zero, decreasing to a minimum value, and increasing to zero again, is called a cycle. (See Figure 1–1.) Since the instantaneous value of the wave is changing, some period of time is required for a complete sequence of changes (a full cycle) to occur. The time period is universally referenced in seconds.

The number of cycles that occur in one second can be called *cycles per second,* which was the international standard reference for frequency for a very long time. The term "cycle" has been replaced with the term *hertz,* named for the German physicist, Heinrich Hertz. The hertz is identical to and interchangeable with the older cycle, so that cycles per second is now stated as hertz per second.

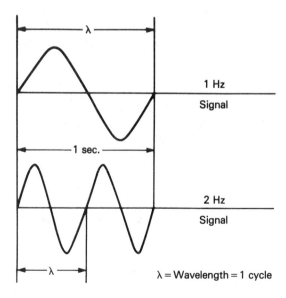

FIGURE 1–1
Wavelength

Figure 1–1 shows both a 1-Hz and a 2-Hz signal. Note that the time period is the same for both signals but two full cycles have occurred in the 2-Hz signal. As a point of interest, in the United States most commercial power is alternating current (AC) and is at a 60 Hz rate; that is, 60 full cycles are occurring every second.

ELECTROMAGNETIC SPECTRUM

The electromagnetic spectrum is shown in Figure 1–2, with all notations in units of frequency (cycles or hertz). The relationship of lightwave frequencies to other, perhaps more familiar, transmission disciplines is clearly shown. This notation becomes quite cumbersome when we approach higher frequencies. As we address the lightwave region of the spectrum, the frequencies involved are very high indeed and must be designated in terahertz (10^{12} hertz), which is a very large unit.

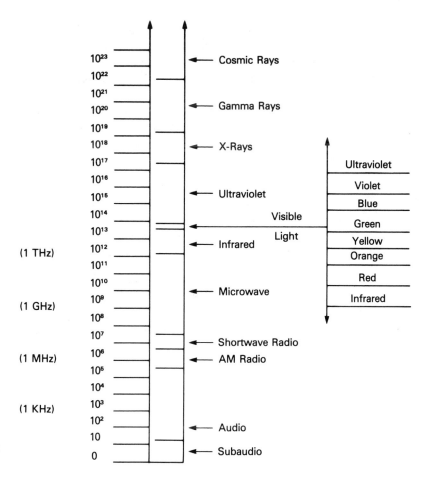

FIGURE 1–2
Electromagnetic
Spectrum

Many optical mechanisms are more directly associated with or related to wavelength rather than frequency. Thus it is much more convenient and practical to use wavelength when working with light, and indeed, even at some higher microwave ranges also. At higher radio frequencies, devices such as antennas, transmission lines, and tuned cavities involve physical dimensions directly related to wavelength, and this is even more pronounced in lightwave technology.

FREQUENCY-WAVELENGTH RELATIONSHIP

Electromagnetic waves, including light, travel in a vacuum at a speed of 186,280 miles per second. It is often more convenient to state this velocity in metric terms, and the figure 3×10^8 meters per second is often used for the speed of light in rough calculations, although 299,793,000 meters per second is more precise.

The distance that the wave would travel while experiencing one full cycle of change is called the *wavelength,* and is generally denoted by the Greek letter lambda (λ). Wavelength may be stated in units of measure of length, assuming that the velocity of propagation is known.

We can easily calculate the wavelength of a 60 Hz wave traveling in space by dividing the velocity of propagation (186,280 miles per second) by the frequency in hertz (60 Hz). The wavelength (λ) of a 60 Hz signal would thus be 3,104.66 miles.

At very low frequencies, such as 60 Hz, this calculation may be somewhat academic, but at higher frequencies (shorter wavelengths) it certainly is not. At broadcasting radio frequencies on the order of 1,000 kilohertz (1 million hertz, or 1 Megahertz), for example, amplitude modulated (AM) commercial stations often use vertical steel towers as radiating quarter-wavelength antennas. In such cases it is necessary to determine the wavelength in feet or meters to establish the required height of the tower itself.

For the example of the radio tower, for a station broadcasting on 1,000 kHz the calculation would be as follows: convert the velocity of propagation (186,280 miles per second) to feet per second (983,560,000 feet per second), and divide this figure by the frequency in hertz to establish the wavelength in feet. Thus, we have

$$\frac{983,560,000}{1,000,000} = 983.56 \text{ feet (one wavelength at 1,000 kHz)}$$

We then divide this figure by four to determine the length of a quarter-wavelength.

$$\frac{983.56}{4} = 245.89 \text{ feet (one quarter-wavelength at 1,000 kHz)}$$

Hence, the tower must be 245.89 feet in height if it is to function effectively as a quarter-wave antenna.

Some other conversion formulas are the following.

To find wavelength in centimeters given frequency in gigahertz (GHz):

$$\text{Wavelength} = \frac{30}{f} \text{ (GHz)}$$

To find wavelength in meters given frequency in megahertz (MHz):

$$\text{Wavelength} = \frac{300}{f} \text{ (MHz)}$$

In both of these examples, the term for the velocity of propagation of the wave is related to 3×10^8 meters per second, and if greater precision was required, the figure 299,793,000 could be employed.

The velocity of propagation (the speed of transmission) is a factor in determining the wavelength of a wave, but it will be different for a wave of a given frequency in different media. For example, a lightwave of a specific frequency will travel at a different speed in glass than it will in air, and consequently it will have a different wavelength in glass than it will in air, even though the frequency of the lightwave remains unchanged. Within the same medium, waves of different frequencies will travel at different speeds also, and thus they will have different wavelengths.

All electromagnetic waves, including lightwaves, travel at a slightly different speed in air than in a vacuum, but for the purposes of our discussion we can assume that the velocity of propagation is the same in both cases.

Use of the term "wavelength" as a common reference may be unfamiliar to some readers, but it is by no means new. Radio amateurs and others have always referred to sections of the spectrum in terms of wavelength, as for example, the 80 meter band (3.5 to 4.0 MHz) or the 10 meter band (28.0 to 29.7 MHz). Even the general public refers to radio frequencies above the commercial broadcast AM band as shortwave radio. Being higher in frequency, these are in fact shorter in wavelength. At higher radio frequencies—in the microwave range, for example—we find common reference to wavelength: the band 30,000 to 300,000 MHz is often called the millimeter microwave band.

At lightwave frequencies which are designated in terahertz (one terahertz is a trillion hertz per second), it is much more convenient and practical to use wavelengths exclusively, but we must specify the transmission medium involved, since it will affect the velocity of propagation and thus the wavelength.

In working with frequency we use convenient prefixes, such as kilohertz (10^3) for a thousand hertz per second, megahertz (10^6) for a million, and gigahertz (10^9) for a billion. In denoting lightwave wavelength we use prefixes also, but they diminish in scale as the wavelength gets shorter (or the frequency gets higher).

WAVELENGTH UNITS

In length measurements, with the meter (39.37 inches) as a point of reference, we have centimeter for hundredths (10^{-2}) of a meter, and millimeter for thousandths

(10^{-3}). At lightwave frequencies, wavelengths are very short indeed, and we need units of length that are much smaller than these. Some readers may be unfamiliar with these terms.

We use the micron (μ) to denote a millionth (10^{-6}) of a meter, but in more general use for the same dimension is the micrometer (μm). The two terms are interchangeable. The nanometer (nm) is a billionth (10^{-9}) of a meter. Beyond these, but infrequently employed, are the picometer (10^{-12} meter) and the femtometer (10^{-15} meter). The reference section in the back of the book includes charts and tables on units of measure for length.

It is interesting to note, and it gives some perspective to these terms, that there are 25.4 million nanometers, or 25.4 thousand micrometers, in one inch.

Figure 1-3 shows the electromagnetic spectrum divided into regions given in terms of wavelength or frequency, as the case may be. Within this spectrum, bands of frequencies where waves have common characteristics are often referred to in

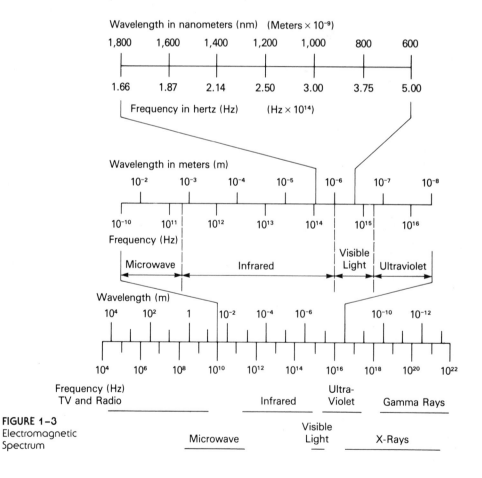

FIGURE 1-3
Electromagnetic
Spectrum

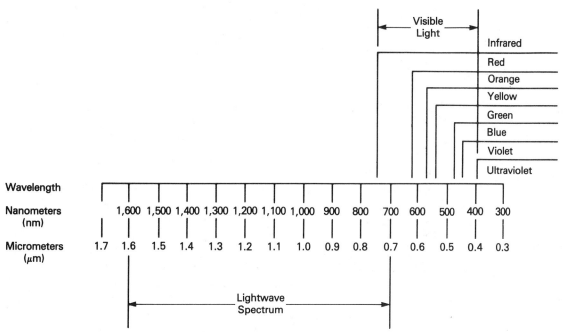

FIGURE 1-4 Lightwave Spectrum

generic terms like the audio spectrum, the radio spectrum, etc. The radio spectrum, for example, is generally accepted to include the frequencies between 8 kHz and 300 GHz. We shall use a similar term, the "lightwave spectrum," shown in Figure 1–4, which also identifies the visible spectrum.

Note that not all light used in lightwave transmission systems is in the visible range and that individual colors, which are determined by individual frequencies of lightwaves, are identified.

DEFINITION OF LIGHT

The definition of light in everyday use refers to the region of the electromagnetic spectrum that can be perceived by human vision. This is designated to be the visible spectrum and nominally covers the wavelength range between 0.4 micrometer (400 nanometers) and 0.7 micrometer (700 nanometers).

In the laser and optical communications field, custom and practice have broadened the usage of the term to include frequencies in those portions of the electromagnetic spectrum that can be processed by basic optical techniques used in working with the visible spectrum, even though such wavelengths may not be discernible to the human eye. This region is not precisely defined, but it is held to extend from the near ultraviolet region (approximately 300 nanometers) through the visible region and into the mid-infrared region (approximately 3,000 nanometers).

PROPERTIES OF LIGHT

Visible light is electromagnetic radiation that can be detected by the human eye, but visible light is only a very small part of the electromagnetic spectrum. Both infrared and ultraviolet radiation are often referred to as light, although they are outside the range that the human eye can detect.

Like all electromagnetic radiation, light is a form of energy and can be converted into other forms of energy. When light strikes certain materials, those materials give off electrons. This transfer of light energy to electrical energy is called the *photoelectric effect*. Light can also produce chemical changes in certain substances.

We can address light purely from the point of view of its illuminating characteristics, or we may address it as a form of energy in applications that utilize light as a medium for transferring energy, such as telecommunications systems.

So called "white light" or sunlight is composed of light waves at all the frequencies (or, alternatively, wavelengths) within the visible light spectrum. If we isolate and examine a single discrete frequency within this spectrum, it will appear to be a specific color. This is readily demonstrated by using a prism, which serves to separate the various wavelengths into component colors called the visible spectrum. Each color in this spectrum corresponds to a particular wavelength of light. A rainbow is simply a visible spectrum that nature produces by separating the different light wavelengths under unique atmospheric conditions.

The wavelengths of light, which encompass one full cycle, are extremely short, and their distance may be measured in a unit called an *angstrom*, named after the Swedish physicist, Anders J. Angstrom. A single angstrom is equal to one ten-billionth of a meter, or, in other words, ten angstroms equal one nanometer. The terms more frequently used in lightwave transmission to denote wavelength, however, are the micrometer and nanometer.

The distance that a wave moves forward during each second is the product of the wavelength of the wave and the number of waves that pass a given point per second, which is the frequency of the wave. Since all light travels at the same speed in air or a vacuum, and each color has a different wavelength, it follows that each color has a discrete frequency.

Light can be emitted from bodies that have been heated until they glow. In the incandescent light bulb, the heating is accomplished by passing an electrical current through the filament of the bulb. The heat transfers more energy to the atoms of the filament than they usually have, and the excess energy is radiated as light.

There are other methods by which atoms can be given extra energy and caused to radiate light. In a gaseous discharge lamp such as a neon lamp, the atoms in the neon gas pick up excess energy when they are struck by electrons passing through the tube of the lamp. In a laser (Light Amplification by Stimulated Emission of

Radiation), atoms are given extra energy that they store until they are caused to emit it.

Reflection

One of the most familiar properties of light is its ability to be reflected. When a ray of light is reflected by a smooth, flat surface, the incident ray, the reflected ray, and the normal (a line perpendicular to the surface) are all in the same plane.

As shown in Figure 1–5, the angle of incidence is formed by the incident ray and the normal, and the angle of reflection is formed by the reflected ray and the normal. Thus, the angle of incidence equals the angle of reflection.

Reflection also occurs from surfaces which are not smooth or polished. The direction of travel of a reflected light beam depends on the roughness of the reflecting surface relative to the wavelength of the reflected light. If the wave is incident on a rough surface, the reflected wave travels in many directions. This is shown in Figure 1–6.

Reflection in many directions is called *diffuse reflection* and occurs if the irregularities in the reflecting surface are about the same size, or are larger than, the wavelength of the lightwave striking the surface. An example of diffuse reflection

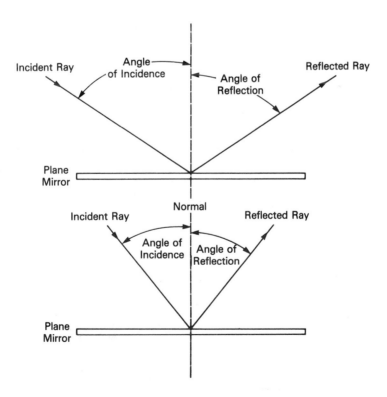

FIGURE 1–5
Reflection

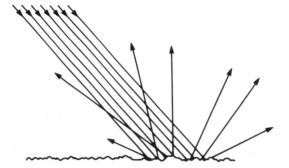

FIGURE 1-6
Diffuse
Reflection

is that occurring on the page you are reading right now. White light that includes all colors, or wavelengths, of visible light is illuminating the page of this book relatively evenly. The surface of the page is rough, and the distribution of the roughness is random. Very tiny portions of this irregular surface are reflective, but the roughness distributes the reflected light in random planes.

The paper has no discrete lightwave-absorbing qualities, so it reflects all wavelengths equally well. Thus, the reflected light appears to be white. The reflected light is widely distributed angularly in a random manner, so white light appears to emanate equally from all parts of the page.

Refraction

A second major property of light is its ability to be refracted. Refraction results from light traveling at different speeds in different media, such as air, water, glass, and other transparent substances. When light passes from air into water, it is abruptly slowed down from a velocity of about 186,000 miles per second to about 140,000 miles per second.

If a ray strikes the surface of a denser substance at any angle other than 90 degrees, it will travel through that substance in a different direction, closer to the normal. This phenomenon is shown in Figure 1-7. The new direction is determined by the index of refraction of the substance. The index of refraction is the ratio of the velocity of light in a vacuum to the velocity of light in that substance.

The index of refraction of air is 1.0003. Thus, the speed of light in air is only slightly (about 50 miles per second) slower than that of light in a vacuum. The refractive indexes of some other common substances are water, 1.33; glass, 1.51; and iodine, 3.34. Clearly light travels quite a bit slower in glass than in a vacuum or in air.

Refraction accounts for the fact that when a person observes an object in water the image of the object is displaced. The light rays are refracted (or bent) at the interface between the two media, which is the surface of the water. Refraction also accounts for why stars appear to twinkle. A star is a point source of light and does not turn "on" and "off," yet it appears to do so. As the light from the star

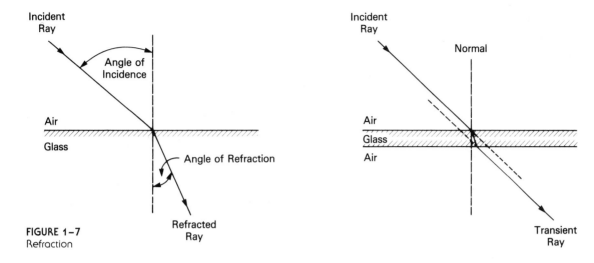

FIGURE 1–7
Refraction

enters earth's atmosphere, it is refracted, since air and a vacuum are different media. Thus, the source of light appears to come from a point other than its actual location. As the atmosphere changes slightly due to wind currents and the like, the refractive index of the atmosphere also changes. The observed point of light appears to move slightly, and so seems to sparkle or twinkle.

The light from the moon and planets is refracted in an identical manner, and it might be said that the moon twinkles also. The moon is observed as a large object, however, and the slight movement is visually imperceptible.

Under certain conditions, light passing from one medium to another will not be refracted at all. Light that is incident normally (along the normal) on the boundary between two media will pass straight through without being refracted.

There is also a critical angle of refraction. Assume that a pencil of light located at point A in Figure 1–8 is incident normally on the boundary between a medium of lower refractive index N_1 and a medium of higher refractive index N_2. Then light passes straight through to point AA.

Next the light is shown incident on the boundary at a larger angle, from point B. The light will then be refracted toward the normal to point BB.

As the angle of incidence approaches 90 degrees—say, the light emanates from point C—the light is bent toward point CC, but when the angle of incidence reaches 90 degrees, no light will be refracted. The area defined roughly as area X in the figure is the area that cannot be reached by refracted incident light.

The situation is different if the light is passing from the denser into the rarer medium. Figure 1–9 shows the refractive indexes of the two materials reversed; that is, the light is now entering the medium with the lower refractive index. Now the light, positioned as shown, is bent away from the normal.

Note that at point D in the figure, the refracted ray emerges along the boundary between the two media. This is the *critical angle*, or minimum angle of total

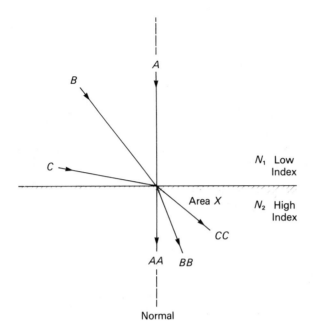

FIGURE 1–8
Refraction

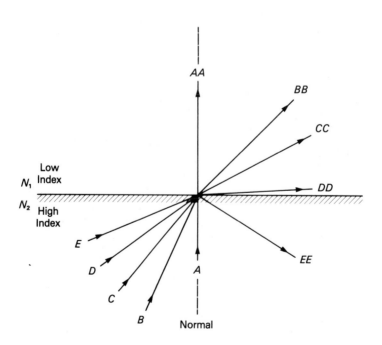

FIGURE 1–9
Refraction

internal reflection. At angles larger than the critical angle the light is totally reflected back into the denser medium, and no rays are refracted into medium N_1 at all.

Dispersion

The process of separating light according to its frequency is called *dispersion*. A prism, for example, disperses white light by refraction, causing the different wavelengths (or colors) to travel in different directions, thus separating them and forming a spectrum, as shown in Figure 1–10.

Any source of light is capable of producing a spectrum, but different sources may produce different spectra. By analyzing the light spectra of individual stars, scientists have been able to determine the elements present in them and the relative abundance of those elements.

Diffraction

Diffraction is the term that describes the bending of the path of waves when they pass close to the edge of an obstacle or through a narrow slit. When a beam of light passes through a narrow slit, it spreads out because of the bending of the light waves around the edges of the slit, as shown in Figure 1–11.

Diffraction can be employed to separate composite light into individual wavelengths by etching or scoring a number of fine parallel lines, closely spaced, on an optical surface. The wavelength that will be diffracted, and thus separated angularly, will be determined by the spacing of the lines and the shape of the etched grooves. Such a device is called a *diffraction grating* and is shown in Figure 1–11.

Light Absorption and Scattering

Absorption is a phenomenon whereby light strikes a substance and gives up some or all of its energy to the atoms of that substance. If the atoms retain the energy, it is converted to heat.

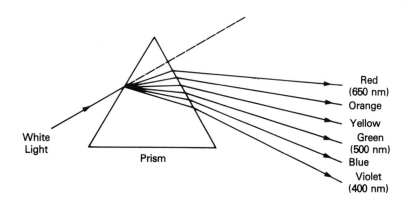

FIGURE 1–10
Dispersion

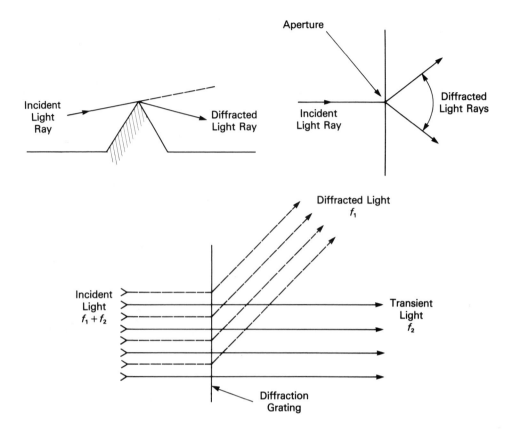

FIGURE 1-11
Diffraction

In some substances, the light energy given to the atoms of the substance raises their energy level to the point where they, in turn, emit light. If this light is emitted as light of a longer wavelength, the process is called *luminescence*. If the light energy is emitted as light at the same frequency as the incident or impinging light was, then the light is said to be *scattered*. Scattering should not be confused with the purely optical phenomenon called diffusion which will be discussed later.

In addressing the propagation of light energy, either in air or in some other transparent material, we recognize that there is some loss, or attenuation, of light energy. This attenuation is largely due to absorption and/or scattering as the light waves propagate through the medium.

Luminescence

The property that enables some materials that have absorbed energy to emit a fraction of that energy as light is called *luminescence*. The process is accomplished in two definite stages, absorption and emission. When energy is absorbed by a material, its atoms are said to be excited, and when the atoms return to their normal state, light is emitted. The form of energy that causes luminescence is denoted by the prefix in such words as electroluminescence (luminescence by electrical action) or chemiluminescence (luminescence by chemical action).

The emission of light by a luminescent process may be by fluorescence or by phosphorescence. *Fluorescence* is a form of luminescence in which the light is emitted instantaneously and in which the emission ceases as soon as the source of excitation has been removed. *Phosphorescence*, or "afterglow," is a form of luminescence in which there is a time delay between the absorption of energy and the emission of light. In phosphorescent materials, the emission persists after the source of excitation has been removed. This phenomenon may be observed in phosphorescent paints that glow long after the activating light has been removed.

ENERGY LEVEL OF LIGHT

The energy level of a single photon of light depends upon its frequency. For example, a photon of red visible light would have an energy level of only 2 or 3 eV. An ultraviolet photon might have an energy level of 300 eV, and an X-ray photon perhaps as much as 40,000 eV.

The energy level of a single photon is quite small, but a great number of photons may be emitted from a light source in a short period of time. An ordinary flashlight, for example, might emit a quintillion photons or more in a second. A quintillion is 10^{18}, while a billion is only 10^9, so that a quintillion is 10^9 billion.

Although the energy level of an individual photon is constant and dependent upon frequency, if the intensity of the light is increased while the frequency remains constant, the number of photons emitted will increase. Consequently, the thermal energy imparted to the atoms and molecules of a body that the light strikes will increase and can become quite substantial in very bright light. On a very sunny day, the sand on a beach can become too hot to walk barefoot on due to this thermal energy.

Groups of atoms such as the sun give off radiation at all frequencies. It is possible to plot the energy distribution in a *Planck curve* (from Max Planck, a German scientist), as shown in Figure 1–12. Note that for the sun, the greatest energy is in the area of the visible spectrum.

The typical response of the human eye to visible spectrum lightwaves is shown in Figure 1–13.

SUMMARY

Light cannot be thought of as either a system of waves alone or a system of particles alone. Sometimes it acts as waves and sometimes as particles, but neither explanation can be considered complete in itself. When light interacts with matter, it exhibits wavelike or particle-like properties, depending upon the nature of the reaction itself.

Light is kinetic energy (energy associated with motion), and the unit of light energy is the photon. Photons exist as such only when they are moving, and when

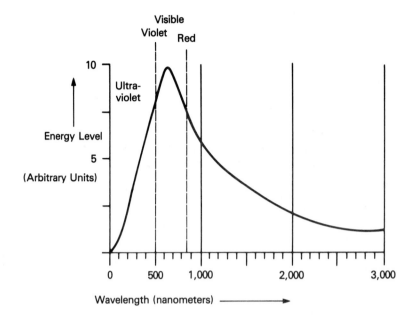

FIGURE 1-12
Planck Curve

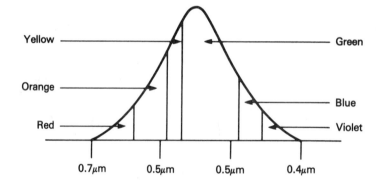

FIGURE 1-13
Typical
Response of
Human Eye

light is absorbed by a material or substance, the photon ceases to exist. The energy of the photon is converted to some other form of energy—heat, for example.

Since light is electromagnetic energy that travels in a wavefront like sound or radio waves, light can be identified within the general electromagnetic spectrum. At lightwave frequencies, it is more convenient to use wavelength instead of frequency. At such frequencies, the wavelengths are very short and are usually denoted in nanometers (nm), a unit of length denoting one billionth (10^{-9}) of a meter.

The spectrum generally referred to as light in laser and communications fields extends from the near ultraviolet region (approximately 300 nanometers) through the visible light spectrum into the mid-infrared region (approximately 3,000 nanometers).

We can address light purely from the point of view of its illuminating characteristics, or as a form of energy, in applications that utilize light as a medium for transferring energy, such as telecommunications systems.

Some of the properties of light are its ability to be reflected, diffused, refracted, dispersed, diffracted, absorbed, and scattered.

With a basic understanding of light as a point of departure, we can usefully discuss some common phenomena of light, the various terms and units of measure which may apply, and those properties of light that may be particularly significant in generating, propagating, and detecting light energy.

REVIEW QUESTIONS _____

True or False?

T F 1. The tactile theory of light explained vision by comparing it to the sense of touch.

T F 2. The corpuscular theory and the wave theory both failed to explain why light did not seem to bend around corners.

T F 3. The Scottish physicist, Maxwell, combined the studies of electricity and magnetism, but did not address the study of light.

T F 4. Light can be completely and satisfactorily explained as a system of particles.

T F 5. A photon has measurable mass.

T F 6. A photon has a measurable level of energy when it is at rest.

T F 7. Light energy travels in a wavefront.

T F 8. The wavelength of an alternating wave is the distance it travels in one half a cycle.

T F 9. We can state the wavelength of a signal in units of measure of length without regard for the medium that the wave is propagating in.

T F 10. The distance that a wave would travel while experiencing a half cycle of change is generally denoted by the Greek letter lambda (λ).

T F 11. A lightwave of a given frequency will travel at the same speed in any medium.

T F 12. Within the same medium, different lightwave frequencies will travel at the same speed.

T F 13. A lightwave signal with a wavelength of 800 nanometers would be a higher frequency signal than a different signal with a wavelength of 1300 nanometers.

T F 14. A nanometer is one billionth of a meter.

T F 15. All lightwave frequencies used in practical transmission systems today are within the visible spectrum.

T   F   16. Light is a form of energy and, as such, can be converted to other forms of energy.

T   F   17. When light impinges on certain materials, those materials give off electrons.

T   F   18. If we isolate and examine a single discrete frequency within the visible spectrum, it will appear to be a specific color.

T   F   19. When impinging light is reflected by a smooth surface, the angle of incidence is twice the value of the angle of reflection.

T   F   20. The attenuation of light energy as it propagates through a medium is primarily due to fluorescence.

CHAPTER 2

Light as a Transmission Carrier

Light is most familiar to us as an illuminating agent, and from this perspective we are little concerned with the energy content of the light rays other than how it affects the intensity of the illumination. We are interested in how the energy is absorbed by materials or substances it may impinge on, since that will determine the color sensation we receive or, in lightwave systems, the amount of power that will be available at the end of long interconnecting optical fibers.

When we attempt to utilize lightwaves as a transmission carrier, we are obliged to address light in a more comprehensive manner. We have to consider the light energy as a level of power in the same sense that we do with electrical energy in copper conductor or microwave transmission systems, and we must calculate the transmission losses of our system to predict the power level of the light that will be presented to the system receiver or detector. Toward these ends, it behooves us to consider in greater detail the illumination-related features of light.

LIGHT AS ILLUMINATION

Lightwave energy can be detected at points distant from the source after propagation. The human eye is a natural detector of lightwave energy, but other devices can be fabricated to detect light also. Applying the photoelectric effect, light meters can be manufactured that measure the electrical energy created by light particles falling on a photoelectric material.

The brightness of a light source indicates only how it appears to the human eye. Brightness is a subjective term that correlates with intensity, which can be measured. The intensity of light being emitted by a source is measured by the flux, which is the rate of passage of light energy. The passage of light energy is given by the number of photons present in any period of time.

The basic unit for measuring luminous flux is the *candela,* or candle. This unit seems archaic perhaps, and was originally defined by the flame of a standard candle, which was somewhat inadequate, as might be expected. It is now more accurately defined in terms of the light coming from a black body at a temperature of 1,773.5 degrees Centigrade (3,224.3 degrees Fahrenheit).

The level of illumination at a surface is determined by the luminous flux on the surface, which is measured in lumens per square foot or per square meter. One lumen is the amount of light falling on one square unit of surface area of which each point is at a distance of one unit from a concentrated light source with an intensity of one candle.

Other units commonly employed are the footcandle and the lux. A footcandle is defined as the unit of illuminance on a surface that is at every point one foot distant from a uniform light source of one candle. One footcandle equals one lumen per square foot, and one lux equals one lumen per square meter.

The Law of Inverse Squares applies in illumination, just as it does in other disciplines: the illumination of a surface by a point source of light varies directly with the intensity of the light source and inversely with the square of the distance from the source.

As shown in Figure 2–1, if a sheet of paper is first held two feet away from a light source and the illumination is measured, and then the paper is moved to a spot one foot away from the source, the measured illumination would now be four times

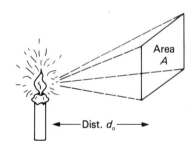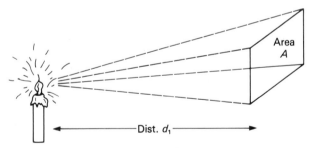

If x_0 = illumination of A at d_0 then $\left(\dfrac{d_0}{d_1}\right)^2 x_0$ is illumination level at d_1

Example: $x_0 = 1$ unit of illumination if $d_0 = 1$ foot

If $d_1 = 2$ feet, then $\left(\dfrac{d_0}{d_1}\right)^2 x_0 = \left(\dfrac{1}{2}\right)^2 \times 1 = \dfrac{1}{4}$ unit

If $d_1 = 3$ feet, then $\left(\dfrac{d_0}{d_1}\right)^2 = x_0 = \left(\dfrac{1}{3}\right)^2 \times 1 = \dfrac{1}{9}$ unit

FIGURE 2–1 Law of Inverse Squares

greater than the original measurement. The figure shows several examples of the application of the Law of Inverse Squares.

Since we are dealing with electromagnetic waves, we can fabricate and apply a variety of devices to focus, concentrate, or narrowly direct the light from the source, rather than simply radiate it in all directions. The effect will be a narrow beam of light that will be higher in intensity and thus may propagate farther or may be more easily detected at a distant point.

Although the terms "footcandle" and "illumination" suggest light energy at wavelengths within the visible spectrum only, i.e., light energy that is detectable by the human eye, the other nonvisible wavelengths discussed earlier can be processed by the same basic optical techniques that are used in working within the visible spectrum. That is, such light can be focused, filtered, etc., just as visible light can.

In a sense, the fact that some wavelengths are visible simplifies the study of light since we can actually perceive many of the phenomena we are discussing. In the study of electricity or radio wavelengths, this facility is not available to us.

In the previous chapter we defined the lightwave spectrum so as to include all wavelengths between 700 and 1,600 nm. Within this range, wavelengths below 900 nm are often referred to as short wavelengths, and those above 1,200 nm as long wavelengths, although this demarcation is by no means precise. The lightwave spectrum was shown in Figure 1-4, which also defined the visible spectrum within its range.

Color is a function of the wavelength of light, and Figure 1-4 showed the relationship between various color wavelengths. The total absence of light energy within the visible spectrum is black, which is also the total absence of color. The presence of all wavelengths within the visible spectrum includes all colors, and this condition is visually perceived as white light.

Note that in Figure 1-4 wavelength was denoted in both nanometers (nm) and micrometers (μm), and the various colors were demarcated so to emphasize their relationship within the spectrum.

COLOR

In Chapter 1 we saw that individual wavelengths of light could be separated or isolated from other wavelengths by using prisms or diffractive structures. Accordingly, it is possible to filter out individual wavelengths at will, just as is done with audio or radio waves. Within the visible spectrum, filtering out specific wavelengths would mean filtering out or attenuating specific colors. But what is color?

To answer this seemingly simple, but in fact quite complicated question, color is not a physical characteristic like weight or mass, nor is it a substance or material. Strictly speaking, color exists only as an observer's sensory perception: it is only visual information. In the physical world there are only matter and energy, and both of these are colorless.

Thus, red is not a property of the fabric of a dress. Rather, the fabric has an individual absorption capacity that enables it to absorb certain wavelengths or spectral elements of the general illumination of white light. Then the portion of the spectrum that is not absorbed by the fabric is reemitted as residual light. But the light rays from this reemission are not a color in themselves; they merely transmit information regarding how the reemission spectrum varies from the spectral make-up of the initial illumination.

Where does the white go when the snow melts? The color white was never a property of the snow, of course; it was only a perception of white which was due to the fact that the snow provided a multiplicity of random-sized surfaces that reflected all visible light wavelengths equally well, thus appearing to be white if illuminated with white light containing all visible wavelengths. If the snow were illuminated with red light, it would appear to be red. When the snow melted, the reflecting surfaces dissolved and there was no longer any uniform reflection effect.

Color materializes only if a color stimulus causes the intact visual system of an observer to produce a color sensation. If the same color stimulus strikes the eyes of a person with defective color vision, an entirely different color sensation or perception will result. If there is no observer at all, then color cannot materialize.

In many optical effects, filtering is achieved by "subtraction"—that is, all undesired wavelengths or colors are absorbed or attenuated. Consider white light, which contains all visible wavelengths, impinging on a material or substance that only allows the wavelength of the color red to pass through, or nearly so. The visual perception of the light after passing through the material would be of red light only. The energy which was present in the form of photons in the other wavelengths that were not passed through will be absorbed as heat in the filtering material itself.

Most of the colors that we see in common objects that reflect light to our light detectors—our eyes—result from the subtraction of selected wavelengths of light. White light falls upon grass. The grass absorbs red and blue wavelengths, or most of them, and reflects only green wavelengths, which our eyes detect. We perceive the grass to be green.

White light falls upon a rose. The rose absorbs green and blue wavelengths, or nearly so, and reflects only red wavelengths, which our eyes detect. We perceive the rose to be red.

The material which produces the color by absorbing particular wavelengths is called a *pigment* and is made of molecules, just as everything else is. The molecules that make up a pigment are of such a nature that they can trap or absorb some wavelengths but not others.

We can paint a wall a particular color by applying material with the particular pigment required to produce that color. If white light falls upon the wall, all wavelengths except that defined by the pigment will be absorbed, or nearly so. The color or wavelength that is not absorbed by the pigment is reradiated from the surface, and our eyes detect this radiation and perceive the wall to be whatever color corresponds to this wavelength.

Since light is energy, its particles, or photons, are present in all the wavelengths that impinge on a surface or material. If only some wavelengths are reflected, the remaining energy must be absorbed by the material or substance, and this energy is converted to heat in the material or substance.

This heat energy is usefully employed by plants in photosynthesis. Of course, it need not be usefully employed at all: when we paint a wall green for esthetic reasons, we do not do so primarily to absorb heat, and the heat actually produced may serve no practical purpose.

COLOR-RELATED TERMS

A number of terms are employed to talk about color which may aid in understanding light, since, as we have seen, color relates directly to different wavelengths of light. Many of these terms are based on the word "chroma," a noun form of the Greek locution "chromo," which is defined as "a quality of color combining hue and saturation."

The word "chromaticity," for example, is defined as "the quality of color characterized by its dominant or complementary wavelength and purity taken together." "Monochromatic" is "having or consisting of one color or hue," or "consisting of radiation of a single wavelength or a very small range of wavelengths." "Dichromatic" is "having or exhibiting two colors." "Dichromatism" is "the state or condition of being dichromatic, that is, partial color blindness in which only two colors are perceptible." "Monochromatism" is "complete color blindness in which all colors appear as shades of gray."

LIGHT-RELATED TERMS

A number of terms are employed to talk directly about light as illumination. "Radiant" is "something that radiates as: a point or object from which light emanates." "Incandescence" is "emission by a hot body of radiation that makes it visible." "Luminescence" is "an emission of light that is not ascribable directly to incandescence, and therefore occurs at low temperatures, and is produced by physiological processes such as chemical action, friction, or electrical action." "Fluorescence" is "emission of, or the property of emitting, electromagnetic radiation, usually as visible light resulting from, and occurring only during, the absorption of radiation from some other source."

"Transparent" is "having the property of transmitting light without appreciable scattering, so that bodies lying beyond are entirely visible." "Translucent" is "permitting the passage of light, but transmitting and diffusing the light so that objects beyond cannot be seen clearly." "Diffusion" is "reflection of light by a rough reflecting surface or transmission of light through a translucent material." "Opacity" is "the quality or state of a body that makes it impervious to the rays of light."

PROPAGATION OF LIGHT

How far can you "see" light? What mechanisms introduce attenuation to light energy in a vacuum, in air, or in any other transmission medium?

Astronomers describe the size of our galaxy by using a measurement of distance known as a light year. A light year is the distance it would take a beam of light a whole year to travel. Since light travels at a speed of 186,300 miles per second in a vacuum (space), in a whole year at this speed light would travel about 6 trillion miles (6×10^{12}).

The human eye can detect light from stars in this galaxy, so obviously, light can travel a very long distance, at least in a vacuum. For a point of reference, the sun is 93 million miles away from earth, and it takes about eight minutes for light from the sun to reach earth. The star nearest to earth is 25 million million miles from earth, and light from this star takes four and a quarter light years to reach us.

Since there is no atmosphere in space, there is no attenuation or decrease in light energy, and light travels these immense distances with relative ease unless an opaque object intervenes. When light enters the earth's atmosphere, or travels in other media such as water or glass, the situation changes drastically. The earth's atmosphere, composed of gases, extends out from the surface about 100 miles or so.

Imagine a theoretical transparent medium that has been carefully freed of all dust and other suspended particles. If a strong beam of light is passed through that medium, observations in a dark room will show that there is still a small amount of bluish color light scattered from the beam. This may be observed when a large searchlight is pointed up toward the sky on a clear night. This light, about 450 nanometers in wavelength, which is of course the wavelength of the color blue, is being scattered by the molecules which make up the medium. No scattering is done by dust or any other particles, since we supposed that they were absent in our hypothesis.

This particle-free transparent medium is exactly what the earth's atmosphere is—at least most of it. There are suspended particles in the lower sections of the atmosphere, but not in the upper regions we are talking about here.

The classical example of such molecular light scattering is the color of the sky. At one time in the past, the blue color of the sky was believed to be due to the scattering of light by dust particles that were suspended in the air. It was Lord Rayleigh who showed that the nitrogen and oxygen molecules in the air are actually responsible for it.

Rayleigh's theory applies to particles that are no larger than about a tenth of the wavelength of light. Rayleigh measured the relative amount of light scattered at different wavelengths in the sky and found that the scattering intensity is inversely proportional to the fourth power of the wavelength. That is,

$$\text{Loss} = \frac{1}{\text{wavelength}^4}$$

Now consider two wavelengths of 0.8 μm and 1.3 μm, respectively. Then

$$\text{Loss}_{0.8} = \frac{1}{0.8^4}$$

$$= \frac{1}{0.41}$$

$$= 2.44$$

and

$$\text{Loss}_{1.3} = \frac{1}{1.3^4}$$

$$= \frac{1}{2.86}$$

$$= 0.35$$

This accounts for the fact that whatever the transmission loss of an optical fiber may be, transmission losses in lightwave systems will always be lower at longer wavelengths. Figure 2-2 shows the relationship between scattering intensity and wavelength.

Since the shorter wavelengths (blue is a short wavelength) are scattered more effectively than the longer wavelengths, a beam of sunlight (white light) passing through the earth's atmosphere will experience more depletion of blue light than it will of yellow or red. Light energy at all light wavelengths will penetrate to the surface, but there will be less blue light than yellow or red at that point.

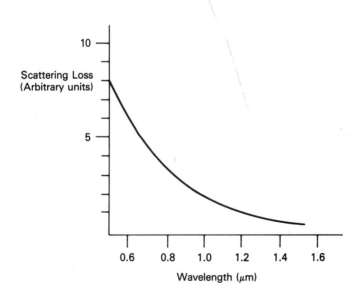

FIGURE 2-2
Rayleigh
Scattering

LIGHT ABSORPTION

Scattering should not be confused with reflection or diffusion. Chapter 1 considered the condition whereby light strikes a substance and then gives up its energy to the molecules of that substance. If the atoms of those molecules retain the energy, it is converted to heat, and the process is called absorption. If, on the other hand, the light energy given to the atoms of the substance raises their energy level to the point where they, in turn, emit light at the same frequency as the incident or excitatory light, the process is called scattering. The emitted or scattered light is polarized in the process, as well as being radiated in all directions. Thus, the sky appears to be of a diffused blue-colored light.

This same mechanism of molecular scattering is experienced in other transmission media, such as glass or fluids, and the nature of the substance and its molecular structure determines the nature of the scattering incurred.

Light scattering of a different kind is experienced in the lower regions of the atmosphere, where suspended particles such as dust or moisture (fog, rain, etc.) are present. At sunrise or sunset, the sun is low in the sky and its light travels through more lower atmosphere air than during the daylight hours. The suspended particles tend to scatter all colors in the light except the red wavelength, which travels directly to the detector, the eye, and the sun appears to be red.

With increasing particle size, the scattering intensity becomes less dependent on wavelength until, for sufficiently large particles, there is no dependency on wavelength at all. Clouds, for example, might scatter other colors than blue if all particles were of a relatively uniform large enough size. In fact, conditions have been observed where the sky appeared red and the sun appeared blue, but since clouds generally have a wide particle size distribution, they scatter all wavelengths equally well. Thus, clouds usually reflect white light and appear to be white or gray, and only rarely colored.

With very large particles, the result is more properly described as diffraction rather than true scattering.

The primary causes of attenuation of light in air or other media are absorption and scattering, and both mechanisms will be encountered when we examine light-wave transmission through optical fibers. The term "extinction" is sometimes encountered and may be erroneously used synonymously with the term "absorption." Extinction of light is the reduction of light energy to a level where it can no longer be detected. Extinction is a broader notion than absorption; indeed, it includes attenuation by both absorption and scattering.

MEASURING LIGHT

The process of measuring electromagnetic energy, regardless of whether or not we can see such energy, is called *radiometry. Photometry* is restricted to measuring that portion of the electromagnetic spectrum that is perceived by the human eye as the sensation of light.

There is a curious, interesting relationship between radiometry and photometry: the fact that electromagnetic energy occurs in the visible region of the spectrum does not define the visual effectiveness of that energy. A blue lamp, for example, may emit the same radiant power in watts in the visible region as a green lamp, but the latter would appear to be brighter because the eye is more sensitive to green light than it is to blue.

Radiometric terms apply anywhere in the electromagnetic spectrum, whereas photometric terms are applicable only in the visible region of the spectrum. Outside of the visible range, there can be no light, by definition; thus, the terms, units, and quantities of photometry do not apply.

RADIANT ENERGY AS POWER

Radiant energy is energy traveling in the form of electromagnetic waves. Radiant energy density is the energy per unit volume, and radiant power, also called radiant flux, is the radiant energy transferred per unit of time.

Radiant energy is measured in the same units as power in general. Since power = work/time, and work = potential energy = force × distance, the unit of power is

$$1 \text{ watt } = \frac{1 \text{ newton-meter}}{1 \text{ second}} = \frac{1 \text{ joule}}{1 \text{ second}}$$

Power is the central term of radiometry, and its unit is the watt. Instruments for measuring radiant energy are called *radiometers*, and the equivalent units for measuring visible light are called *photometers*. Many kinds of devices are made to measure the radiant energy. Some utilize photosensitive material and measure the electron flow that results from the light photons striking the material. In effect, these devices count the number of impinging photons and are calibrated to read directly in power units, usually some fraction or multiple of the watt.

LIGHT AS A TRANSMISSION CARRIER

We know that light can be generated. We know that light will propagate through a vacuum, through air, and through other media as well. We know that light can be detected. These properties of light provide the essential components of lightwave transmission systems that follow the traditional configuration of other transmission systems.

Transmission systems require a modulated source of energy (a transmitter), a transmission medium through which the energy can propagate (copper conductors, air, glass, water, etc.), and a detector to sense the energy at the distant terminal and recover the intelligence which was modulated onto the carrier (a receiver). If light generators, i.e., sources of light, can be modulated with intelligence in some manner, all the necessary functions of a transmission system can be performed.

Modulation is the process whereby the amplitude, frequency, phase, or intensity of a carrier or signal is varied in accordance with these same properties of some other signal which is the intelligence that is to be transmitted. In its simplest form, modulation involves simply turning a carrier on or off. The intelligence could be encoded by varying the intervals during which the carrier is on or off, as in Morse code radio transmission. This technique, which uses light as the carrier, is the "blinker" semaphore used to communicate between ships, for example.

In radio or microwave transmission systems there are a variety of modulation techniques in general use, such as frequency or phase modulation and amplitude modulation. The present state of the art in lightwave transmission systems precludes some of these techniques at this time. In its most commonly applied form, lightwave transmission is restricted today to intensity modulation; that is, the light source is either varied in intensity by an analog input signal, or it is actually turned on or off (sometimes simply made brighter or dimmer) by a digital input signal.

Research into other methods of modulating light are continuing, and alternative modulation techniques will probably be developed.

LIGHTWAVE TRANSMISSION SYSTEMS

In developing practical transmission systems for telecommunication purposes, we are constrained by all the conventional obligations of other transmission technologies. That is to say, we are obliged to design, construct, and operate, over long periods of time, transmission systems that maintain relatively constant transmission quality at all points within the system.

In order to do this, it is necessary to use standard units of measurement, and to be thoroughly familiar with such units and completely comfortable with their use. The two basic considerations in any transmission system are the signal level (the amplitude of the carrier itself) and variations in the signal level, which may be translated to transmission losses or gains as the signal traverses the system. As mentioned earlier, power is the basic quantity used in radiometry and its unit is the watt.

It is possible to do all system engineering and even system maintenance work using actual signal power levels, but this is somewhat awkward. For example, it is sometimes difficult to reference transmission losses directly in power levels. Moreover, if the basic unit of power is the watt, but transmission levels become small increments of a watt, we would find it necessary to make many calculations with small units. This would not only be tedious, but it could introduce error into the calculations.

A more convenient method has evolved over many years of experience in other disciplines. We can adopt units of measure that are equally compatible with signal levels or with transmission losses and gains. Then the development of precise signal levels at any point within a system, at any time, would be a simple matter of addition or subtraction. The basic concept is to deal not with absolute power levels, but with ratios of power. How may this be accomplished?

In lightwave transmission systems we use the familiar term "decibel" (dB) to express ratios of power for purposes of denoting system gain or loss. A negative value of dB denotes transmission loss, a positive value gain.

We use dBm (decibel-milliwatts) as a convenient unit of power for purposes of denoting system transmission levels. The value given to zero decibel-milliwatts is a power level of 1 mw, which may be correctly written as 0 dBm, 0.001 watt, or 10^{-3} watt. A negative value of dBm would denote a power level below 1 mw, and a positive value a level above 1 mw. For example, –3 dBm would denote a precise power level of 1/2 mw since –3 dB would be a 1 to 2 power ratio, or one-half. A value of $+3$ dBm denotes a power level of 2 mw (a power ratio of 2 to 1).

Appendix A presents a detailed discussion on ratios, decibels, logarithms, and decibel-milliwatts for those readers not familiar with these terms and their use.

SUMMARY

Light may be considered either from the point of view of an illuminating agent or as power. Different terms are involved depending on how light is addressed, and even the units of measurement are not the same.

The Law of Inverse Squares applies to light as an illuminating agent, just as it applies in other disciplines. Because we are dealing with electromagnetic waves, we can focus, direct, and concentrate light energy through the use of various devices.

Some wavelengths of light are visible, that is, detectable by the human eye, but many of the wavelengths of light that we utilize in practical transmission systems are not. These nonvisible wavelengths may be treated or processed using conventional optical mechanisms, however.

When light passes through a medium that includes gases or other particulate matter, it may be absorbed or scattered by the molecular nature of the medium it is passing through. If the atoms of the medium retain the light energy, the phenomenon is called absorption. If the light energy raises the energy of the atoms to the point where they, in turn, emit light at the same wavelength as the incident light energy, the phenomenon is called scattering. Scattered light energy is polarized and is radiated equally in all directions.

Both visible and nonvisible electromagnetic energy at light wavelengths can be measured. Photometry is the measurement of visible wavelengths only, and radiometry is the measurement of all light wavelengths. Lightwave energy may be measured as power in the same units used in other disciplines, and the unit of measurement most commonly used is the watt.

Transmission systems employing lightwave energy must consider all the conventional factors involved, such as power attenuation within the transmission medium. Also, just as in other disciplines, transmission calculations employ the extensive use of ratios of power, rather than absolute power levels.

REVIEW QUESTIONS _____

True or False?

T   F   1. All light wavelengths that we employ in practical lightwave transmission systems are visible to the human eye.

T   F   2. Photometry is the measurement of lightwave energy at any wavelength.

T   F   3. The level of illumination of a surface by a light source is constant regardless of how distant the surface is from the source.

T   F   4. Color is a physical characteristic of a substance or material.

T   F   5. The subtraction by absorption or attenuation of specific frequencies of light energy will introduce the perception of a specific color.

T   F   6. When all visible wavelengths of light are present at equal amplitudes, the visible perception is white.

T   F   7. Monochromatic means "having or exhibiting two colors."

T   F   8. Incandescence means "emission by a hot body of radiation that makes it visible."

T   F   9. Opacity means "reflection of light by a rough reflecting surface or transmission of light through a translucent material."

T   F   10. Light scattering is experienced as lightwave energy travels in a vacuum.

T   F   11. Rayleigh's theory applies to particles that are larger than two-tenths of the wavelength of the light energy under consideration.

T   F   12. When the atoms of a material or substance which incident light impinges upon have their energy level raised to the point that they emit lightwave energy at the same frequency as the impinging light, this is called absorption.

T   F   13. The primary causes of the attenuation of light in air or in another medium are incandescence and fluorescence.

T   F   14. The extinction of light includes attenuation due to absorption and light scattering.

T   F   15. Radiant energy is energy traveling in the form of electromagnetic waves.

T   F   16. The unit of measure of radiant energy at light wavelengths is the watt.

T   F   17. Some devices that measure radiant energy at light wavelengths count the number of photons impinging on the photosensitive material in the device.

T   F   18. A ratio is the relation in degree or number between two dissimilar things.

T   F   19. A common term to denote the power ratio of two levels of lightwave energy is the decibel.

T   F   20. We can state the loss of any particular transmission medium of a given length in decibel-milliwatts.

CHAPTER 3

Introduction to Optical Fibers

Lightwave energy can be launched into, and will propagate through, optical transmission lines. Such transmission lines are referred to as *optical waveguides, lightguides,* or *optical fibers.*

The use of these transmission lines presents a spectrum for private use; that is, no transmissions or emissions are launched into the public spectrum, which is defined as free space. Thus, no licensing by any governmental agency is required, and the system operator is entirely free to allocate transmission frequencies or services within the confines of this private spectrum in any manner he or she wishes.

As in transmission lines in other technologies, e.g., metallic conductors or radio frequency (microwave) waveguides, the nature of lightwave transmission lines is to confine the light rays within the boundaries of the transmission line itself. This restricts access to or exit from the transmission line to taps or line terminals that are provided in the design of the facility.

Since basic optical principles apply to the design and function of such transmission lines, the term "optical fibers" is used throughout this text. A number of such fibers may be incorporated into the fabrication of optical cables.

It is possible to launch light energy into optical fibers, and such energy will propagate through the fibers. As in any transmission system, the process will incur transmission losses, and various forms of distortion may be introduced into the lightwave signals themselves.

BASIC OPTICAL MECHANISMS

Figure 3-1 shows a light source (a flashlight) positioned to illuminate the end of a section of pipe which we shall assume has a highly polished inside surface. Note that the light beam is widely diverging, so that some of the light rays will not be coupled into the pipe at all. Under these conditions, less lightwave energy is available at the distant end of the pipe for detection.

In Figure 3-1(b), a lens has been positioned between the light source and the pipe which focuses the light beam more narrowly. Though the inner diameter of the pipe has not been changed, now more light energy is coupled into the pipe, and thus more energy will be available for detection at the distant end.

Quite clearly, any light source that inherently emits a narrower beam of light will couple into a lightguide (the pipe) more efficiently. Some term to define the light-gathering capability of the lightguide will be useful, and we might refer to this capability as the "degree of openness" or the "cone of acceptance" that the end of the pipe presents to incident light.

Recognizing that the light source could be replaced with another section of pipe with its distant end illuminated, as in a long link of pipe sections, then the efficiency of transfer of light between any two pipe ends would, to some extent, depend upon the "cone of acceptance" or "degree of openness" of both pipes. Since the "cone of acceptance" establishes the angle of internal reflections within the pipe, it also determines the divergence of the light beam where it exits the end of the pipe.

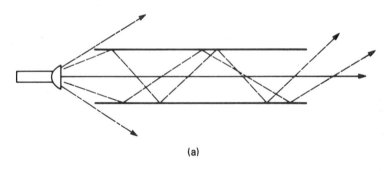

(a)

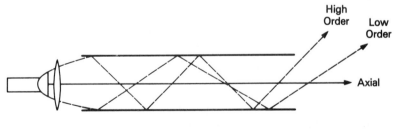

High
Order   Low
Order

Axial

FIGURE 3-1
Reflective
Propagation

(b)

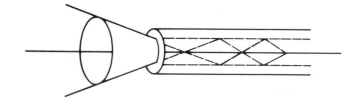

FIGURE 3–2
Cone of
Acceptance

The term describing the light-gathering ability of a lightguide is called *numerical aperture* (NA), which is applied to optical fibers. This is shown in Figure 3–2, and we shall be discussing this term in greater depth later in the text.

PROPAGATION MODES

Figure 3–1 demonstrates another optical phenomenon that will be encountered in lightguiding optical fibers. Whatever the divergence of the incident light beam may be, some light rays propagate directly along the longitudinal axis of the pipe. This propagation path is shown in both parts of the figure as a solid line, and the transmission path length along this line (the optical axis of the pipe) is the shortest possible distance through the pipe. This path represents one mode of light propagation within the pipe.

The figure also shows some light rays entering the pipe at an angle (the dashed lines). If the inner wall of the pipe is highly polished, these rays will be reflected from this surface and will propagate through the pipe in a series of reflective paths. There may be many, perhaps hundreds, of such reflective paths, and each such individual transmission path through the pipe will be called a propagation mode, just as the axial transmission path was.

However many reflective propagation modes there may be, all will be longer in length than the axial transmission mode, due to the multiple reflections involved. In Figure 3–1, reflective modes are identified as being either higher order or lower order. This distinction designates the higher order modes as introducing the longest transmission paths through the pipe, with the lower order modes being shorter paths. The axial mode is the shortest possible transmission path, of course, and is sometimes referred to as the fundamental mode.

If all light rays travel at the same speed—and they do in air—it follows that the reflective propagation modes will take longer to travel to the distant end of the pipe than the fundamental mode. Higher order modes will require the longest transmission time of all.

MODAL DISPERSION

Consider a light source that is simply turned on and off a number of times. This procedure could be a form of modulation, with the intelligence contained in either the duration of the sequence of the on-and-off transitions, or both. The on-and-off functions might be accomplished in response to an electrical input signal. If such a transmission system were in actual use, the objective would be to faithfully reproduce the electrical input signal at the distant end of the system, meaning that the reproduced signal would have the same characteristics of the on-and-off periods of duration, or the same sequence of on-and-off transitions, or both.

If, at a given instant in time, the light source is turned on, all light rays at all lightwave frequencies will be emitted instantaneously. Suppose, then, that many rays propagate through the pipe, as in Figure 3–1. Then the light rays which arrive first at the distant end of the pipe will be those that propagate in the fundamental mode and consequently travel the shortest path along the optical axis of the pipe. If the distant end of the pipe were coupled to a light detector, the detector would sense the presence of these rays and turn on its electrical output signal.

Then, a later instant in time, the other light rays that were reflectively propagating through the pipe would also arrive at the detector input and be sensed, but the detector's electrical output would have already been turned on by the earlier arrival of the axially propagating rays, so no change would occur in the electrical output signal.

Now consider the sequence of events if the electrical input signal at the transmitting terminal of the pipe turned off the light source or light generator. In that case, all light rays at all lightwave frequencies would be extinguished at the same instant. The fastest reaction to this at the distant end would be the loss of light from the fundamental propagation mode first, since that light travels the shortest distance. The detector would still be presented with some light input from the more slowly propagating reflective modes, however, so the detector output electrical signal would not be turned off immediately.

At some later instant in time, after the slowest light rays (the highest order propagation mode) had finished traversing the pipe, the detector would no longer have any light energy input at all, would sense this condition, and would turn off its electrical output signal.

The result would be an unfaithful reproduction of the original electrical input signal, since the electrical output signal would be of longer duration than the initiating signal was. We say that the output pulse was "broadened" or "dispersed," a condition shown in Figure 3–3.

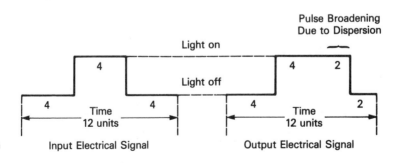

FIGURE 3–3
Pulse Dispersion

Broadening is a form of distortion, of course, and is called *modal dispersion* in lightwave transmission systems. Clearly, it is introduced by the multiplicity of transmission paths or modes supported within the pipe. How pronounced it will be is a function of the number of propagation modes that the pipe supports and, of course, the length of the pipe. This is called *multimode* propagation since a multiplicity of modes are present and sustained in the transmission medium.

INTERSYMBOL INTERFERENCE

Pulse distortion that is evidenced as a longer output pulse period than the input pulse period might be considered academic since the input pulse has been extended by only a very short period of time. This is truly the case at very low transmission speeds, such as when a ship uses blinker semaphore and Morse code. A slight extension of the light-on period in relation to the light-off period would be completely insignificant.

But in lightwave transmission systems, the light may be pulsed at significantly higher rates. In a system operating at 90 megabits per second (Mb/s), for example, the pulse period that the light is turned on is only six to ten nanoseconds or so, and 90 Mb/s is not a particularly fast or uncommonly encountered data rate in today's technology. The pulse distortion introduced can consequently be a significant percentage of the normal pulse period in such applications.

At these higher transmission data rates, more bits (pulses) are transmitted per unit of time. The pulses tend to spread out, as discussed in modal dispersion, and an individual pulse will begin to intrude into a "no pulse present" time slot. This complicates the process of trying to determine, at given instants, whether a pulse is actually present or not. The intrusion of a pulse into an adjacent time period is called *intersymbol interference* and is shown in Figure 3-4. Such distortion can result in errors in data transmitted through a system or link.

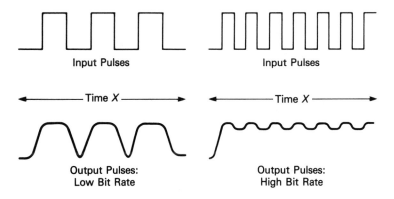

FIGURE 3-4
Intersymbol
Interference

Input Pulses   Input Pulses

Time *X*   Time *X*

Output Pulses:
Low Bit Rate

Output Pulses:
High Bit Rate

SINGLE MODE PROPAGATION

In Figure 3-1, we saw the reflective mechanisms that the pipe provided and the multiplicity of propagation modes that were sustained within the pipe. A significant reduction in modal dispersion (pulse distortion) could be achieved by simply reducing the inside diameter of the pipe, as shown in Figure 3-5.

Although several reflective propagation modes are still supported, since the reflective transmission paths through the pipe are shorter even for the higher order modes, the transit times will be shorter also. It follows that the transit time differential between the fundamental mode (axial mode) and the worst case reflective modes (higher order modes) will be reduced. In effect, all rays that transit the pipe will travel nearly the same distance. Thus, the transit times for all rays propagated will be more nearly the same and the modal dispersion or pulse duration distortion introduced will be less pronounced.

If it were possible to reduce the inside diameter of the pipe even further, perhaps to pinhole size or smaller for example, this might restrict propagation to the axial mode alone, which would effectively eliminate, or at least dramatically reduce, all pulse distortion. This condition would be called single mode propagation, since only one mode (the axial mode), would be supported, or very nearly supported, within the pipe. Figure 3-6 depicts single mode propagation.

Note that with a very small inner diameter it is much more difficult to couple light into the inside of the pipe, and we must use a light source with a narrowly diverging light output or use some focusing lens arrangement as shown in the figure.

Figure 3-7 shows another optical phenomenon, produced by a dent or mechanical irregularity in the pipe wall. Those light rays that previously propagated in an orderly and predictable manner now bounce off the irregularity in a random

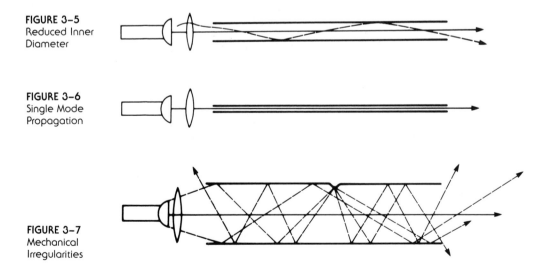

FIGURE 3-5
Reduced Inner
Diameter

FIGURE 3-6
Single Mode
Propagation

FIGURE 3-7
Mechanical
Irregularities

pattern of dispersal angles. Some may continue to propagate through the pipe, but many new modes may be established, possibly introducing new and longer transmission delays, with the consequent effect on modal dispersion. Some of the new rays may even be reflected back toward the entrance end of the pipe, reducing the level of light energy that will be delivered to the distant end for detection.

REFRACTIVE INDEX

Thus far, we have used a pipe as a lightguide and assumed that the pipe was filled with air. As light passes through a vacuum, its velocity is approximately 186,000 miles per second (300 million meters per second). Through the air, the velocity of propagation is slightly slower, 299,739,000 meters per second; but for practical purposes we may assume the 300 million figure to apply in both instances. But optical fibers are solid glass structures and are not filled with air.

When light passes from one medium to another, it changes speed. As discussed earlier, this introduces a deflection of the light rays called refraction. A term which defines this characteristic of media is *refractive index*. The refractive index of a medium is defined as the ratio of the velocity of lightwaves in a vacuum to the velocity of lightwaves propagating in that medium.

The refractive index in a vacuum, then, is 1.0. The refractive index of air is 1.003, often generalized to be 1.0 also. The refractive index of glass is 1.5. Put another way, when light travels in a denser medium, the speed of light (its velocity of propagation) decreases. This slower speed v, when divided into the speed of light in a vacuum, c, yields the index of refraction n. Formally,

$$n = \frac{c}{v}$$

Some typical indexes of refraction are as follows:

Medium	n
Vacuum	1.0
Air	1.0003
Water	1.33
Fused Quartz	1.46
Glass	1.5
Diamond	2.0
Silicon	3.4
Gallium Arsenide	3.6

Indexes of refraction are utilized in the design of optical fibers. To better understand the reflective mechanisms within a fiber, it is useful to review a basic optical principle called Snell's law. Named after the Dutch physicist, Willebrord Snell

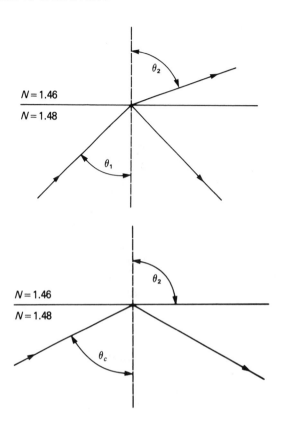

FIGURE 3-8
Snell's Law

van Royen, Snell's law states that at an interface between two transparent media with different indexes of refraction n, light traveling in the higher density medium (higher refractive index) will be partially reflected and partially refracted by the interface between the media when the angle (θ) is small.

The situation is shown in Figure 3-8. As angle θ_1 is increased, the angle θ_2 of the refracted light also increases. At the point where θ_2 becomes 90 degrees, there is total reflection; that is, all light is reflected back into the denser medium, and no light is refracted into the less dense medium.

The point at which θ_2 becomes 90 degrees is called the *critical angle* θ_c of angle θ_2.

STEP INDEX FIBER

It is possible to fabricate optical fibers of silica glass where the center mass of the fiber is a glass material with a given refractive index, but the outer section of the fiber, which is also glass, has a different refractive index. This type of fiber is shown in Figure 3-9 along with a profile of the refractive index of a cross section of the fiber. Since the cross section exhibits only one discrete change in refractive index,

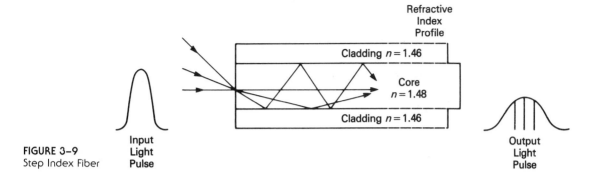

FIGURE 3-9
Step Index Fiber

which is at the line of demarcation between the two glass materials, this fiber is called a *step index* fiber. The refractive indexes shown in the figure are arbitrary and were selected for demonstration purposes only.

The outer layer of glass in a step index fiber has a uniform refractive index and is called the *cladding*. The center material has a different uniform refractive index and is called the *core*. It is possible to manipulate the refractive indexes of both the core and the cladding material by controlling the composition of those materials during fabrication. This is done by judicious introduction of small quantities of other materials, a process called *doping*.

If the cladding material has a significantly lower refractive index than the core material, then Snell's law applies, and it is theoretically possible to have total reflection by introducing light rays into the core that strike the cladding material at suitable incident angles. Then light propagating through the core will be constrained by the cladding material and reflected back into the core throughout the length of the fiber.

Given an optimum incident angle and an optimum difference in refractive indexes, the composite fiber (an integral unit composed of both the core and the cladding materials) will propagate light through the core in the same manner as the light propagated through the interior of the pipe in the earlier example on p. 34. Within the fiber core, both axial and reflective propagation modes will exist, and the mechanisms which introduced modal dispersion (with consequent pulse broadening) in the pipe example will be present in the fiber. Thus, the transmission distortion introduced will be quite similar in both cases.

The ideal condition for efficient propagation within the fiber is not to have reflective angle propagation modes that exceed the critical angle as they impinge on the cladding material. Light rays striking within the critical angle will be totally reflected back into the core, but light rays striking outside the critical angle will be refracted into the glass cladding material. Such rays are called *cladding modes* and the energy they possess will escape from the fiber within a short distance. The mechanism to control the reflective angle within the fiber (at the cladding/core interface) is to restrict the angle at which the light rays are launched or coupled into the fiber initially.

NUMERICAL APERTURE

As mentioned earlier, the term used to define the angle of light entering the fiber is "numerical aperture" (NA). It is a numerical figure that denotes the maximum angle θ_{na} of light entering the end of the fiber that will be propagated within the core material of the fiber. The numerical aperture is shown in Figure 3–10. In accordance with Snell's law, all rays entering the fiber core (refractive index, 1.48) from the air (refractive index, 1.0) are refracted toward the longitudinal axis of the fiber core, which is shown as a dashed line in the figure.

The numerical aperture is a function of the core diameter and the indexes of refraction of the cladding and core materials. In the interest of efficient propagation through the fiber, that is, to ensure total internal reflection, it is desirable to design the fiber with the smallest numerical aperture possible. In multimode fibers this design essentially limits the practical dimensions of the core. For if the core dimension were large, then the light could enter the fiber at angles that exceeded the critical angle at cladding interfaces and would perhaps create new high order propagation modes by reflection. These high order modes would allow light to be refracted into the cladding, of course, and eventually this light would be lost for transmission purposes. These considerations dictate the very small physical dimensions of practical optical fibers.

The same optical principles apply at the distant or receiving end of the fiber. The numerical aperture of the fiber, which is a function of the refractive indexes of the two materials, core and cladding, will establish the angular dispersion of the light that is emitted from the end of the fiber. Just as the term "cone of acceptance" is appropriate in coupling light into a fiber, the term "cone of radiation" is appropriate for the emission of light from the fiber end.

Maximum fiber splicing efficiency, that is, maximum transfer of light between two fiber ends, will depend to some extent on how similar the numerical apertures of the two fibers are. Also, the cone of radiation of light sources that must couple light into fibers and the cone of acceptance of light detectors that must accept the light output from fibers must be considered.

To summarize, light rays impinging on the fiber end at angles greater than the numerical aperture will exceed the critical angle at the cladding/core interface within the fiber and therefore will not be reflected back into the core. Such an errant light

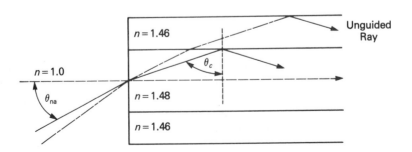

FIGURE 3–10
Numerical
Aperture

ray is shown in Figure 3-10, and rays entering the cladding in this manner are eventually lost for transmission purposes.

GRADED INDEX FIBERS

So far, the discussion has been limited to step index fibers, which, because they support multiple propagation modes, are multimode fibers. The mechanisms of light propagation in such fibers are directly analogous to the example of the pipe as a lightguide. This analogy is useful in understanding modal dispersion, but the deficiencies of multimode (*not* single mode) step index fibers has virtually eliminated their use in telecommunications systems of any length today.

The problem is that in a step index fiber significant modal dispersion occurs because a multiplicity of propagation modes are supported by the fiber. The dispersion is the result of differences in transit time for the different modes, each of which has a different length transmission path. One technique for reducing modal dispersion is, as mentioned, to reduce the fiber core size, and historically the ultimate outcome of this was a very small core which would support only one propagation mode.

There is another remedial measure, however, that depends upon an entirely different optical principle. In the earlier example of a pipe used as a lightguide, the pipe was filled with air, which had a uniform index of refraction. Optical fibers are solid, however, not filled with air; and since we must fabricate such fibers anyway, it is possible to manipulate the refractive index in a variety of ways.

We can, for example, fabricate the fiber in such manner that the refractive index of the core material is not constant throughout the core cross section, but is gradually changing. This can be done by fabricating the core with a series of concentric rings of material wherein each ring or layer has a slightly different refractive index. We can vary the refractive index radially from the axis of the fiber to the outer limits of the core, with the lower refractive indexes being encountered farther away from the axis of the core. Such a fiber, called a *graded index fiber*, is shown in Figure 3-11.

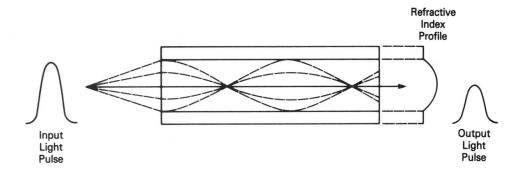

FIGURE 3-11
Graded Index
Fiber

Input
Light
Pulse

Refractive
Index
Profile

Output
Light
Pulse

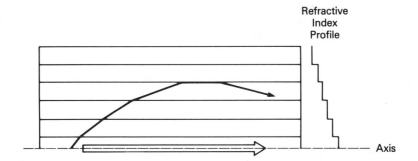

Refractive
Index
Profile

FIGURE 3–12
Graded Index
Fiber

Axis

As light is introduced at an angle into a graded index fiber, the light rays pass through several different refractive indexes within the core material. Since light travels faster in the less refractive material that is located farther from the optical axis, not only is the light refracted back towards the axis, but the time differential between the axial propagation and the refractive propagation is greatly reduced. The refractive mechanism may be seen clearly in Figure 3–12, where the refractive index profile steps have been exaggerated to show the process.

In a graded index multimode fiber, several propagation modes are sustained and different transmission path lengths exist, just as in the multimode step index fiber discussed earlier. The transit time over the longer transmission paths is reduced, however, due to the changing refractive index of the core material. Also, in a graded index fiber, the cladding material has a distinctly lower refractive index than the lowest index in the core material. Thus, a line of demarcation between core and cladding is maintained, and the cladding still acts to confine the light rays within the core material.

Graded index fibers require careful control of the processes whereby they are manufactured. Outside fiber diameters may be 125 μm and the graded index core diameter is 50 μm or so in typical graded index fibers.

Figure 3–12 shows the effect of a varying refractive index on a ray of light in a graded index fiber as compared with the characteristics of a multimode step index fiber, shown in Figure 3–9.

SINGLE MODE FIBER

A single mode fiber is basically a step index fiber since only one demarcation is established between the core and cladding indexes. In a single mode fiber, the reduced core diameter restricts the propagation to a single transmission mode.

If the refractive index differential between the core and cladding material is kept small, the fiber will behave as a true waveguide. It can be shown, using Maxwell's equations, that in a bounding waveguide, electromagnetic energy will propagate and be limited within the confines of the guiding structure. In the case under consideration, the guiding structure is the optical fiber itself, and the light energy is bound within this structure. The light propagates longitudinally through the fiber,

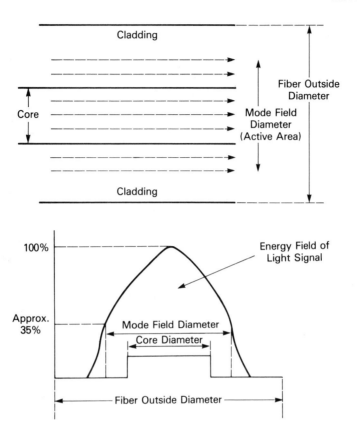

FIGURE 3-13
Mode Field
Diameter

not by means of optical mechanisms such as reflection or refraction. A significant amount of the light energy—as much as 35 percent of the total energy of the light-wave signal—will actually propagate in the cladding material. (See Figure 3–13.)

MODE FIELD DIAMETER

When propagation is limited to the core area alone, the term numerical aperture (NA) is appropriate. Now, however, we must define a larger cross-sectional area of the fiber that includes some of the cladding area. This area, which is larger in diameter than the core but smaller than the cladding outside diameter, is actually propagating light energy, and we might refer to it as the active area of the composite fiber structure.

Since some of the light propagation is effected through the cladding material, a new term, *mode field diameter* (sometimes called *spot size*) is used to define the active or light-propagating cross section of the fiber, which includes all the core area and some of the cladding. Figure 3–13 shows the relationship of the mode field diameter to the fiber core diameter.

Mode field diameter increases with operating wavelength, and thus, at some specific wavelength, a single mode fiber will start to support more than one propagation mode and will function as a multimode step index fiber. The frequency corresponding to this wavelength is referred to as the *cutoff frequency* of the fiber, and is the point in the spectrum below which more than one mode will be supported.

Cutoff frequency will be a function of the fiber geometry and material, the refractive index differential, and the core diameter. Since all these factors can be controlled during fiber fabrication, a fiber can be manufactured to have a specific cutoff frequency characteristic if it is desirable to do so.

If, to have a single mode optical fiber function as a waveguide, we design the fiber for use at a long wavelength, we must manipulate both its core diameter and refractive indexes during fiber fabrication. The core diameter will be slightly increased, and at some shorter wavelength this larger core size will propagate the light energy reflectively as it did in the large-core step index fiber. The cutoff frequency is the frequency corresponding to the wavelength at which more than one mode will propagate through the fiber, and this wavelength will always be shorter than the design operating wavelength of the fiber.

The cutoff frequency may be of interest when more than one wavelength signal will be passing through the fiber, as in a "double window" fiber application or in a system applying wavelength division multiplexing. We shall discuss these applications in greater detail later in the text.

COMPARING TYPES OF FIBER

Both graded index fibers (often referred to simply as multimode fibers) and single mode fibers dramatically reduce modal dispersion problems, but not to the same degree. In the next chapter we shall discuss fiber distortion, loss, and bandwidth characteristics in depth, but we can draw some interesting comparisons even now.

Although large-core, step index fibers are not usually encountered in telecommunication systems of significant length, they do find application in vehicles such as ships, aircraft, and the like, and in other situations where transmission distance or transmission data rate, or both, are nominal. Large-core, step index fibers have the following characteristics:

1. Their large core greatly simplifies fiber connection and splicing.

2. Their large numerical aperture greatly simplifies coupling light into and out of them.

3. Both their modal dispersion and loss of light in cladding modes may be substantial. These penalties may be acceptable if the fiber lengths are limited or data rates are low.

Multimode (graded index) fibers have the following characteristics:

1. Their large core simplifies fiber connection and splicing tasks somewhat.

2. Their large numerical aperture simplifies coupling light into and out of them somewhat.

3. Their modal dispersion introduces some bandwidth restriction, which limits transmission data rates to some degree. A typical usable bandwidth is on the order of 600 MHz per kilometer to perhaps 2,000 MHz per kilometer.

Single mode fibers have the following characteristics:

1. Their typically small core diameter complicates fiber connecting and splicing.

2. Their small numerical aperture (spot size) complicates coupling light into and out of them.

3. Since they have little or no modal dispersion, they present a very wide bandwidth for use, and higher transmission data rates are possible even with simple fiber structures.

SUMMARY

Any source of lightwave energy can couple light energy into an optical fiber. A source emitting a narrower beam of light will couple more energy into a fiber than a source with a widely diverging beam. The term "numerical aperture" (NA) defines the acceptance cone of a fiber or a terminal optical device, such as a photodetector.

Light may propagate directly, reflectively, or refractively through a fiber. If more than one mode of propagation is sustained, the fiber is said to be multimodal. If only one mode is supported, the fiber is called single mode or monomodal.

There are differences in transit time through a fiber for different propagation modes. Thus, some dispersion or pulse broadening is experienced in any fiber supporting more than one mode of propagation. This is called modal dispersion.

In fibers that support only one propagation mode, pulse dispersion may still be introduced, but the mechanism is due to the difference in transit time between two or more lightwave frequencies being propagated if more than one frequency is presented to the fiber. This is called chromatic dispersion.

The cross-sectional area of a fiber may exhibit several changes in refractive index of both the core and cladding materials. If the changes are distinct transitions, the fiber has a step index structure. If the changes in refractive index are gradual, the fiber has a graded index structure.

In a graded index multimode fiber, all light energy propagates through the fiber core. In a single mode step index fiber a significant amount of light energy actually propagates through the fiber cladding material.

REVIEW QUESTIONS

True or False?

T   F   1. A light source that emits a less divergent beam of light will couple less light into a fiber.

T   F   2. The efficiency of transfer of light energy between two fiber ends depends upon the numerical apertures of both fibers.

T   F   3. In a digital system, intelligence can be modulated into a light signal by turning the light on or off, or by turning the intensity of the light higher or lower.

T   F   4. Modal dispersion in a lightwave system is primarily the result of several lightwave frequencies being presented to the system.

T   F   5. Chromatic dispersion is the transmission data rate limitation in multimode, graded index fiber systems.

T   F   6. Pulse dispersion in a digital lightwave system will contribute to the number of errors experienced in the overall transmission performance of the system.

T   F   7. Modal dispersion is more prevalent in a single mode fiber than in a multimode fiber.

T   F   8. All lightwave systems operate with bipolar light energy signals.

T   F   9. When light passes from a less dense medium into a more dense medium, the velocity of transmission decreases; that is, the light travels at a slower speed in the more dense medium.

T   F   10. A step index fiber will have a distinct transition between the core and cladding refractive indexes.

T   F   11. The numerical aperture of a fiber denotes the maximum angle of light entering the end of a fiber that will be propagated within the core material of the fiber.

T   F   12. The diameter of the core of a multimode step index fiber dictates the critical angle within the core, and thus determines whether total internal reflection will occur within the core.

T   F   13. Graded index fibers depend upon the reflective optical mechanisms of the fiber to minimize pulse dispersion.

T   F   14. Light energy propagates through a single mode fiber reflectively.

T   F   15. The mode field diameter of a fiber is the diameter of the core material in the fiber.

T   F   16. The small physical dimensions of multimode fibers used in telecommunication systems are dictated by the need to ensure total internal reflection within the fiber.

T   F   17. Smaller numerical apertures make it more difficult to couple light energy into fibers.

T   F   18. Single mode fibers present more bandwidth for transmission use than do graded index fibers.

T   F   19. The usable transmission bandwidth provided by a fiber has no relationship to the transmission data rate that a fiber can support.

T   F   20. Cutoff frequency applies only to multimodal fibers.

CHAPTER 4

Light Transmission in Optical Fibers

Unwelcome things like distortion and attenuation happen to any signal passed through any transmission system. We must understand these, however, so that we can design systems that provide a usable transmission quality even after the inevitable degradation has occurred. Specifically, we need to learn about what happens to light signals when they are passed through optical fibers. Doing so may help us select fibers that introduce less degradation or design more economical systems that perform well enough that we can accept the degradation and still use the systems.

Just as electromagnetic waves can be injected or launched into transmission lines such as copper wires, so can lightwaves be launched into lightguides, or optical fibers. Transmission of light through optical fibers is, in some ways, analogous to electrical transmission through wires, but there are significant differences in the two technologies, and the principles of physics are quite different for each case.

Thus, whereas electrons are caused to flow through a copper conductor, light is guided through an optical fiber. The optical propagation is much more directly analogous with microwave or radio waveguides. In waveguides the energy does not flow through the material that constitutes the waveguide structure as through an electrically conducting material, but rather it is propagated as a wave through the space enclosed by the conductive waveguide walls.

In lightwave transmission, the fiber functions as a waveguide for the light energy. The manner in which the light is confined within the two basic types of fibers (step index and graded index) involves distinctly different optical principles. In a multimode step index fiber, the light is confined within the core material by total internal reflection that is introduced by design at the core-cladding interface. In a single mode step index fiber, the light propagates longitudinally through both the core and the cladding material. In the graded index fiber, the light is bound not by internal reflection at all, but by refraction, due to the radially varying refractive index of the core material itself.

The two major considerations in addressing the transmission of lightwave energy through an optical fiber are transmission loss and signal distortion. If the signal being transported is analog in nature, then the distortion will be quite similar to that incurred in other analog transmission systems and will be largely related to transmission bandwidth and noise. If the signal being transported is digital, the interference will be evidenced as pulse distortion or dispersion, i.e., pulse broadening.

Figure 4-1 shows the effect of light pulse dispersion on the fidelity of transmission through a digital system. This effect is not experienced in an analog system, where the intensity of the light pulse varies but no rapid transitions of light intensity occur. The problem shown in the figure (exaggerated for clarity) is called *intersymbol interference,* and at high transmission data rates it can render a system unusable. This is because at higher data rates the pulse period is much shorter, and the dispersion becomes relatively large in relation to the pulse period itself. Obviously, such a problem must be addressed and resolved.

LIGHT SCATTERING IN FIBERS

The loss of light energy that occurs as light travels through a fiber can be attributed to several causes. The most pervasive factor, and perhaps the most significant, is light scattering within the fiber itself. The root cause of light scattering was discussed earlier, and the phenomenon, as it occurs in the fiber, is identical.

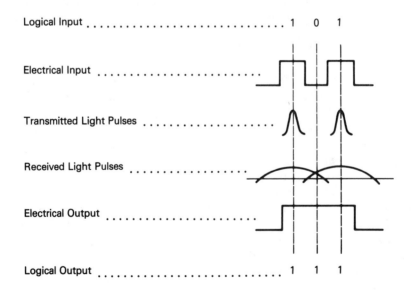

FIGURE 4-1
Intersymbol
Interference

The molecules of the glass themselves absorb photons and reemit light energy. These emissions are random, and the light so emitted does not propagate in an orderly or predictable manner through the fiber. The net effect is a loss of light energy within the fiber, and a lower level of light energy is available for detection at the distant end of the fiber.

The Rayleigh scattering curve for a fiber, presented in Figure 2–2, depends on the material out of which the fiber is fabricated. This does not mean that lower loss fibers cannot be fabricated: if the molecular structure of the silica glass of the fiber itself is changed, then the light scattering losses of the fiber will be changed also.

We can fabricate lower loss fibers by carefully adding different materials in a process called *doping*. Some recent experiments with halides and fluorides have produced very low-loss fibers. But scattering loss is a function of wavelength also, and the Rayleigh curve will still apply, even to the lower loss fiber. We will always experience lower fiber attenuation at longer wavelengths due to the scattering phenomenon.

If fiber composition is altered, the fiber designer must be careful to alter both the cladding and the core material, of course, since the relationship of the refractive indexes of these two materials is critical to the propagation of light within the fiber structure. Losses in fibers manufactured today come close to achieving the theoretical losses of the Rayleigh theory, but variations in fiber density during fabrication do introduce some variations in scattering loss throughout a long length of fiber, and actual losses are higher than those that are theoretically attainable.

LIGHT ABSORPTION IN FIBERS

Impurities in the fabricated glass itself, from which the optical fiber is drawn, introduce absorption of light energy within the fiber. Light absorption was discussed in general in Chapter 1 and is the condition when photons striking a substance give up their energy to the atoms of that substance, but those atoms do not reemit the energy. The phenomenon of absorption in optical fibers is the same as the general one, and the light energy is lost for transmission purposes. Thus, absorption is a form of transmission attenuation in lightwave transmission systems.

Figure 4–2 shows the effects of both scattering and absorption as loss mechanisms in optical fibers. The high attenuation peaks shown at some frequencies are the result of foreign elements or impurities in the glass. Both ferrous and ferric (iron) ions, for example, absorb light in the visible lightwave range. Another common impurity in commercial fibers is water in the form of hydroxyl ions, which introduces the highest loss peak shown in the figure.

The figure shows an early generation multimode fiber, but present-day production should not exhibit the pronounced attenuation peaks seen here since much progress has been made in understanding these phenomena and in producing fibers with much lower levels of contamination.

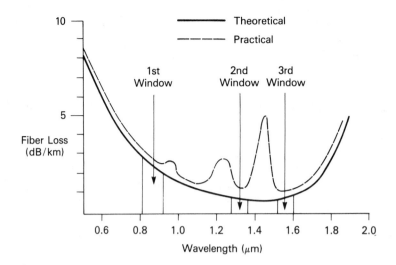

FIGURE 4–2
Fiber
Attenuation

Because the attenuation peaks due to impurities were so pronounced in earlier fibers, transmission designs focused on using frequencies where absorption was less oppressive. The terms "first window" and "second window" came into general usage, referring to those sections of the curve where fiber losses are minimal. The figure identifies the "windows," but as fiber quality improved the terms became somewhat less significant and are now less frequently referenced.

FIBER ATTENUATION MECHANISMS

Another form of scattering loss results from variations in the size of the fiber core, irregularities in the interface between the core and cladding materials, and the microbends which sometimes occur during the process of cabling fibers. Indeed, the fibers may be stressed over small imperfections in the cabling material itself during the cabling process. Figure 4–3 shows the effects of some of these conditions, all of which are due to the limitations of the fabrication or manufacturing processes themselves.

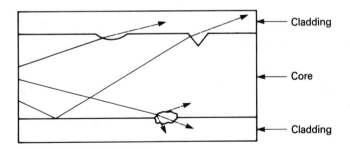

FIGURE 4–3
Microbends and
Attenuation

The attenuation which is intrinsic in fibers can be ascribed to the preceding causes. As a fiber, or a cable made up of fibers, is handled in transportation or placement, other attenuation mechanisms may be introduced. As systems are constructed, optical devices—for example, fiber connectors or splices, or optical splitters and couplers—may be inserted into the fiber for various purposes. In so doing, the optical fibers may be subjected to sharp bends or excessive tension. Any or all of these may attenuate light energy passing through the fiber also. However, such attenuation is extrinsic to the fiber structure and must be addressed in the construction or the design of specific applications or systems.

Loss, or attenuation, may be described as a reduction in the energy level of a transmitted lightwave signal during its passage through the fiber, and it is usually expressed as a power loss ratio in terms of decibels per kilometer (dB/km). Obviously, the stated loss in these terms will refer to a specific wavelength of light. The calculation of loss for a given length of fiber or cable is a simple mathematical process. A fiber specified to introduce 1.5 dB of loss per kilometer would evidence 7.5 dB of loss end-to-end in a five kilometer length (5 × 1.5 = 7.5).

When we address system transmission design, we shall find that some systems will be attenuation limited; that is, the transmission losses will reduce the signal power level below a reliably detectable level, and remedial measures will have to be undertaken. Such measures might be to select a fiber with lower transmission loss, to select a receiver/detector with a lower threshold of detection, to select a light source or transmitter with a higher power output, or to select an operating wavelength where the fiber has evidenced less loss. An attenuation-limited application may require the installation of intermediate regenerative repeaters.

DISTORTION

Distortion may be defined as the unfaithful reproduction, at the device or system output, of the information or signal that was applied at the device or system input. Some form of distortion is introduced in all transmission systems. Since our studies include telecommunications systems, and since such systems frequently involve digital information, we must consider distortion as it is evidenced in digital as well as in other types of signals. It will be helpful to examine digital distortion in other disciplines first.

In metallic conductor systems using, for example, twisted copper pairs, digital transmission is generally bipolar. Since the transmission facility can handle direct current (DC) signal components, a positive-polarity pulse can be employed to denote a digital one for example, and a negative-polarity pulse can be employed to denote a digital zero. A train of such pulses might look like the signal shown in Figure 4-4.

Because of the capacitive or inductive nature of the transmission facility (the copper pair), the recovered pulse train at the end of any significant length of cable would be much less distinct. That is, the rise and decay times of the facility would deform the pulse train at points of transition where the pulse changes polarity. A far-end signal might look something like the pulse train shown in Figure 4–4(c).

Since the capacitive or inductive characteristics of the copper pair do not change, it is reasonable to expect the pulse rise and decay times to have the same effect on positive and negative pulses. The pulse train is deformed or distorted in pulse shape, but positive and negative pulses (ones and zeros) still occupy identical time periods.

In addition to the distortion produced by the rise and decay times introduced by the interconnecting facility, a similar distortion will arise due to the rise and decay times of the electronic terminal devices themselves. Such devices require a finite period of time to switch completely from one state to another, such as from a positive output pulse to a negative output pulse.

The signal at the end of the cable pair might look something like that shown in Figure 4–4(c) as a result of all the contributing distorting mechanisms. The deformation of the original sharply defined pulse train might even be more pronounced than that shown. Note that however distorted the pulse train has become, since the causative rise and decay times were equally applicable to both positive and negative pulses, the time relationship between a positive and a negative pulse has not been altered.

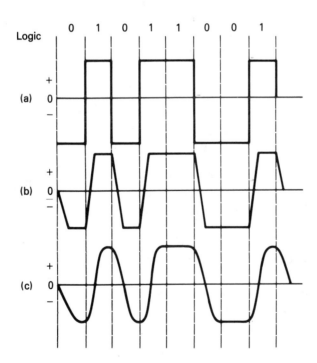

FIGURE 4–4
Bipolar Pulse
Distortion

Carrier System Pulse Distortion

In a carrier-type system such as a digital microwave link, the transmission facility cannot transport any DC signal component. We can turn the carrier on or off, or alter its amplitude or phase, to denote the presence or absence of a pulse or the amplitude of a pulse, but we cannot denote both positive and negative polarities.

The carrier itself is an alternating wave, so both polarities will necessarily always be present, and cannot be employed to represent levels of digital information. There are techniques of bipolar pulse modulation that might be applied, but the majority of carrier-type transmission systems utilize a format that relates the presence of a pulse (a digital one) to a positive voltage amplitude, and the absence of a pulse (a digital zero) to a zero voltage amplitude.

Signal formats may also be "return to zero," "nonreturn to zero," etc., and we shall discuss these at greater length in the next chapter. Note here, however, the relating of digital pulse amplitude to zero pulse amplitude, and the absence of any polarity distinction.

In a carrier-type transmission system, the carrier has only one mode of propagation. In a keyed on-off system, the durations of the detected output signals of a pulse-present period and a no-pulse-present period would be the same. The sequence may change, of course, as the intelligence transported dictates, but any *one* is the same length as any *zero*. In this respect, carrier systems exhibit the same pulse train distortion characteristics as the DC-coupled metallic conductor system did: the pulses may be distorted, but the time relationship between the presence and the absence of a pulse remains undisturbed.

Lightwave System Distortion

As in the cases of the metallic conductor and carrier systems, in lightwave systems we will also be constrained somewhat by the terminal device switching times (rise and decay times), but the mechanism for pulse distortion from the dispersion of light pulses is markedly different than in these other cases. As discussed earlier, pulse broadening in multimode graded index fibers is the direct result of the transit time differential between different modes of propagation, each with a different transmission path length. This modal dispersion may not be the only form of pulse dispersion we have to consider, but before going on, let us make certain that we understand its effects thoroughly.

In a digital lightwave system, if the light is turned on and off, a light pulse will be present for a digital one but no light pulse will be present for a digital zero. If no light is transmitted through the fiber during the off period (the zero), then there will be no modal dispersion of a light pulse during such a period, since no light rays, and consequently no modes of propagation, are present.

During the light-on periods, however, the transmitted light pulse will be broadened, as described earlier, and the duration of the pulse and no-pulse periods in the electrical output signal from the link receiver will not be the same as they were in the other systems. In effect, the output "one" pulse will intrude into the time period allocated to the output "zero" pulse condition. This is a somewhat unique and interesting difference inherent in lightwave transmission, so that the term "pulse broadening" is, in this context, peculiarly appropriate. Note that if we do not actually turn the light off entirely to indicate a no-input-pulse condition, then the phenomenon would not be evidenced: the pulse broadening was the result of light transit time differentials, and had nothing at all to do with the intensity of the light.

Thus, if we simply turn the light intensity higher and lower to denote the presence and absence of a pulse, respectively, the pulse broadening would be evidenced equally in both states of the output electrical pulse train. This technique of system operation might be called intensity modulation.

This interesting difference in lightwave transmission as compared to more conventional technologies is perhaps a bit academic. For unless the transmission data rate (bit rate) were very high, the detection of the presence or absence of a pulse would not be significantly less dependable even if there were a nominal time period differential. But our objective was to understand lightwave transmission, and this mechanism is somewhat peculiar to that technology.

MODAL DISPERSION

As explained earlier, modal dispersion is due to the transit time differential introduced by the propagation of several modes of light rays in a step index fiber structure, each mode having a different transmission path length and thus a different transit time. The mechanisms whereby pulse dispersion is introduced in graded index fibers are quite different, but we still refer to the pulse broadening that such mechanisms produce as modal dispersion.

Of course, in graded index fibers, the same transmission path length differences exist as in the step index structure, and thus the same transit time differentials *could* exist. But the propagation mechanism is refraction rather than reflection, due to the gradual radial changes in the refractive index of the core material. Thus, when light passes through the material with the lower refractive index, its velocity of propagation increases, and the time differential between several or many modes at the end of a graded index fiber is not as pronounced as in the step index structure.

SINGLE MODE FIBER PROPAGATION

As discussed in Chapter 3, a single mode fiber structure essentially limits the propagation of light to the fundamental or axial mode. This eliminates or greatly reduces any modal dispersion of the light pulses. Whatever dispersion there is will be very small and will be the product of minor low-order reflective propagation that may still occur in some fibers, but it will not introduce any pronounced transit time differential. For all practical purposes, we can say that no modal dispersion is evident in a single mode fiber. This is not, of course, to say that no other forms of distortion are introduced.

Figure 4–5 shows the structure of a single mode fiber. This structure will effectively limit the fiber to supporting only one mode of propagation, the axial mode, and it accomplishes this by a drastic reduction in core diameter. Note that the output light pulse is not substantially broader than the input pulse.

As is so often the case with technological advances, when the things that were previously limiting or restrictive get resolved, new limitations become apparent and require remedial attention. When single mode fiber first evolved, it was a dramatic improvement over multimode structures in that there was, for all practical purposes, no pulse dispersion. This situation encouraged and accelerated the application of higher transmission data rates on lightwave systems, since pulse dispersion was eliminated.

However, as the transmission bit rates got higher and higher, even the nominal pulse dispersion that previously was insignificant eventually became a limiting factor. It then became necessary to define the mechanisms of dispersion in greater detail and to reduce even further the dispersion that had previously been, compared to multimode dispersion, quite acceptable. The higher transmission data rates demanded even better performance from single mode fibers than they inherently offered at that time.

Single mode fibers take on the true function of waveguides, and the mechanisms of distortion are more complex than those shown in the previous examples. Since only a single demarcation is created between cladding and core materials in a simple single mode structure, and since each material has a different index of refraction, single mode fibers can still be defined as step index structures. But in a single mode fiber acting as a waveguide, a significant amount of the light energy actually does propagate within the cladding material. Of course, it does not propagate reflectively or refractively; rather, it propagates longitudinally, as a wave front. Figure 4–6 sketches this propagation characteristic.

Since some of the propagation is effected through the cladding material, the notion of numerical aperture (NA) is not directly applicable to single mode fibers. Instead, mode field diameter (MFD), sometimes called the spot size, is used to de-

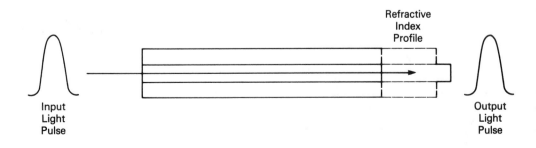

FIGURE 4–5
Single Mode
Fiber

Input
Light
Pulse

Refractive
Index
Profile

Output
Light
Pulse

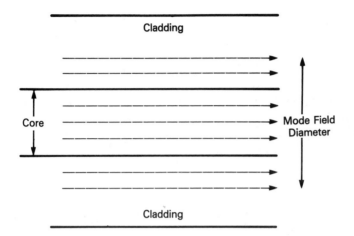

FIGURE 4–6
Single Mode
Propagation

fine the "active," or light-propagating, cross-sectional area of the fiber. Figure 4–7 depicts the relationship between mode field diameter and fiber core diameter.

Mode field diameter increases with operating wavelength, and a single mode fiber will begin to support more than one mode of propagation at some specific wavelength. This cutoff wavelength is the point in the spectrum at which and below which more than one mode will be supported. Cutoff frequency is a function of fiber geometry and material, refractive index differentials, and core diameter.

WAVEGUIDE DISPERSION

Another law of physics also comes into play at this point. Light energy of a specific wavelength will travel at a different speed through different media, that is, through media with different refractive indexes. Since, in a single mode fiber, some light is propagating through the cladding, and since the cladding does have a different refractive index than the core, then even if a single-wavelength signal were presented to the fiber, some dispersion would be incurred due to the transit time differentials through the two dissimilar materials (the core and the cladding). This phenomenon is referred to as *waveguide dispersion.*

Waveguide dispersion is wavelength dependent and may well be more pro-

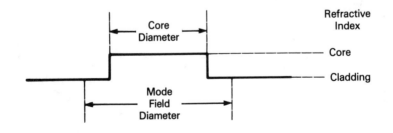

FIGURE 4–7
Mode Field
Diameter

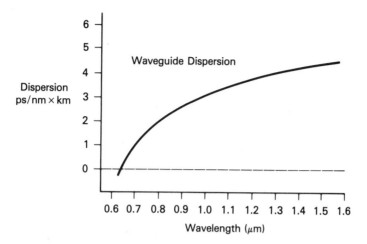

FIGURE 4-8
Waveguide
Dispersion

nounced at one wavelength than at another in a particular fiber structure. At some wavelength, the fiber may even exhibit zero dispersion. By manipulating the fiber geometry (that portion of the fiber that is propagating the light), it is possible to shift the point of minimal dispersion in a fiber to a different place within the light-wave spectrum or to flatten the dispersion curve so that less dispersion is presented across a significant range of wavelengths. Figure 4-8 shows a typical curve of wave-guide dispersion within a fiber. Note that near 6.5 micrometers the fiber introduces no dispersion at all.

Fibers whose dispersion characteristics have been altered are called *dispersion shifted* or *dispersion flattened* fibers. Obviously, we could select operating wave-lengths where a particular fiber dispersion would be least disruptive, or we could shift the dispersion in the manner discussed to accommodate the operating wave-length. In either case, a system can be optimized for minimum dispersion and thus can be conditioned for best performance at extremely high transmission data rates.

Figure 4-9 shows the geometry of a simple single mode fiber structure that has been altered by depressing the cladding index across the mode field diameter. By changing the refractive index of the cladding in the precise radial area where light will be propagated, the zero point of waveguide dispersion can be manipulated. A wide range of fiber geometries is possible, and the one shown is purely for demon-strative purposes. Basically, all such changes would involve manipulation of the cladding material index within the physical confines of the mode field diameter. This delineates the waveguide structure of the fiber.

MATERIAL DISPERSION

A second form of distortion that involves yet another law of physics is evidenced in single mode fibers. At different wavelengths, light travels at different speeds, even in a constant or uniform index medium. Thus, even in a constant medium such as

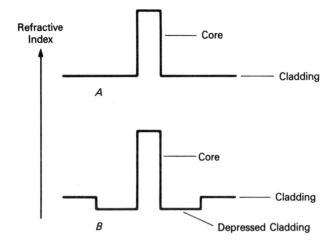

FIGURE 4–9
Dispersion-
Shifted Fiber

that presented by the fiber core material, and even if only a single mode of propagation exists, the speed of light will be different for different wavelengths.

Most light sources produce light output that contains energy at a number of
wavelengths around the designed "center" wavelength. Figure 4–10 shows the spectral output of a typical laser light source. Although the output is centered at 1.3
μm, it contains some lower level energy at wavelengths both above and below that
value. The device is said to have a finite *spectral width*. Laser diodes have spectral
widths typically on the order of 4 nm, while LED (light emitting diode) sources have
a significantly greater spectral width, on the order of 40 to 100 nanometers.

Although the fiber supports only one mode of propagation, the energy
launched into the fiber includes the several wavelengths shown in the figure. Despite
the uniformity of the core material refractive index, these different wavelengths
transit the fiber at different speeds; thus, a light pulse at the far end of the fiber

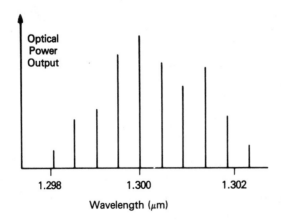

FIGURE 4–10
Spectral Width

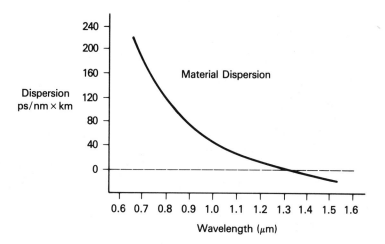

FIGURE 4–11
Material
Dispersion

will be dispersed in time, or broadened, just as the multiple reflection paths broadened the pulse in a multimode fiber. This phenomenon is referred to as *material dispersion.*

Note also that lightwave sources are not perfectly stable. A source designed to output a light pulse with a wavelength of 1,300 nm will emit other wavelength pulses part of the time. When pulse wavelengths are shorter or longer than 1,300 nm, chromatic (material) dispersion may impose significant increases or decreases in light pulse transit time. There can be as much as 100 picoseconds per nanometer of wavelength variation for each kilometer of fiber length.

Figure 4–11 shows a typical curve of material dispersion in a single mode fiber. As with waveguide dispersion, material dispersion will not be the same at all wavelengths, and it is possible to manipulate the refractive indexes of the cladding and core materials somewhat to minimize the dispersion. It is also possible to select operating wavelengths that coincide with the minimum dispersion wavelength of a particular fiber.

Material dispersion will be dramatically reduced (and transmission bit rates will increase) when single-frequency light sources are developed that emit a single wavelength of lightwave energy that is very stable in frequency.

SINGLE MODE FIBER DISPERSION

The two forms of pulse dispersion that may be introduced in a single mode fiber are material and waveguide dispersion. Common usage sometimes combines both these effects in the single term *chromatic dispersion,* which at least clearly denotes the wavelength dependency of both phenomena. Modal dispersion, which is a function of transit time differentials between propagation modes (be they reflective or refractive), is not predominantly wavelength dependent, but chromatic dispersion is.

Within the optical fiber alone, the bandwidth-limiting restrictions that are related to pulse dispersion mechanisms are modal dispersion, waveguide dispersion, and material dispersion. The last two are sometimes referred to as chromatic dispersion.

When lightwave transmission is applied in systems that are shorter in length or have lower transmission data rate requirements, then either single mode or multimode fibers may be considered, from a strictly technical point of view. Such applications might often be encountered in distribution or exchange level telephone plant, or in limited density rural telephone network trunking.

For higher bit rate systems (higher traffic density applications) over long distances between two discrete points, single mode fiber is almost mandatory. If very long, repeaterless fiber spans are to be achieved, then the most refined techniques of dispersion shifting may be required. Such applications might be found in submarine cables connecting continents or islands, or in systems between major population centers.

As fiber attenuation is reduced through research and development efforts, the limiting factor in system design tends to shift from transmission levels (fiber loss) to distortion (fiber dispersion), and this is more pronounced in higher bit rate systems. The overall quality of system transmission is less dependent upon merely sensing the presence or absence of light at the receiver input, being much more affected by the broadening of the light pulses as they traverse the system. It becomes increasingly less certain at higher data rates that the absence or presence of a pulse can be reliably detected. This uncertainty may be evidenced in higher error rates in the electrical output of the receiver.

SUMMARY

In waveguides, the energy does not flow through the material that constitutes the waveguide structure as through an electrically conducting material; rather, it is propagated as a wave through the space enclosed by the conductive waveguide walls.

In lightwave transmission, the fiber functions as a waveguide for the light energy. The manner in which the light is confined within the two basic types of fibers (step index and graded index) involves distinct optical principles.

In a multimode step index fiber, the light is confined within the core material by total internal reflection introduced by design at the core-cladding interface. In a single mode step index fiber, the light propagates longitudinally through both the core and the cladding material. In the graded index fiber, the light is bound not by internal reflection at all, but by refraction. This is introduced by the radially varying refractive index of the core material itself.

The two major considerations in addressing the transmission of lightwave energy through an optical fiber are transmission loss and signal distortion. Impurities in the fabricated glass from which the optical fiber is drawn introduce absorption of light energy within the fiber. Other sources of attenuation may be variations in the size of the fiber core, irregularities in the interface between the core and cladding materials, and the microbends which sometimes occur during the process of cabling fibers. The fibers may be stressed over small imperfections in the cabling material itself during the cabling process.

In a single mode fiber acting as a waveguide, light energy propagates within the cladding material. It does not propagate reflectively or refractively, however, but propagates longitudinally as a wave front. Since some of the propagation is effected through the cladding material, the notion of numerical aperture (NA) is not directly applicable to single mode fibers. Instead, the mode field diameter (MFD), sometimes called the spot size, is used to define the ''active,'' or light-propagating, cross-sectional area of the fiber.

Waveguide dispersion is wavelength dependent and may well be more pronounced at one wavelength than at another in a particular fiber structure. At some wavelength the fiber may even exhibit zero dispersion. By manipulating the fiber geometry (that portion of the fiber that is propagating the light), it is possible to shift the point of minimal dispersion in a fiber to a different place within the lightwave spectrum or to flatten the dispersion curve so that less dispersion is presented across a significant range of wavelengths.

REVIEW QUESTIONS _____

True or False?

T   F   1. In all commercial optical fibers, the core material has a higher refractive index than the cladding material.

T   F   2. One form of attenuation in optical fibers is due to absorption of light energy in the fiber material itself.

T   F   3. All fibers exhibit pronounced peaks of attenuation, and these cannot be reduced in the fiber fabrication processes.

T   F   4. Mechanical stress imparted to the fiber during the cabling processes will reduce the tensile strength of the fiber, but will not introduce any additional optical attenuation.

T   F   5. Fiber attenuation is generally specified as a power loss ratio in units of dBm/km.

T   F   6. When the fiber attenuation due to the physical fiber length is excessive, a regenerative intermediate repeater may be required.

T   F   7. In most commercial lightwave transmission systems, bipolar pulse transmission is utilized just as it is in metallic conductor systems.

T   F   8. Pulse broadening is a form of distortion which produces an unwanted extension of the pulse duration.

T   F   9. Modal dispersion is introduced by the transit time differential of different wavelength optical signals passing through the same optical fiber.

T   F   10. A graded index fiber will generally exhibit a cutoff frequency which is approximately twice the frequency of the system's operational wavelength.

T   F   11. A significant amount of modal dispersion is introduced to signals passed through a single mode fiber.

T   F   12. In optical fibers employed in telecommunications systems, no light energy propagates through the cladding material of a single mode or a multimode fiber.

T   F   13. The optical mechanism most important for supporting the propagation of light energy in a single mode fiber is refraction.

T   F   14. Dispersion-shifted fibers will evidence significantly less optical attenuation, and this is why they are usually used in very long fiber links.

T   F   15. The velocity of propagation of light through any specific fiber is the same for all lightwave frequencies.

T   F   16. The spectral width of a light source is the light energy present in the device output that is at any frequency other than the center or design operating frequency of the device.

T   F   17. A single mode fiber will pass lightwave energy of only one specific frequency.

T   F   18. Chromatic dispersion is a phenomenon unique to multimode fiber systems.

T   F   19. Higher transmission bit rate systems require that pulse dispersion from any cause be strictly controlled and limited.

T   F   20. Higher transmission bit rates require more transmission system bandwidth.

CHAPTER 5

Lightwave Terms and Factors

Any transmission system is composed of several discrete elements. In lightwave transmission systems, these can be identified as terminal devices, a transmitter at one end and a receiver at the other, interconnected by an optical fiber. The nature of each of these individual elements is quite different. For example, terminal devices such as transmitters and receivers are power-consuming active units, whereas the interconnecting fiber is a passive transmission medium. It follows that the manner in which we define the characteristics and performance of these various system elements will vary also.

The *American Heritage Dictionary* (New College Edition) defines the word *system* as "a group of interacting, interrelated, or independent elements forming, or regarded as forming, a collective entity." In designing lightwave systems, we are obliged to assemble the various elements into a system in a manner that ensures overall performance that is responsive to our requirements, as well as cost effective.

To do this successfully, we must use a number of definitive terms, some of which may appear to be unrelated. We must also understand these terms and their relationship to each other. Finally, we must state our transmission requirements and all equipment or material specifications in terms that are unambiguous and that are correctly related to all other terms that we employ.

SIGNAL-TO-NOISE RATIO AND BIT ERROR RATE

An optical fiber is immune to many of the noise influences that metallic conductors suffer from, such as electromagnetic coupling and transient signals. Wideband white noise can be introduced into lightwave systems through the transmitters and receivers, however. Because the lightwave receiver is usually working with very low-level input signals, it is the principal source of noise in lightwave systems. Careful circuit board design and judicious selection of the photodetector are essential in minimizing the noise the receiver will introduce into the system.

In speaking of lightwave systems, it is necessary to establish a term that relates system noise contribution to the transported intelligence itself; in analog systems the term employed is *signal-to-noise ratio* (SNR). This has the same meaning in any analog system, whatever the transmission medium may be. It is the difference between the amplitude of a signal (before modulation or after detection of a modulated carrier) and the noise present in the spectrum occupied by the signal, when both are measured at the same point in a system.

In digital systems, we use the term *bit error rate* (BER). This is defined as the ratio between the transmitted bits detected in error and the total number of bits transmitted. Thus, a BER of 10^{-9} simply means that for every billion bits received, one bit was detected incorrectly.

Note that where an SNR relates primarily to the white noise contribution of the terminal equipment, the system BER is not a function of noise alone, but also reflects pulse distortion (dispersion). This distortion may be due to several causes, including, in lightwave systems, modal dispersion and chromatic dispersion (consisting of material and waveguide dispersion) of the light energy pulses as they transit the optical fiber.

As a general rule, lightwave receivers are specified with input optical signal level referenced to the electrical output SNR in the case of analog systems, and BER in digital applications. For example, a manufacturer might state that a specific receiver will produce a BER of 10^{-9} with a minimum optical input signal level of -40.5 dBm. If the system designer is prepared to accept a BER of 10^{-9}, he or she must assure through the design that the optical input level to the receiver does not fall below -40.5 dBm. If the input level should fall below this level, we expect that the output electrical signal quality may be poorer than a 10^{-9} BER.

It is also possible to overdrive a receiver with too high an optical signal input level, resulting in some degradation of quality, but we shall discuss this possibility when we examine lightwave terminal equipment later in the text.

DATA RATE AND DIGITAL SIGNAL FORMAT

The *bit*, the fundamental unit of information in a digital system, may represent either a logic one or a logic zero. In digital electronics, bits are represented by pulses, so that a description of an individual bit must include rise and fall times, pulse width or period, and some format designating when a single pulse is associated with other pulses to form digital information.

Data rate, sometimes referred to as *bit rate*, is the number of bits processed in a given period of time. Since each individual bit contains information, and bits can be formatted to represent intelligence, the data rate relates to how much intelligence can be processed or transported in a given period of time.

Data rates are typically expressed in bits per second (b/s), kilobits per second (kb/s), megabits per second (Mb/s), or gigabits per second (Gb/s). Higher level bit rates mean faster processing or transmission, that is, more intelligence handled per unit of time.

Bit rate is influenced by format, which is the particular method of encoding bits for serial transmission. In Figure 5–1, three popular formats are shown: non-return to zero, return to zero, and Manchester code. Note that the bit period, i.e., the time interval allotted for each bit, is given to be 0.5 μs. This period corresponds to a data rate of 2 Mb/s, since a full second could accommodate two million such bits. Note that the minimum pulse period is not necessarily the same as the bit period for every format.

Examine the non-return to zero (NRZ) signal in the figure. The signal level changes only when the bit logic level shown at the bottom changes. The signal voltage goes high for a logic one and low for a logic zero. If a series of identical bits is transmitted (all ones or all zeros), the signal voltage remains at the same level, high for ones and low for zeros. Thus, for NRZ data, the bit period and the minimum pulse period are the same. As the figure shows, both periods are 0.5 μs, which corresponds to a 2 Mb/s data rate.

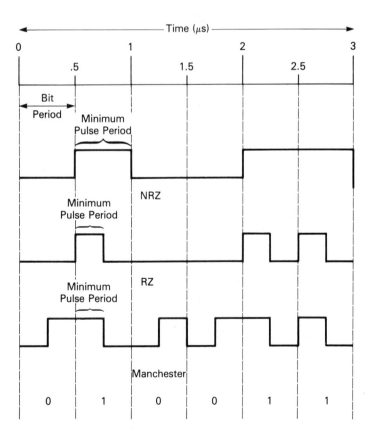

FIGURE 5–1
Digital Signal Format

The transmission bandwidth required to accommodate a 2 Mb/s signal is equal to the reciprocal of the minimum pulse period, but we must specify how bandwidth is measured. Transmission bandwidth is generally accepted to be the spectrum of the device or facility within which range the amplitude of a signal is no lower than half the amplitude of the maximum signal level present within that same spectrum. Since half-power is denoted as 3 dB, we often refer to the 3-dB bandwidth of a device or system.

In the case of the NRZ signal shown in Figure 5–1, the bandwidth required to accommodate a 2 Mb/s signal may be calculated as

$$B = \frac{1}{t} = \frac{1}{0.5} = 2 \text{ MHz}$$

where

B = Bandwidth in megahertz
t = Pulse period in microseconds

For the RZ formatted signal in the same figure, the signal voltage goes to a high level and returns to a low level for each logic one bit, and the minimum pulse period is 0.25 ms. Thus, a system using RZ encoded signals must be able to operate twice as fast as one using NRZ signals, since it must respond correctly to a pulse period that is one-half as long as the pulse period of the latter to handle the same data rate. Applying the previous formula, a pulse period of 0.25 ms requires a bandwidth of 4 MHz:

$$B = \frac{1}{t} = \frac{1}{0.25} = 4 \text{ MHz}$$

Note that, as with NRZ signals, a series of identical bits transmitted (all ones or all zeros) would produce no change in signal voltage level.

The Manchester code shown in Figure 5–1 represents a logic one bit by a transition from high signal voltage to low signal voltage, and a logic zero bit by a transition from low voltage to high voltage. Hence, since two voltage levels must be handled for every bit regardless of its logic value, the minimum pulse period for this signal is 0.25 ms, as it was for the RZ format. The required bandwidth is again 4 MHz.

Both RZ and NRZ codes require a separate synchronizing clock signal since the format allows signal voltage levels to remain constant through several bits in some cases—strings of consecutive ones or zeros, for example. If the signal voltage level remains constant for any extended period, then the periodic change in level that constitutes the synchronizing characteristic of the signal would not be present, and the receiver could not recover timing directly from the data received.

Because the Manchester code requires two voltage levels within each and every bit period, regardless of whether the bit logic value is one or zero, synchronization (timing) is apparent in the coded signal. The signal is said to be *self-clocking,* i.e., timing can be recovered from the signal itself. This simplifies receiver design somewhat.

PULSE BIPOLAR CODING

Another coding technique for lightwave transmission systems is *pulse bipolar* coding. This is not quite the same thing as the more familiar bipolar signals in metallic conductor systems. In pulse bipolar systems the light source (the transmitter) is biased to about half its peak-to-peak drive current, which establishes a quiescent level of light energy output. For each one bit, a positive level of light energy is generated relative to the quiescent level. For each zero bit, a negative level is generated, again relative to the quiescent level. Figure 5–2 shows an NRZ signal encoded in a pulse bipolar format.

A three-state scheme of this type offers a couple of advantages: the receiver design is simplified because of the constant mid-level quiescent signal, and clocking recovery is relatively easy. On the other hand, the technique suffers some disadvantages if the duty rate varies substantially from 50 percent, which is the case when an equal number of ones and zeros are being transmitted. Also, systems with burst-mode transmission, i.e., in which data transmission occurs in bursts separated by

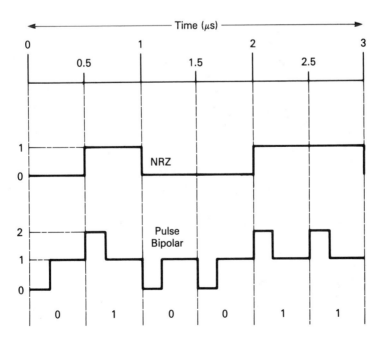

FIGURE 5–2
Pulse Bipolar
Format

periods of inactivity, may present problems, since such transmission requires some sustained data input to reestablish the mid-level reference signal.

OTHER CONSIDERATIONS

Other considerations are involved in ensuring a compatible interface with digital terminal equipment—recovering data timing in the electrical output signal, for example, or pulse stuffing and destuffing. But these functions and others are not directly associated with the generation, modulation, propagation, and recovery of the lightwave signal itself, and as such, are beyond the scope of this text. Before leaving the subject, however, two basic observations are worthwhile.

First, we can expect to find different encoding formats used in different systems, and even within sections of a single system. At link terminals, for example, some sophistication may be required in the equipment to convert the recovered electrical output signal from the light receiver to some other, more compatible electrical signal format, such as Binary Three Zero Substitution (B3ZS) or Binary Six Zero Substitution (B6ZS), that the digital terminal equipment demands.

Second, different formats are commonly employed in digital transmission systems to protect against the problems introduced by strings of consecutive ones or zeros in the transmitted information. These formats are beyond the scope of our discussion, but the interested reader may pursue them further in any text addressing digital transmission or digital telephony.

RISE TIME

Rise time may be defined as the time period required for the instantaneous amplitude of a pulse to go from 10 to 90 percent of peak pulse amplitude. In Figure 5–1 the pulses are shown as clean and sharply defined, a somewhat ideal condition. In actual systems, depending upon how rapidly a device can be turned on and off or effect a transitional level change, the output pulses, be they electrical or lightwave, will be distorted to some extent.

The speed of operation of a light source or detector is often specified in terms of rise time. For example, laser sources have very short rise time characteristics— on the order of 1 ns or less, typically. Rise time for light emitting diodes (LEDs), on the other hand, might range from 3 ns to over 100 ns. These figures refer to the lengthening of the electrical or optical output pulse in relation to the length of the electrical or optical input pulse. In a lightwave transmitter, the input is electrical and the output optical; in a receiver, the opposite is true.

The electrical pulse of Figure 5–1 had a period of 0.5 ms, which is 500 ns. If this were the electrical input pulse to a transmitter with a rise time of 50 ns (such as an LED), the light (optical) output pulse would be 550 ns.

In Figure 5–3(a), we see just such an input pulse with the lengthened pulse (550 ns) shown in Figure 5–3(b). The scale is exaggerated to emphasize the time

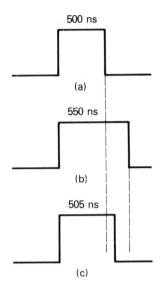

500 ns

(a)

550 ns

(b)

505 ns

FIGURE 5–3
Pulse Dispersion   (c)

relationship, and the drawing does not show the pulse distortion or rounding that would also be introduced. In Figure 5–3(c), we see the light output pulse from a source whose rise time is only 5 ns, perhaps a laser source.

Similar pulse distortion would be evidenced in a receiver, except that the input would be optical and the output electrical. Keep in mind that as the bit rate gets higher, the pulse period gets shorter. However, the device rise times do not change, since they are fixed by device design and geometry. Thus, as the bit rate increases, the pulse dispersion (broadening) becomes a more significant portion of the shorter pulse duration time.

Whereas rise time is the time required for a device to change the state or level of its output signal in response to a change in its input signal, dispersion is the direct broadening in time of a pulse period. The two phenomena are closely related and sometimes confused, but they are not identical.

Some pulse broadening or dispersion experienced in lightwave signal pulses that pass through an optical fiber is entirely independent of the transmission bit rate or pulse period. Recall that material and waveguide dispersion (chromatic dispersion) are strictly functions of wavelength alone in single-mode fibers. Although dispersion is measured in units of time as a lengthening of the pulse period, it cannot be strictly defined as rise time, which is also measured in units of time.

When the electrical interfaces with the lightwave system or link are electrical pulses, as in a digital application, it is common practice to use the term "rise time" to denote all pulse distortion as it is presented in the electrical output pulse. We may also encounter the term "System Rise Time Budget," which is used to help characterize the transmission performance of a link, but we should recognize the subtle distinction between dispersion and rise time.

We must be concerned with the system overall (end-to-end) distortion, since we must detect and recover the initial input pulses at the system output with as few errors as possible. We are obliged to make a selection of terminal equipment and fiber types; in order to do this, we must establish, for digital systems, the overall dispersion we are prepared to accept as usable.

Due to rise time, the terminal equipment will contribute some distortion to the overall system performance, and the passage of lightwave energy through the interconnecting fiber itself will introduce pulse dispersion, thus contributing to the overall system distortion. Also, the BER we are prepared to accept for system performance has an impact on the bandwidth and dispersion requirements. A narrower bandwidth, which might be equated to longer rise times, would introduce more pulse distortion. In such cases, the positive detection of the presence or absence of a pulse would be less certain, i.e., the bit error rate may be degraded. Conversely, if we could accept a higher number of errors, we might relax the bandwidth or rise time/dispersion requirements.

Thus, we must establish a system overall rise time requirement taking BER into account. We must determine the rise time contribution from each system transmitter and receiver, and then determine the dispersion contributed by the fiber. Finally, we must combine these rise times in some rational manner to determine the overall (end-to-end electrical transmission path) rise time that will be introduced. If this figure is compatible with the previously established requirements, then we should have an adequate, acceptable system.

Calculating Rise Time

For NRZ electrical pulses, experience has shown that for acceptable transmission performance, pulse rise time should be no more than 70 percent of the bit period (pulse period) itself. Reflecting this, the actual rise time requirement in nanoseconds for a given NRZ bit rate can be determined by dividing 700 by the transmission bit rate. In Figure 5–1, for example, the NRZ bit period was given to be 0.5 microseconds. This corresponds to a data rate of 2 Mb/s from

$$\text{BR} = \frac{1}{t} = \frac{1}{0.5} = 2 \text{ Mb/s}$$

where BR is the transmission data rate (Mb/s) and t is the pulse period in microseconds.

The acceptable system rise time for a 2 Mb/s NRZ signal can be determined from

$$\text{RT} = \frac{700}{\text{BR}} = \frac{700}{2} = 350 \text{ ns}$$

where RT is the rise time in nanoseconds and BR is the bit rate in Mb/s. Thus, if the end-to-end system rise time does not exceed 350 ns, then the system will satisfactorily transmit data at a 2 Mb/s rate.

For a higher data rate, say 45 Mb/s, the acceptable rise time would be much shorter for an NRZ signal, from

$$\frac{700}{45} = 15.5 \text{ ns}$$

While 70 percent is an acceptable factor for NRZ signals, for RZ signals experience indicates that the rise time should be no greater than 35 percent of the bit period, because of the shorter pulse period. For RZ applications, we divide 350 by the transmission bit rate in Mb/s to establish the required rise time in nanoseconds.

For a 2 Mb/s RZ signal, the acceptable rise time would be

$$\frac{350}{2} = 175 \text{ ns}$$

For a 45 Mb/s RZ signal, the acceptable rise time would be

$$\frac{350}{45} = 7.77 \text{ ns}$$

This shows quite clearly that the rise time requirements are more severe, i.e., a shorter rise time is required for higher transmission data rates.

From this discussion, it is plain that if the transmission bit rate can be held to a lower level, the bandwidth or rise time requirements will be less severe. This might translate into lower cost fiber or less sophisticated, less expensive terminal equipment. We might even consider system configurations that permit lower transmission bit rates if possible, and we shall explore this idea further in a subsequent chapter.

BANDWIDTH

Since terminal equipment consists of discrete units, we can define and measure the electrical bandwidth of such units. The electrical bandwidth is usually specified in Hertz and generally defined in both equipment and systems to be the spectrum provided for use to the half-power points. This means that no signal at any frequency within the bandwidth of the unit or system will be attenuated more than 3 dB (such reduction representing half the original input power) more than the signal within that bandwidth that is attenuated the least, or is the highest in amplitude. The signal least attenuated is usually at a frequency in the center of the spectrum identified as the bandwidth.

The optical bandwidth of the fiber, on the other hand, cannot be defined without taking into account the physical length of the fiber. We define the optical bandwidth as the spectrum presented for use by the physical interconnecting facility—the fiber alone—separate and apart from the lightwave terminal devices. For example, a particular fiber of indeterminate length may be specified to present an optical bandwidth of 100 MHz end-to-end. This simply means that no high frequency components of any information carried through that fiber, up to 100 MHz in frequency, will be lower in amplitude than 3 dB below the highest amplitude components of the transported information that is recovered at the end of the fiber.

As a point of interest, since the fiber is only one of three system elements, the other two being the transmitter and receiver, and since all three elements may affect the system bandwidth, the optical (fiber) bandwidth of a system may not be the same as the information bandwidth the system presents for use.

It is obviously necessary to establish the required system bandwidth. Once this requirement is known, we have only to assure that no single system element presents less than this required bandwidth. Each discrete element—the transmitter, the receiver, and the fiber—must meet or exceed the system requirement.

Calculating Bandwidth

The required fiber optical bandwidth for a given transmission bit rate (assuming NRZ format) may be calculated approximately by using the factor 0.9 as

$$\text{Bandwidth (optical)} = \text{Bit Rate} \times 0.9$$

For example, if the required transmission bit rate for an NRZ signal is 45 Mb/s, the optical bandwidth required of the fiber itself (end to end) is $45 \times 0.9 = 40.5$ MHz.

Conversely, if the optical bandwidth of a given length of a fiber is known, then the pulse-dispersion-limited transmission bit rate through that particular fiber may be calculated approximately as

$$\text{Transmission Bit Rate (Mb/s)} = \frac{\text{Optical Bandwidth (MHz)}}{0.9}$$

Since the reader may encounter this notation in some material reviewed, it is helpful to understand this relationship, but in general practice the calculation is seldom required or performed. It is commonly accepted that the optical bandwidth relates directly to transmission bit rate for NRZ signals, and the required bandwidth is simply taken to be the required bit rate. For example, transportation of a 90 Mb/s signal (NRZ) is taken to require 90 MHz of usable information bandwidth and 90 MHz of optical bandwidth end to end through the interconnecting fiber. Although this is admittedly a conservative approach, it does not penalize the fiber unduly either technically or economically, and it does simplify system design.

BANDWIDTH AND RISE TIME

System calculations can be developed using either bandwidth or rise times, and some confusion can be introduced if the relationship between the two terms is not understood. It is not uncommon to find bandwidth used in fiber specifications, but rise time is more often used in terminal equipment descriptions. We can define systems and specify requirements for fibers and terminals using either term.

If all system elements (transmitter, receiver, and fiber) are specified in rise time and dispersion, we can calculate the system rise time by squaring all the individual element rise times and adding the squared quantities together. The square root of the sum of the squares will then be the root mean square (RMS) system rise time. This RMS value is often multiplied by 1.1 as an arbitrary system degradation factor.

For example, given a transmitter rise time of 1.5 ns, this would be squared to be 2.25, and a receiver rise time of 4.76 ns squares to be 22.66. Similarly, if the total fiber dispersion were known to be 5.92 ns, this would be squared to give 35.05. The system rise time is then calculated as

$$RT_{Sys} = 1.1 \sqrt{RT_{Trans}^2 + RT_{Recvr}^2 + RT_{Fiber}^2}$$

$$= 1.1 \sqrt{1.5^2 + 4.76^2 + 5.92^2}$$

Sum of the squares = 2.25 + 22.66 + 35.05 = 59.96

The square root of 59.96 = 7.57

7.57 × 1.1 = 8.30 ns

For some period of time it was generally accepted industry practice to prepare, for every digital lightwave link or system, a rise time budget in the manner just described. The end result was a predicted overall system performance to be expected. But some system elements are specified by manufacturers in bandwidth units alone, and bandwidth is also a more familiar term to many people in the telecommunications industry. So the preparation of rise time budgets for applications has fallen into disuse, and it is now common practice to simply call out bandwidth for all system elements and for overall system performance.

Engineers engaged in the design of terminal equipment such as lightwave transmitters and receivers must address rise time in developing their designs. However, those people designing systems or links are not so constrained and may work entirely in bandwidth if it simplifies the process. The end result, applying both techniques correctly, will produce acceptable system performance predictions.

It may be useful, on occasion, to convert one term to another. Dispersion can be converted to an approximate 3 dB optical bandwidth in a single mode fiber, assuming Gaussian response, by applying the formula

$$B = \frac{440,000}{T \text{ (ps)}}$$

or

$$B = \frac{440}{T \text{ (ns)}}$$

where

B = Link bandwidth (MHz)

T = Link dispersion (ns or ps)

Table 5–1 shows both the optical bandwidth required and acceptable rise times for RZ and NRZ signals for several transmission bit rates that are frequently encountered.

SYSTEM OR LINK DISPERSION

We can calculate the dispersion that will be introduced into a particular link, but we must first introduce a new factor. Earlier, in Figures 4–8 and 4–11, we showed material and waveguide dispersion characteristics for a typical single mode fiber. Recall that dispersion is linearly sensitive not only to fiber length, but also to light source spectral line width. Thus, to calculate the expected dispersion from a particular installation, the laser spectral line width must be known.

If we wish to specify the spectral output of a light source, we must establish a limiting optical power level of the output lightwave energy that will be considered to be significant as spectral output. The term usually employed is full width half maximum (FWHM), which refers to the difference between the wavelengths at which the magnitude of the lightwave energy has dropped to half its maximum value.

Consider Figure 5–4, which shows a light source spectral output that has lightwave energy at 1,298 nm and 1,302 nm that is in each case 50 percent of the maximum amplitude lightwave energy output, at the design wavelength of 1,300 nm. The

TABLE 5–1
Rise Time
and Optical
Bandwidth

Digital Signal	Bit Rate (Mb/s)	Return to Zero (RZ)		Non-return to Zero (NRZ)	
		Acceptable Rise Time (ns)	Required Bandwidth (MHz)	Acceptable Rise Time (ns)	Required Bandwidth (MHz)
DS 1	1.544	226	1.544	453	0.77
DS 1C	3.152	111	3.152	222	1.58
DS 2	6.312	55.4	6.312	110	3.16
DS 3	44.736	7.82	44.736	15.6	22.38
DS 3C	90.148	3.88	90.148	7.76	45.07
DS 3D	135.532	2.58	135.532	5.16	67.77
DS 4	274.176	1.28	274.176	2.55	137.09

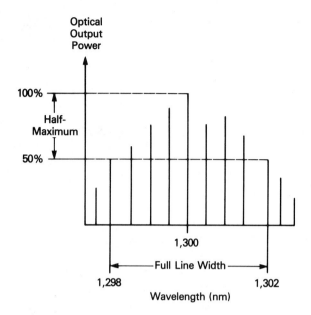

FIGURE 5-4
Full Width Half
Maximum
(FWHM)

full width half maximum (FWHM) spectral line width of this source would be said to be 4 nm since that is the length of the spectrum between the half-power points.

Note that although the term "FWHM" implies a light source that is not monochromatic, that is, a source that has significant lightwave energy output at several wavelengths, even a monochromatic source with a single wavelength output may experience chromatic dispersion. If the source operating wavelength drifts or changes, then the end result is several different wavelengths of light propagating through the fiber at different times, each incurring a different dispersion. Laser line width is specified for all sources, even though they may exhibit monochromatic or nearly monochromatic output.

OPTICAL FIBER BANDWIDTH

The bandwidth presented for use by an optical fiber is a function of the physical length, as well as the characteristics of the fiber. Thus, the units of fiber bandwidth are Megahertz (MHz) times kilometers (km), or MHz-km.

In single mode fiber calculations, bandwidth is a linear function of length. Thus, a fiber specified to provide 250 MHz-km would provide 50 MHz of usable bandwidth through a 5 km length of fiber, calculated as follows:

$$BW_{(system)} = \frac{250}{5} = 50 \text{ MHz}$$

Conversely, if we required 50 MHz of usable bandwidth in a single mode fiber 5 km long, the bandwidth requirement for the fiber would be

$$BW_{(fiber)} = 50 \times 5 = 250 \text{ MHz–km}$$

But bandwidth is not a strictly linear function of length in multimode optical fibers. When a fiber is propagating many modes, each mode behaves as a relatively independent transmission channel. The measured system bandwidth (or pulse broadening) is just a measure of the lack of equal delay times among these independent channels.

Fibers of equal bandwidth usually do not have identical modal delay characteristics. In constructing systems, several different fibers are often connected together end to end. Joining fibers together in this manner connects the independent transmission channels in one fiber to the channels in another. The resultant system modal delay characteristics will depend upon the modal characteristics of each fiber. It is entirely possible for one fiber to almost or completely cancel the delay characteristic of another fiber in one or more transmission channels.

If it actually did so in all channels, and the cancellation was 100 percent, then we would have no overall modal dispersion in the system at all. Of course, this doesn't happen, but the statistical nature of the fiber characteristics does result in a length dependence of bandwidth somewhere between linear and square root, depending upon many factors. A generally accepted length dependence factor that reflects this phenomenon is $L = 0.8$.

If bandwidth were truly a linear function of fiber length, then the bandwidth requirement for a 5 km length of fiber carrying a signal requiring 50 MHz of usable bandwidth would be 5 km × 50 MHz = 250 MHz-km, the same as for a single mode fiber. Because of the statistical nature of multimode propagation, however, the length dependence is L. Thus, the effective length of the 5 km system is 5 km raised to the 0.8 power, which becomes 3.62 km. Since the effective length is shorter than the actual length, the bandwidth required of the fiber to carry the 50 MHz signal would be lower also.

We can establish the required fiber bandwidth by multiplying the effective length by the required system bandwidth as follows:

$$BW \text{ (fiber)} = BW \text{ (system)} \times \text{Effective fiber length}$$

$$BW \text{ (fiber)} = 50 \text{ MHz} \times 3.62 \text{ km}$$

$$BW \text{ (fiber)} = 181 \text{ MHz-km}$$

Note that the actual physical length, which was 5 km, produces an effective length of 3.62 km from $5^{0.8}$. Note also that the 181 MHz-km fiber specification is less stringent than the 250 MHz-km required earlier. In multimode fibers, because of the length dependence factor L, the system bandwidth is effectively increased, and thus the transmission capacity is also. The length dependence factor applies only to multimode fibers.

HIGH BIT RATE SYSTEMS

The introduction of lightwave transmission into very high traffic density situations, such as those typical in trunking between major population centers, has encouraged the development of very high transmission bit rate systems on the order of hundreds of megabits per second and even up to gigabit rates. At these very high bit rates, even high grade single mode fibers can introduce significant pulse dispersion. These levels of dispersion could be, and were, ignored at lower bit rates, but even nominal dispersion can be cumulative to the point of significance at high bit rates.

This situation has led to the use of more sophisticated fiber specifications. In lower bit rate systems, the rise times addressed were adequately expressed in nanoseconds per kilometer (ns/km). Using single mode fibers at higher bit rates or in very long systems, we must consider chromatic dispersion, and with high grade fiber the dispersion will be relatively small, but it still becomes a significant part of the pulse period. The achievable rise time in good single mode fiber can be very short, on the order of 10 picoseconds or so, so the units generally employed in these applications are picoseconds per kilometer (ps/km) rather than nanoseconds per kilometer (ns/km).

Another problem is encountered at these higher bit rates as well. In our earlier discussions on waveguide and material dispersion (chromatic dispersion) in single mode fibers, we saw that both forms of distortion were frequency dependent. This was clearly evident in Figures 4–8 and 4–11. But then, any calculation of fiber dispersion must incorporate some consideration of the light source (transmitter) output spectral width, since the source will be launching light energy at more than one wavelength. The result, from a source with greater spectral width, that is, more wavelengths emitted, will be more pronounced dispersion of the transmitted pulses.

A great deal of research work is being done to develop monochromatic light sources which have a single—or nearly single—wavelength output. Systems using such sources would incur less dispersion, thereby permitting even higher transmission bit rates or wider bandwidth as it were.

For greater precision in calculating bandwidth, a unit that references the spectral bandwidth of the light source itself, as well as the wavelength at which it is operating, is required. The unit currently employed is picoseconds per nanometer-kilometer (ps/nm-km), which denotes the rise time per fiber kilometer for a particular wavelength. Use of this unit reflects the fact that several wavelengths may be propagating through the fiber, and that each individual wavelength may introduce its own dispersion per kilometer of fiber length. Thus, we must consider the dispersion contribution from several signals of different wavelength.

The unit is supported by a specification that identifies the spectral bandwidth of the transmitting light source. It is inadequate to simply call out the source output as being a signal at 1,300 nanometers; rather we need to define the source output in terms of a range, as, for example, a 1300 ± 20 nanometer signal. This specification, in nanometers, is, as discussed earlier, the full width half-magnitude (FWHM) spectral width, which is somewhat analogous to half power bandwidth.

One method of calculating single mode end-to-end system dispersion incorporating all these factors is

$$T = \frac{t \times \text{FWHM} \times L}{1,000}$$

where

T = Rise time end-to-end (ns)

t = Fiber dispersion (ps/nm-km)

FWHM = Laser line width (nm)

L = Link length (km)

As an example, assume an operating wavelength of 1,300 nm with a fiber specification of 5 ps/nm-km, a laser line width of 4 nm, and a link length of 40 km. The dispersion we would incur in such a link would be

$$T = \frac{t \times \text{FWHM} \times L}{1,000}$$

$$= \frac{5 \times 4 \times 40}{1,000}$$

$$= 0.8 \text{ ns or } 800 \text{ ps}$$

Since we are discussing single mode fibers, by selection and mixture of the raw materials of the fiber during fabrication, we can manipulate the chromatic (material) dispersion of the fiber for a selected wavelength. In this manner, refractive index, and hence the transit time, and hence the dispersion will be different for any other wavelength than the one optimized. So the material dispersion will also be different at other wavelengths.

This phenomenon does not preclude using the fiber to transport several lightwave signals of different wavelengths simultaneously, as might be done by employing wavelength division multiplexing. Although the different wavelengths will experience different dispersions, as they may also have different attenuations, we could calculate the performance for each individual signal (each wavelength).

OPTICAL TRANSMISSION LEVELS

Since we are obliged to select both lightwave transmitters and receivers for use in specific applications, we must be able to establish and specify the optical transmission signal levels with which these devices must generate or operate satisfactorily.

TABLE 5–2
Power Levels

Milliwatts	dBm
10.00	+10
3.98	+6
1.99	+3
1.00	0
0.50	−3
0.25	−6
0.10	−10
0.01	−20
0.001	−30
0.0001	−40

A lightwave transmitter outputs lightwave energy which is coupled into the interconnecting optical fiber. The transmitter output signal levels are usually specified in terms of optical dBm where decibels are referenced to 1 milliwatt (0 dBm = 1 mw).

Referencing transmission levels in this manner rather than expressing them as power in watts or milliwatts greatly facilitates system calculations, since transmission losses are specified in dB. Thus, all calculations of level can be accomplished by simple addition and subtraction.

Transmission losses are specified in negative decibels (-dB), of course, but transmitter output may also be specified as a negative value of dBm. This would indicate a transmitter whose lightwave energy output was some level below the 1 mw power level. Some manufacturers may specify transmitter output as so many milliwatts of power. This may be awkward or confusing in calculating system levels, but the conversion of one term to another is straightforward. Given power in milliwatts, the equivalent in dBm is given by the formula

$$\text{Pwr (dBm)} = 10 \log \text{Pwr (mw)}$$

Table 5–2 gives some representative signal levels in milliwatts that might be encountered in lightwave systems, along with their equivalent values in dBm.

Readers less familiar with decibels and decibel-milliwatts may find Appendix A in the back of the text helpful.

SUMMARY

To define the quality of transmission that a particular facility will produce, we require two quantities, one for digital applications and the other more appropriate for analog systems. In analog systems we use the signal-to-noise ratio (SNR), which is the power ratio between a desired signal (the unmodulated information that is presented to the system for transmission) and the electrical noise that is present in the spectrum occupied by the desired signal. The unit of measure is the decibel (dB).

In digital systems we use the bit error rate (BER), the ratio of the number of bits detected in error to the number of total bits transmitted and received. A BER of 10^{-9} would indicate that for one billion bits transmitted and received, only one bit would is detected or read out incorrectly.

In digital systems transmitting voice, such as telephone trunking systems connecting switching centers, a BER of 10^{-6} might be considered adequate. A voice circuit so derived might experience barely audible clicks every ten seconds or so, which might be considered satisfactory for telephone trunk service. The requirements for data transmission are less forgiving, and a BER of 10^{-8} or better is generally required. Since data transmission is pervading the public telephone switched network, a transmission quality standard of 10^{-9} BER is generally used in designing transmission systems without regard to their utilization, that is, to whether such a facility is initially intended for voice or data transmission service.

Transmission data rate, sometimes referred to simply as bit rate, is the number of bits transported or processed through a system in a given time period. Data rates are expressed in bits per second (bp/s), kilobits per second (kb/s), megabits per second (Mb/s), or gigabits per second (Gb/s).

Transmission data rate is influenced by format, which is the particular method employed to encode bits for serial transmission through a system or facility. Popular formats used in lightwave transmission systems are return to zero (RZ), non-return to zero (NRZ), and Manchester code.

Some encoding formats are appropriate for transmission purposes, but less satisfactory for detection, timing recovery, and digital multiplex-demultiplex functions. Within sections of any system, we frequently find transitions from one format to another more suitable format, particularly when the electrical signal is extended to other terminal equipment such as a digital carrier or multiplexer.

A contributing factor to erroneous bit detection in a system can be the rise time of the terminal devices themselves. In a lightwave system, these devices are the light sources and the photodetectors employed. Rise time requirements are more severe for higher transmission data rates.

System bandwidth relates directly to transmission data rate and rise time. Since rise time requirements depend on the encoding format employed, bandwidth requirements do also.

The bandwidth presented for use by an optical fiber is a function of the fiber length as well as the fiber characteristics. At higher transmission data rates, even single mode fibers introduce significant pulse dispersion. It then becomes necessary to specify fiber performance in a more sophisticated manner, referencing the spectral output line width of the light source. As a consequence, calculations of system end-to-end dispersion are possible that incorporate all the variables involved, and the end-to-end dispersion can be directly related to system bandwidth.

Transmission power levels (light energy levels) may be specified in the standard units of power, the watt or milliwatt, but a more generally employed term is the decibel-milliwatt. Zero decibels-milliwatt (0 dBm) equals one milliwatt.

REVIEW QUESTIONS _____

True or False?

T	F	1. A power level of 2 mW is the same as +6 dBm.
T	F	2. A power level of 0.01 milliwatt is the same as –20 dBm.
T	F	3. The signal-to-noise ratio of a digital lightwave system is most generally used to define the quality of transmission through the system.
T	F	4. The BER of a lightwave system is a function of system noise alone.
T	F	5. Lightwave receivers are generally specified with their optical input signal levels referenced to the electrical output signal BER.
T	F	6. A BER of 10^{-9} denotes a more superior transmission performance than a BER of 10^{-8}.
T	F	7. In a digital transmission system, the bit is the fundamental unit of information.
T	F	8. Data rate is the total number of bits that are transmitted or processed independently of the time such transmission or processing may require.
T	F	9. Format is the method employed to encode bits for serial transmission.
T	F	10. In digital electronics, electrical pulses are used as bits.
T	F	11. If the bit period of a signal is known to be 1 μs, the data rate is 2 Mb/s.
T	F	12. In an NRZ formatted signal, the signal level changes only when the bit logic level changes.
T	F	13. The transmission bandwidth required to accommodate a 2 Mb/s signal is equal to the minimum pulse period of the signal.
T	F	14. In an NRZ formatted signal, the signal voltage goes to a high level and returns to a low level for each logic one bit.
T	F	15. A lightwave system transmitting RZ formatted signals requires half as much bandwidth as a system transmitting NRZ formatted signals.
T	F	16. At the receiving location, timing can be recovered directly from a signal received that is formatted using the Manchester code.
T	F	17. In a pulse bipolar lightwave system, the lightwave signal is actually turned on and off.
T	F	18. As transmission data rate increases, pulse period gets longer.
T	F	19. Rise time is the time period required for the instantaneous amplitude of a pulse to go from 20 to 80 percent of peak pulse amplitude.
T	F	20. Typical laser light source rise times are on the order of one nanosecond or so.

CHAPTER 6

Lightwave System Terminal Equipment

Terminal equipment, as presented in this text, is equipment that is installed at points within the lightwave system where the optical transmission terminates. Such points would normally be at the system extremities or at intermediate service locations where a transition is required from lightwave signals to electrical signals—for example, at an intermediate repeater location.

Terminal equipment can be logically subdivided into two categories, those devices that are electro-optical in nature, and those that are entirely electronic or electrical. The electro-optical category includes all devices that accept electrical input signals and generate or emit lightwave output signals as a result of that input, and those devices that accept lightwave input signals and produce electrical output signals as a result of that input. Typical electro-optical devices are laser diode or light emitting diode light sources and photodetectors of all varieties. Electronic terminal equipment includes devices that process electrical signals only, and typical units might be digital multiplexing equipment or digital regenerative repeaters.

We shall address both categories. For discussion purposes or block diagram presentations, it may be convenient to assume that electro-optical and electronic equipment are discrete and separate entities. In actuality, the capabilities and functions of both are often provided within a single product configuration. The situation is similar to equipment used in other kinds of transmission. For example, in digital microwave equipments, many units are available that provide some level of digital multiplexing in the same product package that generates or receives the microwave frequency carrier itself.

LIGHT SOURCES

In lightwave technology, transmitters are transducers that convert electrical input signals to lightwave energy output signals. This requires a source of light energy. At present, such sources are limited to light-emitting diodes (LEDs) or laser diodes. Both are very small semiconductor chips of gallium arsenide (GaAs) or some other material. The major considerations in the design or selection of light sources are as follows:

Operating Speed. The source must be able to turn on and off fast enough to satisfy the transmission requirements of the system. The speed at which a source can operate is usually specified in terms of rise time, defined as the time required for a source to affect a transition from 10 to 90 percent of the peak output power.

Power Output. The source must couple enough optical power into an optical fiber to operate a detector at the distant end of the fiber. The power level must be high enough to overcome fiber and fiber splicing losses and still provide sufficient light input to the detector to satisfy system signal-to-noise or bit error rate specifications.

Peak Output Wavelength. The wavelength emitted by the source should be stable and should coincide with the low loss bands of the fiber to which it will be connected.

Spectral Width. Ideally, a light source should produce a single-wavelength output to eliminate material dispersion in the optical fiber. Most sources emit a number of wavelengths with spectral width usually taken to be 50 percent of the maximum output.

Light Emitting Diodes (LEDs)

An LED is essentially a p-n junction that spontaneously emits light when current is passed through it. By introducing materials such as indium, aluminum, or phosphorus, the wavelength of the emitted light can be controlled.

The cone of radiation from LED devices is considerably larger than the cone of acceptance of optical fibers. Thus, efficient coupling of light output into small-core fibers with low numerical apertures is difficult unless special packaging techniques are employed. Recent developments have resulted in the manufacture of some LEDs that can couple usable levels of light energy into even small-core single mode fibers.

Although a variety of structures can be fabricated, two types of LEDs predominate today: the *well emitter* and the *edge emitter,* shown in Figure 6-1.

Well emitters produce light from a surface area fabricated in an open well in the LED. This type of structure generates a widely diverging light beam. Edge emitters produce light from a narrow strip fabricated in the LED material and generate a light beam considerably narrower than a well emitter.

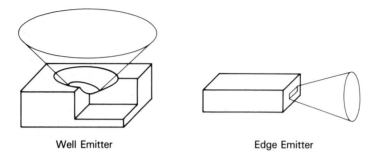

Well Emitter   Edge Emitter

LED light sources have the following characteristics:

1. Output power is relatively insensitive to device temperature, rendering devices less complex and requiring little or no temperature compensation.
2. LEDs have wide spectral output, typically 25 to 50 nm or greater.
3. Emission pattern is quite broad, and coupling losses into fibers may be on the order of 10 dB and higher.
4. LEDs are generally considered to have long operating life.
5. LEDs are generally inexpensive.
6. LEDs have a rather large active area where light is generated and have characteristically high capacitance. This makes them difficult to digitally modulate at higher transmission bit rates.
7. Since LED light power output is relatively linear over a rather wide range of input drive current, LEDs are particularly well suited for analog modulation systems.
8. LEDs typically have lower light power output than lasers.

For all of the preceding reasons, LED sources are best suited for lower bit-rate applications or shorter length systems where many of these deficiencies are not particularly significant. The use of larger core multimode fiber may offer some advantages in plant construction and maintenance, and in such cases LED sources may find practical application, since the broad emission pattern typical of the LED structure would not be inhibitive.

Laser Diodes

The term "laser" is derived from Light Amplification by Stimulated Emission and Radiation. The laser structure provides an "optical cavity." A laser functions much like an LED until a threshold drive current level is reached. At this point incident photons created in the recombination process (the same as in the LED function) stimulate the emission of additional photons within the optical cavity in a process called *lasing*. A typical laser structure is shown in Figure 6–2.

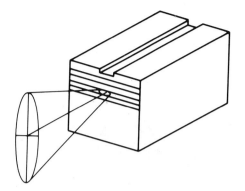

FIGURE 6–2
Laser Diode

As the device temperature increases, the threshold current increases and the optical output power changes. The wavelength of the optical output is temperature dependent also. Thus, lasers are significantly more complex than LEDs in that some form of temperature sensing and compensation is required.

One technique commonly employed involves *thermoelectric cooling*. A thermoelectric cooler (TEC) is not a simple heat sink. When electrical current is passed through it, one side of it becomes cooler and the other side becomes warmer. By sensing the temperature of the assembly with a thermistor, the laser temperature can be controlled and stabilized as shown in Figure 6–3.

In addition to temperature stabilization, most lasers incorporate some laser bias control to stabilize the threshold current itself, which otherwise might change with temperature and device aging. Bias control functions are shown in Figure 6–4.

It is plain that these stabilizing requirements add to the sophistication and complexity of laser diode light sources.

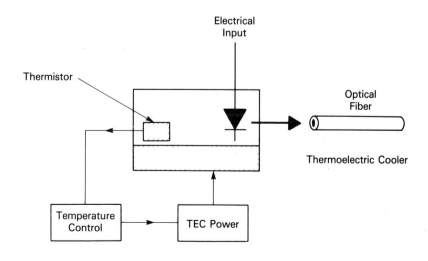

FIGURE 6–3
Thermoelectric
Cooling

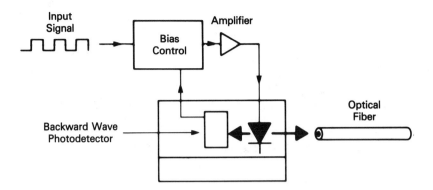

FIGURE 6-4
Laser Biasing

Laser light sources have the following characteristics:

1. Output power is temperature dependent and requires that sophisticated temperature compensation be built into the device.

2. Output wavelength is temperature dependent.

3. Threshold current must be controlled by monitoring light energy output power.

4. Lasers are maintained at a bias above the lasing threshold current to avoid having to operate in the threshold current region. Thus, lasers typically produce some optical output all the time.

5. Lasers typically have higher output power than LEDs.

6. Lasers have very narrow beam width output that facilitates coupling output energy into small-core fibers.

7. Lasers have very narrow, almost monochromatic spectral width output, so all of the light output is at the same wavelength or nearly so. Material dispersion in fibers is minimal due to this feature.

8. Lasers are more expensive than LEDs.

9. Lasers are generally considered to have shorter operating lives than LEDs.

10. Lasers have short rise times, making it easier to modulate them at higher transmission bit rates.

For the preceding reasons, lasers are the most suitable choice for high-bit-rate systems and for long length facilities where single mode fiber is employed. Research to restrict the spectral output of laser sources to a single wavelength is continuing and, if successful, will improve laser performance even further.

LIGHTWAVE RECEIVERS

In lightwave technology, receivers are transducers that convert light energy input to electrical energy output signals. Receivers process the electrical output of an internal photodetector by amplifying, pulse reshaping, or effecting current-to-voltage conversions such that the electrical output is compatible with the system in which the receiver is employed. At the present time photodetectors are largely limited to positive-intrinsic-negative (PIN) diodes or avalanche photodiodes (APDs). The major considerations in the design or selection of light receivers and photodetectors are as follows:

Responsivity. Responsivity is a measure of a detector's output current or output voltage in relation to optical input power. Since responsivity varies with wavelength, it is specified either at the wavelength of greatest responsivity or at a discrete wavelength of interest.

Quantum Efficiency. Quantum efficiency expresses a diode's sensitivity as the ratio of photons impinging on, to the number of electrons flowing in, an external circuit. An efficiency of 1, or 100 percent, would mean that every photon sets one electron flowing.

Dark Current. Even in the absence of incident light, some current flows because electron-hole pairs can be generated thermally. Dark current is device temperature dependent and is also called leakage current.

Noise Equivalent Power (NEP). NEP is the RMS signal power required to produce a signal-to-noise ratio of unity. It is the minimum incident optical power required to create a current equal to the inherent RMS noise current, and is analogous to the thermal threshold of detection in receivers in other technologies.

Response Time. The response time of a detector is the time it takes the electrical output to rise from 10 percent of peak output to 90 percent of peak output. Such time might be on the order of 1 ns for an APD and perhaps 3 or 4 ns for a PIN detector, and in either case is dependent upon bias voltage.

Bias Voltage. Operating with a current, a detector requires a bias voltage. PIN devices typically require less than 100 volts, while APDs might require several hundred volts. Since bias voltage influences response time, dark current, and responsivity, and can increase device temperature, it can significantly affect detector operation.

PIN Diodes

A PIN diode consists of a negative and a positive terminal separated by a "depletion region," created by reverse biasing so that very little current flows through the device. Electrons tend to migrate from the negative terminal and to form holes near the positive terminal, depleting the area around the junction of carriers. Thus the reference to the depletion region.

When light impinges on the diode, the photons absorbed create electron-hole pairs in the depletion region. Then carriers separate and drift toward their respective terminals, and for each electron-hole pair created, an electron is set flowing in the external circuitry.

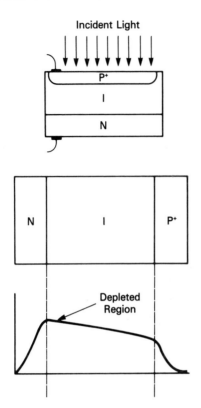

FIGURE 6–5
PIN Diode

In an ideal PIN diode, each incident photon creates one electron-hole pair, which in turn sets one electron flowing in the external circuit. If the light flux impinging on the diode is weak, the electrical current generated may not be strong enough to be detectable in the presence of the noise that is inherent in the PIN diode and receiver circuitry. Figure 6–5 shows a simple PIN diode structure.

PIN diodes have the following characteristics:

1. They possess a relatively simple structure compared to APDs.

2. They are relatively insensitive to device temperature.

3. Quantum efficiency, i.e. photon-to-electron conversion gain, is typically 1 or less.

4. They have a relatively limited dynamic range.

5. They have high reliability and a long operating life.

6. They are inexpensive.

7. Compared to APDs, they have low responsivity for a given signal-to-noise ratio or BER.

APDs

When the light flux impinging on the detector is weak, it is desirable to increase the photodetector output before any processing or amplification is attempted within the electrical section of the receiver. This detector gain can be accomplished through the use of an APD. The structure of this device is shown in Figure 6-6.

The APD establishes a strong electrical field within a portion of the depletion region, as shown by the depletion region profile. Primary carriers created by incident photons (as in the PIN diode) that enter the strong field can gain several electron volts of energy. In a collision with the lattice structure, the primary carrier loses enough energy to promote an electron from the valence band to the conduction band in a process known as collision ionization. The secondary carriers may, in turn, create even more carriers. The result is a phenomenon known as avalanche breakdown, evidenced as internal gain.

The number of electrons set flowing in the external circuit of an APD is equal to the number of incident photons times the device's multiplication factor. An APD may produce a photon-to-electron conversion gain on the order of 3 or 5, although it may also produce a higher noise output.

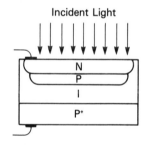

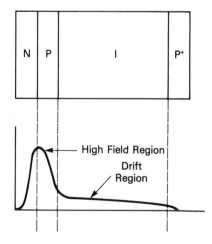

FIGURE 6-6
APD

APDs are sensitive to device temperature, so automatic gain control (AGC) is usually included to maintain a stable bias voltage as the temperature varies. Thus, receivers using APDs are generally more complex and expensive.

APDs have the following characteristics:

1. They are more complex in structure than PIN diodes.
2. Their responsivity is sensitive to device temperature.
3. Their quantum efficiency, and hence photon-to-electron conversion gain, is typically 3 to 5.
4. They usually present an extended dynamic range.
5. They are highly reliable and have a long operating life.
6. They are more expensive than the simpler PIN diodes.
7. Their responsivity is better than that of PIN diode detectors. A typical improvement over the PIN device is 5 or 6 dB.

Lightwave Receiver Design

To ensure adequate detection, generally the optical input signal level should be at least twice the inherent receiver noise current. However, more optical power input may be necessary to produce a desired signal-to-noise ratio or bit error rate.

To reduce the effects of noise, a transimpedance amplifier is included in some receiver packages. Such receivers using a PIN diode detector are sometimes called *PIN-FET units*. "FET" refers to a field effect transistor employed to amplify the detector electrical output signal.

The detector output also requires other processing, e.g., pulse reshaping. Such functions are provided in the receiver, in sufficient amount to produce an electrical output signal fully compatible with the system requirements. In digital systems, an electrical pulse stream receiver output might be applied directly to a conventional digital demultiplex equipment input. The receiver electrical output must be fully compatible with such a device.

Because detector active surface areas are relatively large, efficient coupling of light from an optical fiber into the detector is not difficult. Sometimes fibers with core area larger than the core size used in the link are utilized as detector "pigtail" fiber inputs, to minimize receiver coupling losses.

Usually, PIN diode packaging is simpler than APD packaging. The latter can be quite complex, particularly if a device cooler (TEC) is included.

TERMINAL EQUIPMENT APPLICATIONS

Because a laser couples the most light into a fiber, and because an APD detector is most sensitive to weak optical signals, APDs offer optimum performance in high-speed, long distance applications. The higher cost and complexity of such devices may be more than offset by the elimination of an expensive intermediate repeater.

Their use is not mandatory, however, and their unique characteristics do not automatically qualify them to be the correct solution for every application.

An analogy might be drawn with microwave transmitters. Some units employ a klystron tube, producing significantly higher radio frequency (RF) output than completely solid-state units. Clearly, the higher RF output signal provides more transmission path "fade margin," and this might be considered to be inherently superior. Consider the possibility that the microwave path length is only nominal, however.

Link design might provide an adequate fade margin in other ways, perhaps in antenna size and gain. If this were possible, the reduced cost and complexity of a completely solid-state microwave transmitter might be very attractive. The application must be considered in the decision-making process, and this is equally true in the design of lightwave systems.

It can be shown that even using multimode fiber with a significant contribution of modal dispersion, a transmission system using a data rate of 45 Mb/s would not be bandwidth limited until the cable length was on the order of 40 km or so. The rise time characteristics of the light source and detector become somewhat academic in such an application. Stated another way, an LED source and a PIN detector might function quite satisfactorily in such a link.

This kind of installation would be primarily attenuation limited. It might be necessary to specify a laser light source to provide adequate light input to the receiver, and it might be necessary to specify and use an APD detector as well, if the interconnecting fiber length and loss so dictates. It might also be possible to select a lower loss fiber or a different operating wavelength, or to consider the combinations of a laser source and PIN detector instead of an LED source and APD detector. The link design problem resolves itself to transmission losses and levels alone.

Transmission capacity at 45 Mb/s might be more than adequate for many applications within the distribution or exchange sections of telecommunications networks. It also seems reasonable to assume that a great many rural network requirements could be served with fiber lengths of 40 km or less. If these assumptions are acceptable, then the design of lightwave links for such service, and the selection of terminal equipment also, can actually be limited to consideration of transmission levels alone.

The process may have to be more sophisticated for higher transmission rates or for longer link lengths. We shall explore this in greater depth in a later chapter. Here, it is useful to briefly review those terminal equipment specifications that will be of primary concern in the design and selection process.

TERMINAL EQUIPMENT SPECIFICATIONS

It is necessary to establish and specify the optical signal levels that electro-optical terminal devices will generate or with which they will operate satisfactorily. A lightwave transmitter outputs light energy which is coupled into the interconnecting optical fiber. Optical transmitter output signal levels are usually specified in terms of dBm, where dB is referenced to one milliwatt, that is, 0 dBm = 1 mw.

Referencing transmission levels in this manner facilitates system calculations, since transmission losses can be specified in decibels and all calculations of signal levels can be reduced to simple addition and subtraction. Transmission losses would be stated as negative values of dB, but transmitter output levels could also be negative values of dBm. The latter would denote a transmitter whose light energy output was at some level below one milliwatt.

Some manufacturers may specify transmitter output as milliwatts of power. Given a discrete output power in milliwatts, the equivalent power level in dBm can be determined by the formula

$$Pwr \ (dBm) = 10 \log Pwr \ (mw)$$

It is necessary to identify the point at which the transmitter output is specified. In some configurations an optical fiber "pigtail" may be provided as part of the transmitter assembly, and the light output power is specified at the end of this pigtail. In other configurations the unit may deliver its specified output into an optical connector on the rear panel of the transmitter assembly. In such a case, the link calculations must include the optical transmission loss of the connector.

A lightwave receiver accepts light energy input from the interconnecting optical fiber, but specifying receiver performance is a bit more sophisticated than in the case of the transmitter. Most receivers are specified by the manufacturer to provide a stated quality of electrical output signal at a specific optical input signal level. This might be termed the threshold of the receiver, at least for the stated quality of electrical output signal. In digital systems this quality is generally related to bit error rate, whereas in analog applications it is stated as a signal-to-noise ratio.

A manufacturer might specify a receiver to produce a BER of 10^{-9} when the optical input signal level is –39 dBm. Such a unit should consistently deliver a BER of 10^{-9} or better at input levels of –39 dBm or higher. At input levels below –39 dBm, the BER may be poorer than 10^{-9}. The system designer is obliged to assure, through his or her design, that receiver input levels are maintained above this threshold, or else accept the fact that the system transmission quality may be degraded.

There is a condition of excessive optical input signal that can degrade the electrical output signal. The manufacturer will generally define this point of receiver overdrive by specifying the receiver dynamic range in dB. The dynamic range of a receiver is a function of the device design, the type of light detector employed, and the range of automatic gain control (AGC) the receiver has. The dynamic range specification states that at some known input level above the previously defined receiver threshold, electrical output signal quality will be degraded.

Consider again a receiver that produces a BER of 10^{-9} at an optical input level of –39 dBm. If this unit has a specified dynamic range of 30 dB, then at optical input levels between –9 dBm and –39 dBm, the output electrical signal quality should be 10^{-9} BER or better. At input signals below or above these levels, the electrical

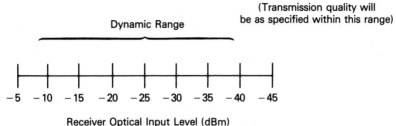

FIGURE 6-7
Receiver
Dynamic Range

output quality may be degraded. Receiver dynamic range is illustrated in Figure 6-7.

Given a clear definition of the transmitter optical output and the receiver optical input requirements, it is a relatively simple matter to calculate all intermediate link transmission losses and predict overall transmission performance with confidence. We shall perform such functions and make such calculations in addressing transmission engineering later in the text.

ELECTRONIC TERMINAL EQUIPMENT

Earlier, it was mentioned that at the light transmitter location an electrical signal from external equipment was utilized to drive or pulse the light transmitter. The light receiver shown in Figure 6-8 is designed to reproduce the original, initiating electrical signal, which would be a pulse train in a digital system, and to extend that reproduced signal to some external equipment again.

There is a clearly defined demarcation point at the electrical interface with either a light transmitter or receiver. The nature of the electronic terminal equipment external to the light transmitter and receiver may determine the functions required in these units. It may be necessary to convert the electrical output of the receiver shown in the figure to some more suitable electrical format, such as B3ZS, for example, before extending the signal to any external electronic equipment.

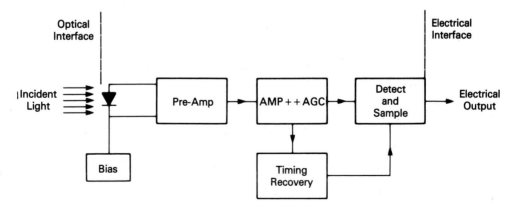

FIGURE 6-8
Lightwave
Receiver

If the lightwave transmission link is considered between the electrical interfaces of the transmitter and receiver alone, it is quite evident that such a link can be designed, constructed, and put in service as either an analog or a digital facility. In telecommunications applications the link would most probably be digital, and the electronic terminal equipment would likely be conventional digital multiplex-demultiplex units.

A terminal of this nature is shown in block diagram as an unprotected facility in Figure 6-9. The electrical and optical transmission bit rates are identified. Although such an electro-optical terminal is shown consisting of discrete elements of equipment, in most commercially available products today the complete terminal is usually more consolidated. Flexibility in use is somewhat restricted due to this consolidation, however.

In telephone networks, it is usually necessary to design and construct systems that are capable of more transmission capacity than is initially required for use. In some instances the subsequent heavier traffic requirement may never actually develop or may be delayed for an indeterminate period of time, but planning for such potential growth must be a part of the initial system engineering process.

One technique to increase a transmission system's capacity in digital applications is to consolidate or multiplex several digital signals into a single higher bit-rate digital signal. Equipment is commercially available to do this in various stages.

The basic digital signal in the North American hierarchy is DS 1, a pulse stream at a 1.544 Mb/s rate. This signal is often denoted as a T–1 system, a generic term that refers to a digital carrier system capable of supporting 24 equivalent voice circuits. The standard North American digital hierarchy is shown in Table 6–1.

Table 6–2 lists some of the available digital multiplex equipment.

When digital multiplexing is employed to increase a system's traffic-carrying capacity, the resultant higher transmission rate imposes wider bandwidth or faster

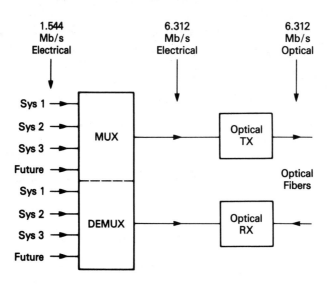

FIGURE 6-9
M 12 Digital
Multiplexer

TABLE 6–1
North American
Digital
Hierarchy

Digital Signal Level	Number of Equivalent Voice Circuits	Transmission Rate (Mb/s)
DS 1	24	1.544
DS 1C	48	3.152
DS 2	96	6.312
DS 3	672	44.736
DS 3C	1344	90.148
DS 3D	2016	135.532
DS 4	4032	274.176

TABLE 6–2
Digital
Multiplex
Equipment

Multiplexer Designation	Acceptable Inputs	Output Signal	Output (Mb/s)
M 12	4 DS 1	DS 2	6.312
M 13	28 DS 1	DS 3	44.736
M 23	7 DS 2	DS 3	44.736

rise time system requirements. This may introduce higher system costs by requiring higher performance fiber or more sophisticated electro-optical terminal equipment.

Consider an application where the initial traffic requirement is 6 DS 1 signals that can support a total of 144 equivalent voice circuits. Suppose that the long-term traffic projection anticipates an eventual requirement for 17 DS 1 signals, a capacity of 408 voice circuits. The initial requirements could be met with the terminal configuration shown in Figure 6–10. For clarity, only one system terminal is shown.

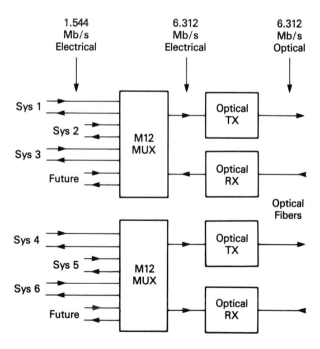

FIGURE 6–10
Digital
Multiplexing

As shown in Figure 6–11, the terminal could be expanded to accommodate the growth requirement given in the traffic study. The projection was for 17 DS 1 signals to support 17 T–1 systems, and the figure presents a capacity of 28 such systems.

Note that here the optical transmission bit rate is 44.736 Mb/s initially, and the M 13 multiplex equipment is also provided in the initial installation. This arrangement permits a graceful growth, since DS 1 signals can be incrementally added by circuit board additions to the in-place M 13 multiplexer. Such a configuration does increase the initial installation cost, however, and the transmission capacity provided is initially underutilized. If the traffic growth that was projected does not actually develop, then the configuration might not be cost effective.

Figure 6–12 shows an M 23 digital multiplexer. This unit accepts 7 DS 2 inputs, each at a 6.312 Mb/s rate, and consolidates them into a DS 3 signal at a bit rate of 44.736 Mb/s. The unit can be useful in the graceful growth of a system from lower traffic requirements to higher levels at a later date.

As a general rule, for terminal equipment available today, the demarcation between electro-optical and electronic devices is somewhat academic. The demarcation that is shown in some of the figures presented is technically correct, but packaging of terminals by manufacturers often obscures the interface by consolidating all terminal equipment into a single discrete product.

The effect is to provide complete terminal assemblies that perhaps are quite suitable for individual system applications, but that do not lend themselves to sys-

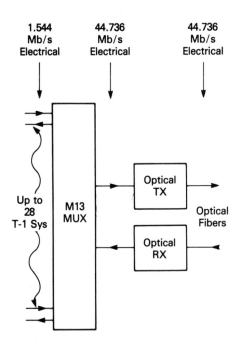

FIGURE 6–11
M 13 Digital
Multiplexer

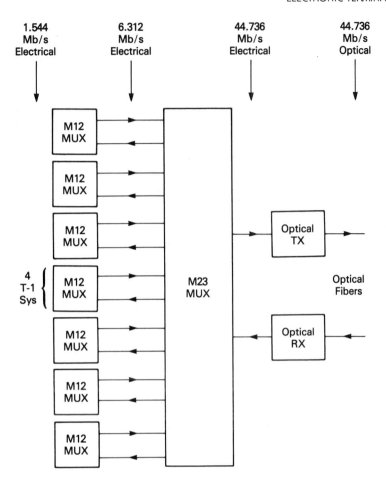

FIGURE 6-12
M 23 Digital
Multiplexer

tem evolution. If the anticipated future requirement is a 45 Mb/s data rate, then the system designer may be influenced to provide a complete terminal at this level even though the initial requirements might be adequately satisfied with lower level equipment, and even though the future necessity for higher level equipment is uncertain.

If the demarcation between electro-optical and electronic equipment were preserved in equipment configurations, the system designer would have more freedom of choice. If light transmitters and receivers were widely offered as stand-alone products, independent of the electrical terminal units, then the lightwave link of the earlier example could have been designed differently. For instance, it could have been equipped for 45 Mb/s service, but a lower level electrical terminal could have been installed initially. A lightwave link capable of 45 Mb/s service can certainly be keyed satisfactorily at any lower transmission rate. Thus, the designer might have elected to provide a 45 Mb/s light link but stay with the M 12 multiplex level in the electrical terminal, as was shown in Figure 6–10.

The trend in consolidated product packaging probably reflects the heavy influence on lightwave technology evolution that has been due to the bulk of applications being predominantly in major, high-traffic-density installations. As the technology is more widely applied, particularly at lower levels of network plant such as in exchange or distribution, perhaps the products will be offered in a more flexible manner.

Lower revenue base support is implicit in lower traffic density applications, and a more flexible selection of products as system "building blocks," rather than as predefined transmission systems, would be useful.

NONSTANDARD DIGITAL SIGNALS

Products are offered for lightwave systems that neither conform to the North American digital signal hierarchy nor utilize standard increments of digital multiplexing. If applications develop where traffic density is limited, as, perhaps, for distribution or exchange plant in rural telephone networks, then a lower level of digital multiplexing may be technically acceptable, and may have merit. Reducing the transmission bit rate to 12 or 16 Mb/s or so, a purely arbitrary figure used here for discussion purposes only, may reduce the complexity, and thus the cost, of the multiplex equipment itself. Lower transmission bit rates would certainly relax the specifications for both fibers and electro-optical terminal equipment.

Offsetting the potential advantages of such a product philosophy may be the economics of both fiber and terminal product fabrication. There appear to be, at this time, no clearcut economies to be gained by relaxing fiber performance requirements. It is doubtful that a fiber specification for 45 Mb/s operation would actually impose any cost penalty at all over a fiber specification to support 16 Mb/s operation.

As for the circuit cards or modules themselves, the production volume of such items has a substantial impact on the end cost, and this may nullify any economic advantage that technical simplification might offer. In the long run, it may prove to be less expensive to standardize on a minimum transmission data rate, such as 45 Mb/s, regardless of what level of electronic terminal equipment is actually installed, but the market forces that will determine this have yet to clearly emerge.

NEW DEVELOPMENTS IN LIGHTWAVE TECHNOLOGY

Lightwave technology is by no means static, and new development efforts are extensive and continual. It is likely that fiber transmission losses will be reduced in the near future, and efforts to improve fiber performance at higher transmission rates have already been translated to field installations in the form of dispersion-shifted or flattened fibers. Operating at wavelengths in the region of 1.5 μm rather than 1.3

μm allows transmission over longer fiber lengths, but higher transmission bit rates will incur greater dispersion distortion and special measures will be required.

One promising effort is directed at producing improved single mode lasers with spectral widths on the order of 0.05 nm or so. This would dramatically reduce pulse dispersion. Such efforts are largely in product development or improvement areas, and as better devices become available, system and applications engineers will undoubtedly incorporate them in new installations.

One interesting approach is known as a *coherent system.* In such a system, the light source (laser) output would be held at a constant level rather than amplitude modulated in an on-off fashion. The information would be modulated onto the lightwave carrier signal by externally varying the phase of the light. At the receiving location a local light source would be phase and frequency locked to the incoming carrier, but at the present time this is very difficult to accomplish. The lightwave signal received is combined with the phase-locked local lightwave signal, and the result is an output signal whose amplitude varies in proportion to the phase variations of the signal received. Coherent systems of this type would permit extremely high transmission data rates over very long fiber lengths.

Although the industry emphasis is clearly on increasing fiber transmission capacity, and the improvements achieved to date are truly remarkable, these improvements become somewhat academic in the environments presented by the exchange plant and the rural telephone network. The requirements for increased system capacity are not persuasive here, and the practical transmission distances involved do not demand higher level technology either.

Recognizing these facts, together with the smaller supporting revenue bases which these levels of plant present, it appears that such applications may be better served by pragmatically applying lightwave technology rather than simply opting for "state-of-the-art" or "leading edge" technology for each and every application.

ANALOG TERMINAL EQUIPMENT

It is possible to multiplex several analog type signals on a lightwave system. For example, in cable television applications, it might be desirable or even necessary to transport several television signals through a lightwave facility. If the facility presents sufficient bandwidth for use, then all that is necessary is to combine several carriers, each individually modulated with a different television signal, and apply the composite signal composed of all these carriers as the modulating baseband information for the lightwave facility.

One technique used employs frequency modulation (FM) of individual carriers in much the same fashion that carriers can be modulated for transmission over coaxial cables. The carrier frequencies are selected so as to be compatible with each other (adequately separated in frequency), and so that all carriers fall well within the transmission bandwidth of the lightwave system. At the distant system terminal the electrical output of the lightwave receiver is presented to a number of companion demodulators, each tuned to an individual carrier frequency, and thus the individual television signals may be recovered. The present state of the art can accommodate perhaps eight television signals on a single optical fiber in this manner, but this depends on the lightwave facility bandwidth. More capacity is certainly possible.

Note that this technique does not involve digital encoding of the television signals and is really basically not different from the frequency division multiplexing techniques widely used in coaxial cable systems. Although most analog television lightwave applications to date seem to have used FM modulation, AM modulation is also possible.

TERMINAL EQUIPMENT COSTS

Although terminal equipment costs are changing rapidly in today's market, the cost differential between different types of terminals may be expected to remain relatively constant. The cost data offered here are only representative, but they do provide a reasonable basis for evaluating alternative system designs and configurations. The reader is advised to solicit current cost quotations before finalizing any system design. A fully protected (1 × 1) 45 Mb/s lightwave terminal, including the digital multiplexing equipment necessary to accept up to 28 DS 1 inputs at 1.544 Mb/s each and produce a 45 Mb/s optical output, will cost approximately $18,000. Two such terminals are required, one for each end of the lightwave link; thus, a fully protected 45 Mb/s link will cost about $36,000.

An unprotected 45 Mb/s lightwave terminal, including the digital multiplexing equipment necessary to accept up to 28 DS 1 input signals, each a T-1 system at 1.544 Mb/s, and produce a 45 Mb/s optical output, will cost approximately $11,000. A link requires two such terminals and will cost approximately $22,000.

A fully protected (1 × 1) 90 Mb/s lightwave terminal capable of accepting up to two DS 3 digital signals at 45 Mb/s each and producing a 90 Mb/s optical output will cost approximately $24,000. A link requires two such terminals and will cost about $48,000. Note that this quotation does not include the digital multiplexing equipment required to consolidate 28 DS 1 signals at 1.544 Mb/s each into a single DS 3 45 Mb/s signal. A protected (1 × 1) multiplexer of this type (28 T-1 inputs with a 45 Mb/s electrical output, designated an M 13 multiplexer) would cost approximately $10,000 per terminal, or $20,000 if both terminal ends of a link require such a unit.

A fully protected (1 × 1) 140 Mb/s lightwave terminal capable of accepting from one to three DS 3 45 Mb/s inputs and producing a 140 Mb/s optical output will cost approximately $16,000 equipped to handle only one DS 3 signal. Equipped to handle two DS 3 inputs, the unit would cost approximately $19,000, and for three DS 3 inputs about $23,000. As before, this cost would be incurred at each end of a lightwave link, and these costs do not include any multiplexing equipment to produce the DS 3 level signals from lower digital level signals.

A fully protected (1 × 1) 565 Mb/s lightwave terminal capable of accepting up to 12 DS 3 inputs and providing a 565 Mb/s optical output would cost approximately $45,000 per terminal end, and twice this amount for a complete link. These units can be implemented at a lower capacity, and a rule of thumb concerning cost might be $1,000 per terminal end for each DS 3 signal added. For example, a terminal implemented for only four DS 3 signals might only cost $37,000 initially, since eight DS 3 signals would not be equipped. Each new implementation of a terminal would cost about an additional $1,000.

Keep in mind that a lightwave terminal operating at data rates higher than 45 Mb/s will usually require some digital multiplexing equipment if the derived services are to be accessed at that location. A terminal may accept and present DS 3 signals, but terminal traffic is rarely available as a DS 3 signal, nor can it be accepted by switches or toll terminating equipment at that rate.

In cost estimating any application at a 45 Mb/s data rate or lower, the costs given will usually include the digital multiplexing capability, and entry or exit out of the system can be made directly at the DS 1 data rate (1.544 Mb/s). In cost estimating an application at a transmission data rate higher than 45 Mb/s, digital multiplexing requirements must be established and their cost included. An example is illustrative.

Figure 6–13 shows a simple network providing trunking between Station *A* and each of three other locations, identified as Stations *B, C,* and *D*. All interconnecting lightwave links are operating at a 135 Mb/s data rate, and the optical terminal equipment would cost $16,000 at each end of each individual lightwave link. Six optical terminals are shown to be required in the figure, so optical terminal costs would be $6 \times 16,000 = \$96,000$ total for the network.

If a DS 3 signal were allocated to serve each of the stations, then each of stations *B, C,* and *D* would require an M 13 digital multiplexer, and three digital multiplexers would be required at Station *A,* one for each service link. Each multiplexer would cost approximately $10,000, so that $60,000 would be required for the network.

The reader is again cautioned that these costs are only representative and undoubtedly will change, perhaps drastically and often, as time passes.

SUMMARY

At the present time, lightwave transmitters are limited to using either a light emitting diode (LED) or a laser diode as a light source. LED sources are best suited for low-data-rate, shorter length systems, while lasers are best employed in high-data-rate, longer length applications using single mode fibers.

Currently, lightwave receivers are limited to PIN diodes or APD (avalanche) diodes as detectors. PIN diodes are more suitable for shorter length systems while APDs are generally used in higher-bit-rate, longer length installations.

Transmitter performance and receiver performance are usually specified in terms of optical signal power. Transmitter output is generally specified in decibel-milliwatts (dBm), and receiver sensitivity as signal-to-noise ratio (SNR) or BER, both expressed in terms of receiver input power level in dBm.

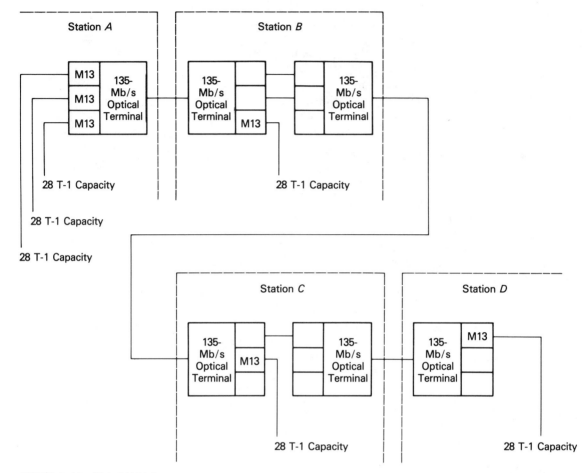

FIGURE 6-13 Digital Multiplexers

A clearly identifiable demarcation exists between optical terminal equipment and those units required to further process the recovered electrical digital signal which is the lightwave receivers' output. Lightwave receivers are often required to convert the recovered electrical signal to a different format, in addition to detecting and demodulating the optical input signal. For example, a common required interface conversion would be to change the signal received to a bipolar electrical output signal even though the format of the input optical signal was not bipolar.

A standard hierarchy of digital signals has been established but there are transmission data rates that are not consistent with this hierarchy. This type of equipment is sometimes used for lower traffic density applications where lower data rates are adequate to serve the requirements.

Digital multiplexing to a higher transmission data rate increases a facility's traffic-carrying capacity, but requires wider transmission bandwidth. Traffic-carrying capacity in analog systems may be increased by frequency multiplexing additional carriers onto the lightwave system baseband. Each such carrier would be a different frequency and would be modulated with different information. Increasing capacity in this manner also requires greater system transmission bandwidth.

REVIEW QUESTIONS

True or False?

T   F   1. Digital multiplexers are electro-optical devices.

T   F   2. LED light sources typically couple their light output into small core fibers very efficiently.

T   F   3. Characteristically, LED light sources have wide spectral output.

T   F   4. LED light sources are well suited for use in analog modulated systems.

T   F   5. LED light sources are usually difficult to digitally modulate at higher transmission data rates.

T   F   6. Laser output power changes as device temperature increases.

T   F   7. A thermoelectric coupler functions the same way a heat sink does.

T   F   8. Laser output wavelength is not temperature sensitive.

T   F   9. Laser optical output is usually monitored to control laser threshold current.

T   F   10. LED sources generally produce higher light energy output than lasers.

T   F   11. Lasers are typically less expensive than LEDs.

T   F   12. Avalanche photodiodes (APDs) are widely employed as light sources in lower data-rate lightwave systems.

T   F   13. PIN diodes are generally less expensive than APDs.

T   F   14. APDs are typically more temperature sensitive than PIN diodes.

T   F   15. Lightwave transmitter output power is generally specified in decibels (dB).

T   F   16. Lightwave receiver sensitivity generally references S/N Ratio or BER-to-receiver input light signal power level.

T   F   17. The basic digital signal in the North American digital hierarchy is 6.312 Mb/s.

T   F   18. Digital multiplexing is a technique commonly employed to increase the traffic-carrying capacity of lightwave transmission systems.

T   F   19. Higher level digital multiplexing requires wider system transmission bandwidth.

T   F   20. A coherent lightwave system eliminates the necessity for phase locking the local receiver light source to the incoming light signal.

CHAPTER 7

Optical Devices

This chapter discusses a variety of ancillary devices that can be fabricated for use and employed in lightwave transmission systems. These devices permit us to configure systems in various ways to best accommodate different applications.

As used here, the generic term "optical devices" denotes devices whose basic functions derive from optical principles and techniques. Such devices directly process light rays by focusing, filtering, collimating, or otherwise handling the light rays themselves. Their use in lightwave systems will thus be confined to the optical transmission path that is established between two or more terminal points at which electro-optical devices are located to generate or detect the lightwave signals.

Many of these optical devices present the same functional capabilities of comparable units employed in other kinds of transmission that we may be more familiar with, such as filters, power splitters or dividers, or service point taps used in multipair metallic or coaxial cable systems. Most of these devices do not require operating power to perform their various functions, and we generally refer to such units as passive, as opposed to active power-consuming devices.

Although lightwave energy is electromagnetic energy that travels in a wave front just as radio or microwave energy does, the much higher frequencies (shorter wavelengths) require that optical techniques and mechanisms be employed. Some of the more common phenomena we may have become familiar with in earlier experiences will be less applicable at these wavelengths. For example, in conventional metallic cable systems, impedance matching at cable and device interfaces must be considered to assure the efficient transfer of energy and to minimize possibly disruptive energy reflections from the interface. This is true even of cable ends, which must be terminated in the cable characteristic impedance to inhibit reflections.

In theory, the same constraints apply to lightwave transmission, but the mechanisms involved are less restrictive. Although lightwave energy reflections do occur from an open end of an optical fiber, the amplitude and time delay of the reflected energy is such that no specific protective action is normally necessary. Later we shall even see where the phenomenon of reflected energy is actually employed in a test technique.

Conversely, the connection of two fiber ends (as in a fiber splice) presents more severe problems than an equivalent conductor splice in a multipair metallic cable. With fibers, the interface at a splice becomes largely a mechanical problem involving fiber core size, alignment of the two ends, ellipticity of the cores, etc., to ensure a minimum of reflection and a maximum transfer of light energy. In optical interfaces of this type, irregularities are more seriously evidenced as a loss of lightwave energy than as a disruptive reflected signal.

In optical devices, additional and possibly disruptive optical interfaces are frequently and unavoidably introduced. For example, when mirrors, prisms, or diffraction gratings are employed in such devices, additional glass-to-air or glass-to-glass interfaces cannot always be avoided. Each of these individually can introduce Fresnel reflection losses and other problems. In such devices, the mechanical construction and geometry can have a significant effect on the transmission characteristics of the unit.

OPTICAL COMPONENTS

Prisms have properties that refract light rays, and so they can be used to separate different wavelengths of optical energy. Thus, they can be employed in optical devices to filter out signals of different wavelength. It is also possible to utilize several prisms in an optical switch that could direct lightwave energy into one or more optical fibers selectively by mechanically altering the physical relationship of two or more prisms. Or, an optical switch can also be fabricated to mechanically reposition the ends of fibers to selectively direct the light into particular fibers. Optical switches of either variety will require some electrical energy to reposition the optical elements, but not to effect the actual transfer of the light energy.

Mirrors are an obvious optical component and are fabricated in many forms, including elliptical and parabolic varieties. Thus, light rays may be focused and directed as necessary.

Another optical technique employs dichroic filters. *Dichroism* is the property a surface has of reflecting light rays of one wavelength while passing light rays of a different wavelength. Figure 7–1 shows how a dichroic filter may be employed in practical lightwave devices.

It is also possible to etch or score a grating or number of parallel lines on an optical surface that will diffract light rays of a particular wavelength while passing other wavelengths. Figure 7–2 shows a diffraction grating employed as a filter, and this mechanism is often encountered in practical optical devices in use today.

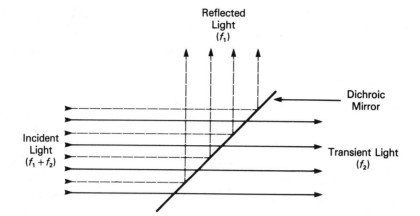

FIGURE 7–1
Dichroic Filter

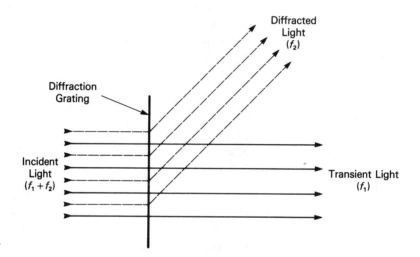

FIGURE 7–2
Diffraction Filter

In applying optical principles, any number of lenses can be fabricated to focus, collimate, or otherwise process light rays. Combinations of filters and lenses are also possible, as shown in Figure 7–3. In the figure, a glass rod, essentially a section of a graded index fiber structure with optical lensing properties, is combined with a partially reflective surface (a dichroic mirror) that allows one wavelength to pass through it while reflecting a different wavelength.

The geometry of this unit is such that the reflected rays are focused on a point where a separate optical fiber is positioned. The light entering the glass rod at any point other than along its axis is refracted back toward the axis at the middle of the rod.

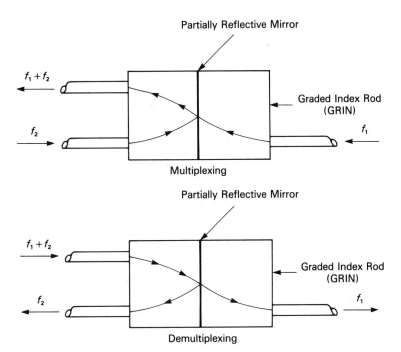

Partially Reflective Mirror

Graded Index Rod (GRIN)

$f_1 + f_2$

f_2

f_1

Multiplexing

Partially Reflective Mirror

Graded Index Rod (GRIN)

$f_1 + f_2$

f_2

f_1

Demultiplexing

FIGURE 7–3
GRIN Filter

FABRICATED OPTICAL DEVICES

Instead of the discrete optical elements described, optical devices such as couplers and splitters can be made and employed using optical fibers alone. The technique is to twist the fibers together, taper them somewhat by elongation, and then fuse the glass into a single, solid structure. Such units are referred to as *biconical* or *tapered fused* units.

Figure 7–4 shows a four-port biconical device. Note that along the fused cross section, the cladding materials have been fused together and the two fiber cores are separated by only one common cladding. Light energy within this common cladding is shared between the two fiber cores by power splitting or coupling and is recaptured through the common cladding by combining.

The use of biconical devices can be extended to a relatively large number of fibers permitting so-called star type network configurations to be built up. As Figure 7–5 shows, light can be coupled from one fiber into many, or from many into one.

Star type couplers may also be constructed by using reflective mirrors in discrete components, as shown in the three-port device of Figure 7–6. Any light rays injected into the coupler from any port will reflectively propagate and appear on any optical fiber connected to any of the other ports.

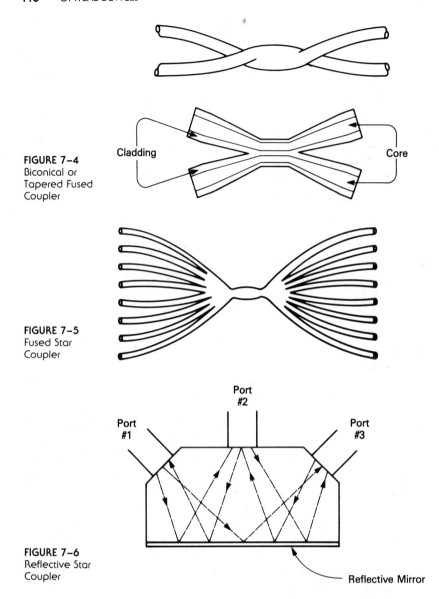

FIGURE 7–4
Biconical or
Tapered Fused
Coupler

Cladding

Core

FIGURE 7–5
Fused Star
Coupler

Port
#2

Port
#1

Port
#3

FIGURE 7–6
Reflective Star
Coupler

Reflective Mirror

Optical Taps

Taps are devices that permit service points to be established along the length of an optical fiber. Taps may couple light energy out of the fiber, into the fiber, or both. The optical fiber itself can be utilized in fabricating a tap, as shown in Figure 7–7.

In the figure, the thickness of the deposited film or the air-glass interface can be varied during fabrication to select the amount of light energy that is actually

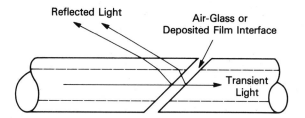

FIGURE 7-7
Passive Tap

coupled into the output tap port. The latter will determine the amount of energy that remains available at the "through" output port. Stated another way, the device insertion loss—the loss apparent as signal level difference between the input and the "through" output ports—will depend upon how much input energy is diverted to the tap port.

Many of the devices fabricated are inherently frequency independent; that is, they do not manifest any frequency discrimination to any wavelengths of light passed through them. For example, a biconical tapered fused unit would evidence the same frequency characteristics as the fiber from which it was fabricated. Frequency discrimination can be introduced almost at will into many of these devices. For example, in the tap shown in Figure 7-7, if a dichroic coating were applied to the glass-glass interface or a diffraction grating were inserted, the device would exhibit frequency-discriminating characteristics similar to those shown in Figures 7-1 and 7-2.

Active Taps

It is possible to fabricate a variety of active devices. In Figure 7-8 a tap is shown that incorporates a photodetector within the tap configuration itself. The unit shown employs a beam splitter and would have no inherent frequency-discriminating characteristics, but a dichroic coating could be easily introduced.

Since the detector is associated directly with the optical tap, it is not necessary to divert a large amount of energy to the tap port. Consequently, only nominal "through" loss will be introduced to transmissions that pass through the tap to the main output port.

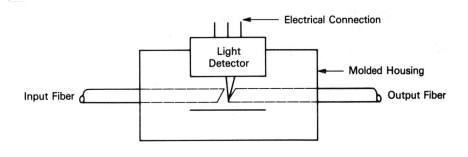

FIGURE 7-8
Active Tap

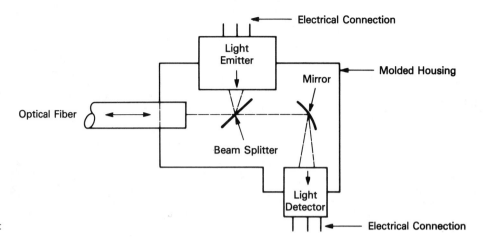

FIGURE 7–9
Bidirectional Unit

The detector mounting and the optical elements can all be assembled as a single mechanical unit with electrical leads to the detector. Some manufacturers use a high-precision plastic molded body that assures precise alignment of the optical elements. The optics are an integral part of the molded body itself in such structures.

This type of structure can be expanded to two-way, or bidirectional devices, also, as shown in Figure 7–9. The unit shown, fabricated in a single molded structure, incorporates all optical functions and mounts both active elements, the light energy emitter, and the photodector. Note that in such a structure, although frequency discrimination is not inherent, it could be easily provided.

It is evident that many of the devices we are discussing are directly related to other units that provide similar functions in other transmission disciplines. For example, the couplers and splitters offer the same functionality that line taps do in coaxial cable systems. And the use of dichroic mirrors and diffraction gratings present the same basic frequency discrimination that filters effect at other transmission frequencies.

The problems of impedance matching that are associated with other media are not encountered in lightwave transmission, but the mechanical constraints of precision mating of fibers, and concern with fiber compatibility in core size and numerical aperture, might be considered analogous. If we understand the nature of the optical devices that are available, then we can address the possible application of such units in systems.

CONFIGURING LIGHTWAVE SYSTEMS

Two concepts commonly encountered in telecommunications work today, but perhaps a bit obscure to individuals with a background in telephone communications, are local and wide area networking, usually denoted by the acronyms LAN and

WAN respectively. The nature of most of these installations is quite different from telephone interoffice trunking, and this has influenced the development of optical devices to some extent.

An example is the optical star type coupler. This type of coupler permits a multiplicity of stations or terminals to be interconnected without any circuit switching. Such an application is not directly related to telephone trunking, which provides nondedicated transmission capacity between discrete points, such as two switching centers or offices. The facility is not dedicated in the sense that traffic routing or circuit usage is flexible, is assigned on demand, and varies from call to call.

Given these basic differences, we can expect that some optical devices will be less directly applicable in telephone-type networks and will be infrequently employed. Since this text is primarily directed at telecommunications transmission, we shall not pursue the development of local or wide area networks in great depth, but rather shall focus in on telephone trunking instead.

In telephone trunking, we can clearly identify applications where optical devices may be usefully employed. An excellent example is the frequency discriminating devices that permit the use of two or more lightwave signals of different wavelength to be carried on the same optical fiber. The technique of using these devices in this way is called *wavelength division multiplexing* (WDM) in lightwave technology, and this is directly analogous to frequency division multiplexing in other disciplines. Multiple channel telephone carrier systems, for example, or CATV systems using coaxial cable both employ frequency separation to carry a number of channels of different information. The commercial broadcasting field, i.e., radio and television, does the same thing of course, the only difference being that such transmissions use free space, the atmosphere, as the medium.

If we can use two or more light wavelengths on a single fiber, then obviously we can increase the transmission capacity of the fiber. If the transmission data rates (bit rates) of the different lightwave signals are the same and relatively nominal, we might relax the specifications for fiber and terminal equipment also, while still deriving a substantial increase in transmission capacity.

For example, if a system were designed and constructed for operation at a 45 Mb/s data rate, it would present a finite transmission capacity for use. If we can add to that system a second or third lightwave signal, also at a 45 Mb/s bit rate, then we can double or triple the transmission capacity of the basic facility.

Keep in mind that we have not increased the sophistication of the facility; that is, it is not necessary to increase the transmission bit rate, so we need not install a higher grade fiber or more sophisticated terminal units.

Figure 7–10 shows a two-frequency WDM system capable of carrying two DS 2 signals, each at a 6.312 Mb/s bit rate, on a pair of optical fibers. The two lightwave frequencies are denoted as f_1 and f_2. The transmission capacity of the link might have been increased by digitally multiplexing several signals up to a higher digital level or bit rate, of course, but this option might not be available in a situation where the fibers are already in place, having been initially designed for a lower

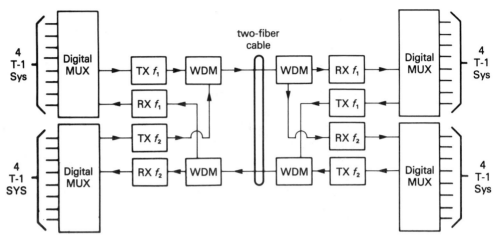

FIGURE 7-10 Unidirectional WDM

transmission data rate. A facility initially designed for a 45 Mb/s bit rate may not function adequately if the bit rate is subsequently raised to 90 Mb/s.

BIDIRECTIONAL OPERATION

There is nothing inherently unidirectional about an optical fiber. Just as we can transmit in both directions over a coaxial cable, so can we transmit information in both directions through an optical fiber. If light energy can be launched at one end of a fiber and recovered at the distant end, then it can be simultaneously launched at the distant end and recovered at the near end. Of course, it may be necessary to discriminate between two lightwave signals of different wavelength, but that is entirely possible and practical using the optical devices just discussed.

Figure 7-11 shows a bidirectional system using WDM. This is essentially the same installation as that shown earlier in Figure 7-10, except that each individual DS 2 signal or subsystem is now independent of the other, since the two do not share any common equipment or travel over the same optical fiber. As a result, a failure of any terminal equipment unit or optical fiber can disrupt only one-half of the traffic or circuits provided by the system. In Figure 7-10, a failure of either optical fiber will impose a 100 percent traffic disruption. The service reliability is due entirely to the use of the optical devices (the WDM couplers) and can be achieved at a relatively small cost.

In lightwave technology it is difficult to fabricate a receiver that will respond to only one discrete wavelength of light energy. Frequency discrimination must generally be provided externally to the receiver itself, and this involves filtering in the optical transmission path. The filters discussed here can certainly be used in this manner, but their characteristics must be compatible with the system requirements in regard to bandwidth, transmission loss, and the like.

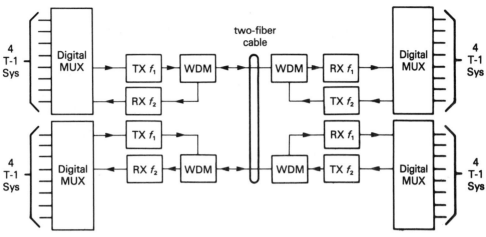

FIGURE 7-11 Bidirectional WDM

FREQUENCY DISCRIMINATION REQUIREMENTS

The frequency discrimination or isolation required in an optical device is determined not only by the separation between the wavelengths of the several signals involved, but also by the amplitude of these signals. In some situations, there may be substantial differences in the signal levels involved.

In a unidirectional WDM application, both signals will be generated at the same terminal location. Since both have to overcome the same transmission link losses, such as those due to splicing and the fiber, then both signals would be launched into the fiber at the same relative amplitude. At the receiving terminal, both signals will be at, or very nearly at, the same amplitude again. In such applications, the design of an optical device or filter to separate the two signals is rather straightforward.

Consider, however, the problem in a bidirectional system. We have two signals at one terminal that must be isolated to some degree, but one of these signals is the relatively high-amplitude output of the lightwave transmitter while the second signal is at a much lower level since it has traversed the interconnecting facility and been attenuated by the fiber, splices, etc., that are associated with that transmission facility. Thus, a WDM device would have to provide a significantly higher degree of isolation between the two signals in the bidirectional application than it would in the simpler unidirectional link.

OPTICAL DEVICES IN DISTRIBUTION PLANT

In telephone trunking the bulk of operating lightwave systems have already been installed. Logically, the next step is the extension of lightwave transmission out into the telephone distribution or exchange level plant, including, perhaps, the individual subscriber loop. The distribution application is quite different, both as a technical problem and from the economic view of supportive revenues, than the trunking requirements were. If trunking philosophy is applied without consideration of these differences, it is possible that distribution applications will be inhibited by the costs alone.

The distribution or exchange plant has the following characteristics:

1. It involves relatively short transmission distances.

2. Traffic requirements do not demand high bit rates.

3. Service growth may be uncertain, both as to traffic volume and as to geographical distribution.

4. The plant does not demand, and usually cannot support, sophistication or cost in redundant facilities or equipment.

Where trunking applications have clearly identifiable service points, most generally the two ends of a link, the economic analysis resolves itself quite simply into three questions: how much will the facility cost, how many circuits or services can it accommodate, and what is the prognosis for traffic and revenue for the facility?

In distribution plant, unfortunately, neither the service requirements nor the revenue potential are so certain for the long term. Traffic studies and projections only present the "best guess" prognosis—an educated guess perhaps—but still considerably uncertain.

Capital investments in distribution plant might be much more effectively applied if the initial facility could provide some flexibility in future use, that is, if it could be tapped for service anywhere along its length as subsequent requirements developed. This flexibility can be provided by the judicious use of optical devices.

Figure 7–12 shows a fiber pair installed initially to provide trunking service between two switching offices, labeled *A* and *B*. In distribution or exchange level plant, the majority of such applications usually do not require a large number of derived circuits, certainly not in rural areas, for example. In any event, for purposes of discussion, let us assume that the facility shown was initially predicated on a 45 Mb/s data rate. Fully implemented, this facility could provide 672 individual voice circuits.

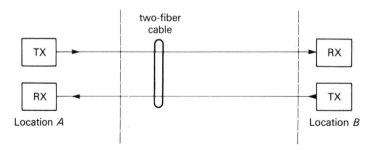

FIGURE 7–12
Sample System

The present state of the art in fiber losses and terminal equipment is such that additional transmission losses can often be accommodated in many installations without compromising the quality of the initial installation at all. Keep in mind that, by definition, distribution plant fibers would be limited in length also. Under these conditions, if fiber facilities are designed and placed strictly in response to the perceived requirements, both initial and projected, then these facilities will be underutilized in many cases. Inherently, they offer a great deal more transmission capacity than is actually required. WDM might, then, be applied to take advantage of this circumstance.

Consider, in Figure 7–12, a subsequent requirement for service that develops at a point *C*, which might be located anywhere along the route of the initial facility. In Figure 7–13 we have responded to this new requirement by installing, at point *C*, WDM couplers to create a new service point. Of course, companion WDM devices were installed at office *A* at the same time.

Using a new wavelength signal, at or lower than the original data rate, on the same original optical fibers, an independent transmission facility is provided between points *A* and *C*. This new facility has, or could have, the same 45 Mb/s transmission capacity that the initial system had. Indeed, it may be possible in many distribution plant applications, to initially design links that could accommodate, at a later date, two or three separate wavelengths in this manner, simply by incorporating the WDM coupling losses into the initial design of the facility.

The points of future service may not be identifiable at the time the initial facility is constructed, but they do not have to be. Allowing for the coupler insertion losses in the initial design makes the geographical service location purely academic. Wherever such a point was subsequently located, it cannot compromise the end-to-end system performance.

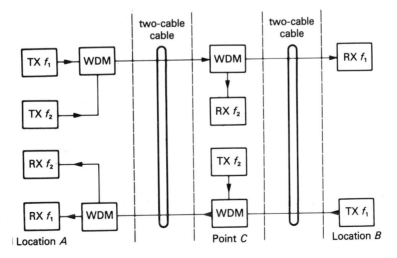

FIGURE 7–13
WDM Terminal
Addition

Whether service points are developed along a fiber facility for remote switches, grouped carrier terminals, private branch exchange service, or what have you is irrelevant. It does not even matter if the derived services are for trunking or for subscriber loops: from a transmission point of view, there are no significant distinctions among digital signals for any level of service.

It is important to avoid strict preconceptions of facilities as either purely trunking or purely for exchange service. The inherent flexibility of lightwave facilities can be employed, in the manner discussed, to drastically improve the cost effectiveness and utilization of such facilities, regardless of how we conceive them. It matters little if the cost is distributed across trunk or distribution services; the cost per circuit or per service, and per location serviced, is reduced, and that is the ultimate objective.

OPTICAL DEVICE PARAMETERS

There are a number of basic parameters that are of primary concern in the design and use of optical devices. Some of them require definition as follows:

Coupling Ratio: The coupling ratio is the ratio of lightwave energy level between the input port of a device and the output port or ports. For example, a device that splits power from a single input port equally between two output ports (a 1×2 port device) would be said to present a 50/50 coupling ratio. Such a unit would theoretically produce an output signal level at each of the output ports that was 3 dB down from the input signal power level, since –3 dB would be one-half of the input power.

It is possible to fabricate a 1×2 port device where the coupling ratio would be 90/10. In such a device, 10 percent of the input energy would be presented to the tap port and 90 percent to the "through" port. The tap port in this case would theoretically produce a –10 dB output, since –10 dB is one tenth of the input power.

Excess Loss: As applied to optical devices, excess loss refers to energy losses intrinsic to the device itself. These losses are in addition to the theoretical loss that coupling ratio alone introduces. For example, a 50/50 coupler would introduce more than the 3 dB theoretical loss to the tap ports because the device has intrinsic losses. The difference is referred to as excess loss in optical devices and typically ranges from less than 1 dB to perhaps as much as 3 dB or more.

Intrinsic losses of this type are not unique to optical devices alone. For example, in a coaxial power splitter, which divides a single input equally across two outputs, the theoretical loss is 3 dB, but the practical losses incurred are on the order of 3.5 to 4.5 dB or so.

Other optical device parameters of interest are uniformity of coupler output and coupler crosstalk. Ideally, all outputs in the same device have the same level of energy if they all have the same coupling ratio. However, this is difficult to achieve in practice. A measure of the quality of a device is how closely this uniform output level is approximated.

Optical crosstalk is analogous to the crosstalk we experience in other transmission devices, such as trans-hybrid crosstalk at four-wire/two-wire demarcation points in a telephone circuit. It denotes the level of discrimination that a device provides between signals present at one port and signals present at another. Crosstalk values typically range from as little as 15 to as much as 50 dB in optical devices.

In considering device losses, the wavelength of the signals passed through a device must be considered, but in most commercially available nonfrequency-discriminating units losses show little sensitivity to wavelength. Factors such as the numerical aperture of the fibers that the unit is connected to may influence how efficient a particular device is at a particular wavelength, however.

The manufacturers of optical devices will usually specify the following performance characteristics:

Operating Wavelength	Excess Loss
Fiber Numerical Aperture	Coupling Ratio
Fiber Dimensions	Directivity
Cutoff Frequency (single mode)	Thermal Characteristics
Number of Ports	Insertion Loss

Devices are available for both multimode and single mode fibers, and the mechanical dimensions of optical devices are very small. Thus, the units can usually be easily installed in all the environments we would expect to encounter in telecommunications installations, including splice cases or splicing enclosures.

Optical devices are purely optical in nature and do not, as a rule, consume any operating power. As a consequence, such units should be extremely reliable and should not introduce any unusual concern in this regard.

ECONOMIC CONSIDERATIONS

The economic tradeoff that is most important in evaluating optical device utilization in systems is generally the cost of such devices compared to the cost of additional fibers. There are, of course, the more subtle factors such as the flexibility for use that the constructed plant may provide, as discussed earlier, but most often what is involved is a cost comparison against more fibers.

Optical device development is not static, and unit costs will surely change as manufacturing costs are reduced or as volume of use throughout the industry increases. The cost data presented here should be considered representative rather than authoritative, but the information may be useful in initial economic analyses of applications.

At this writing, the cost of a three-port or four-port coupler for use with either single mode or multimode fiber ranges from $100 to $300, depending upon the coupling ratio and other characteristics required. WDM couplers that are frequency discriminating are available at costs ranging from $250 to $800, depending upon characteristics and complexity.

Active devices, such as the active, bidirectional unit shown in Figure 7–9, are indicative of the reduced costs that may be anticipated through advanced production techniques. These units are precision-molded plastic. Including the light emitter and photodetector, they are available today for $200 each. In reasonable production quantities, the manufacturer predicts a $10 or so price may be possible, including the electro-optical units. The unit is presently a subassembly produced for sale to other manufacturers as a component for use in other product lines.

SUMMARY

A variety of optical devices are available which, by applying basic optical principles, can process lightwave signals in a number of useful ways. Some of these devices can be employed as filters, power dividers, or power combiners in configuring lightwave transmission systems.

Many optical devices are inherently passive in nature, that is, they require and consume no operating power to perform their functions. Devices of this type are extremely reliable since they are completely mechanical and optical by nature.

It is possible to create hybrid devices that combine passive optical qualities with active power-consuming opto-electronic functions such as light generation or detection. An example of such a hybrid device would be an active tap, which can tap lightwave energy out of or into an optical fiber with only a minimal introduction of transmission loss into the fiber.

By inserting various optical devices such as diffraction gratings or dichroic mirrors into a lightwave system, we can, almost at will, introduce frequency discrimination into the system. Such a capability permits wavelength division multiplexing (WDM), where more than one lightwave signal can be applied to and transmitted over a single optical fiber. WDM techniques can also provide bidirectional transmission on a single optical fiber.

As optical devices come into more general use, and familiarity with their application becomes more widespread, they may contribute to more flexible and cost-effective system designs.

REVIEW QUESTIONS

True or False?

T   F   1. A major disruptive factor in lightwave transmission systems is reflected lightwave energy from the ends of unterminated fibers or from other optical interfaces.

T   F   2. Prisms can be used to separate lightwave signals of different wavelength.

T   F   3. A dichroic mirror will reflect light signals of all wavelengths equally well.

T   F   4. A diffraction grating is often used to filter out unwanted noise that may be present in lightwave signals.

T   F   5. A short section of a graded index fiber rod exhibits the properties of an optical lens and can be utilized in a variety of optical devices because of these properties.

T   F   6. Some optical devices can be fabricated by fusing existing optical fibers together.

T   F   7. All optical taps require the application of operating power to function.

T   F   8. Bidirectional optical devices must provide high isolation between ports because they are required to discriminate between and separate lightwave signals of substantially different amplitude.

T   F   9. The star-type coupler is particularly well suited for use in telephone network trunking applications.

T   F   10. Wavelength division multiplexing can only be applied in lightwave systems transmitting digital signals.

T   F   11. Increasing transmission capacity by wavelength division multiplexing requires that more transmission bandwidth be provided by the system.

T   F   12. Lightwave detectors, as employed in receivers, are inherently single-frequency, narrow-bandwidth devices, and will only respond to a very narrow spectrum of lightwave signals.

T   F   13. Excess loss is a term that denotes the losses that are intrinsic to optical devices, that is, those losses that occur over and above the theoretical losses that a particular coupling ratio should introduce.

T   F   14. Optical functions can be incorporated into some hybrid devices by precision molding optical capabilities like lenses or mirrors directly into unit structures.

T   F   15. In fabricating optical devices, not all potentially disruptive optical interfaces can be completely avoided.

T   F   16. Fresnel reflection losses can be introduced at glass-to-glass, glass-to-air, or air-to-glass interfaces.

T   F   17. In an etched diffraction grating, the spacing between the grating lines will influence the frequency-discriminating characteristics of the device.

T   F   18. A system that employs WDM to increase its traffic-carrying capacity cannot use the same transmission data rate for more than one of the lightwave signals carried.

T   F   19. When a second lightwave carrier is added to an existing system, as when WDM is applied to an older system, the system bandwidth must be increased to accommodate the additional transmission capacity.

T   F   20. Although WDM can be effectively employed to increase a system's capacity, it cannot be used to serve additional points along a system cable route.

CHAPTER 8

Configuring Lightwave Systems

The evolution of any transmission facility follows a logical sequence of operations. First, a traffic study establishes the long term and immediate transmission quality and capacity that are required. The study may include some analysis of the service reliability requirements.

The next phase might be referred to as applications engineering. It is directed at developing system configurations that satisfy the technical performance requirements and provide sufficient capacity for both immediate and projected traffic. The effort must also address the possible provisions for improving system reliability through redundant equipment, spare cable pairs or fibers, and protection switches. When applications engineering efforts have developed all practical alternatives, an economic analysis can be made to select the most efficient approaches, and then transmission engineering can be pursued.

Lightwave facilities are quite different in nature from multiple-pair cables, a fact that influences the process of lightwave applications engineering. Any lightwave facility presents a substantial transmission bandwidth for use. Where additional transmission capacity for traffic growth in paired-cable plant can be directly translated to higher pair counts, even lower sophistication lightwave plant (such as those using multimode fibers or LED sources) will often present bandwidth in excess of requirements. This presents the engineer with both new options and new obligations in developing system configurations for review.

The factors determining the transmission capacity of a lightwave link are the electro-optical terminal devices and the fiber characteristics. If a fiber inherently provides at least the necessary bandwidth, then the factors are reduced to the terminal devices or system configurations alone.

If, in such a situation, the designer is unduly influenced by previous experience with paired cables, where the pair count is the ultimate determinant of cable capacity even when cable carrier is applied, then it is possible that the design may underutilize the fiber, resorting to additional fibers as the only rational solution to more transmission capacity. Thus, the designer must be knowledgeable to effectively apply this new technology and not simply substitute one transmission medium (glass) for the other (copper).

APPLICATIONS ENGINEERING CONSIDERATIONS

Obviously, the system under design must be fully responsive to all immediate transmission capability requirements, and the applications engineer is granted no latitude whatever in this area. Less obviously perhaps, but equally restrictive, is the provision of sufficient additional capacity to accommodate all projected future requirements. Here also, no compromise with the requirements can be allowed, since such projections are the result of an earlier traffic growth study, and presumably a management judgment has already been made on the question. Within these constraints, however, and with the separate issue of facility reliability noted, the applications engineer has considerable latitude. If one considers that the expanded capacity requirements are by no means assured but simply represent the best judgment possible at the time, then the responsibility that the applications engineer bears is very real and quite substantial.

The initial development phase of system design is reflected not only in initial facility cost but in subsequent (and somewhat uncertain) facility utilization. If applications engineering is casually pursued, then the capability may easily be over-built, with substantial economic penalty for providing in-place capacity that may never be utilized at all. On the other hand, an inflexible design could involve massive equipment obsolescence if the projected upgrade in capacity actually does develop.

If the end-result system design is to be efficient and economical, that objective must be addressed in this phase of the work. Transmission engineering processes are unavoidable and essential for any system configuration, but such processes will have relatively little impact on cost as compared to the applications engineering effort.

It may appear to be somewhat incongruous, but even if a single fiber presents sufficient bandwidth for use to satisfy the transmission requirements—even perhaps enough bandwidth to accommodate the bidirectional bandwidth requirements—there may still be compelling reasons to provide additional fibers in a facility. To reconcile this seeming incongruity, it may be helpful to review the factors that determine the transmission capacity of a link, and thus identify any options or alternatives that might be available. In digital lightwave telecommunications applications, there are three techniques that can be employed to increase transmission capacity: higher level digital multiplexing, WDM, and the provision of additional fibers.

HIGHER LEVEL DIGITAL MULTIPLEXING

Higher level digital multiplexing consolidates several lower bit rate digital signals into a single, higher bit rate pulse stream. For example, in the North American digital hierarchy, a DS 1 signal at 1.544 Mb/s can provide 24 equivalent voice channels, as in T–1 carrier systems. If four such DS 1 signals are digitally multiplexed together, the resultant pulse stream at 6.312 Mb/s is called a DS 2 signal.

Tables 8–1 and 8–2 list the standard North American digital signals and some standard digital multiplex equipment that is available, which was presented earlier in Chapter 6.

When digital multiplexing is employed to increase the traffic-carrying capacity of a system, the resultant higher transmission data rate requires a wider bandwidth characteristic in the system. This may translate to higher fiber performance requirements or terminal equipment specifications and thus to increased system costs. Digital multiplexing also introduces a higher degree of commonality in terminal equipment, possibly requiring the installation of redundant, automatically substitutable units to protect traffic capacity continuity.

An example will perhaps best demonstrate digital multiplexing. The basic requirement for purposes of illustration will be taken to be 96 voice circuits initially, with expansion capability to 192 circuits at a later date.

It is possible to apply a DS 1 signal (24 voice channel capacity) directly to a pair of fibers, using one for each direction of transmission. To satisfy the initial 96 channel requirement, four DS 1 signals must be transported, requiring four fiber pairs (eight fibers) as shown in Figure 8–1. Note that a respectable level of service protection has been provided due to the absence of common equipment. If any single fiber or any optical terminal equipment fails, only 25 percent of the traffic

	Line Code	Data Rate	Capacity
TABLE 8–1 Standard North American Digital Hierarchy	T-1 (DS 1)	1.544 Mb/s	24 Chans.
	T-1C (DS 1C, 2-DS 1)	3.152 Mb/s	48 Chans.
	T-2 (DS 2, 4-DS 1)	6.312 Mb/s	96 Chans.
	T-3 (DS 3, 28-DS 1)	44.736 Mb/s	672 Chans.
	T-3C (DS 3C, 2-DS 3)	90 + Mb/s	1344 Chans.
	T-3D (DS 3D, 3-DS 3)	135 + Mb/s	2016 Chans.
	T-3E (DS 3E, 4-DS 3)	180 + Mb/s	2688 Chans.
	TX (9-DS 3)	405 + Mb/s	6048 Chans.
	TX (12-DS 3)	565 + Mb/s	8064 Chans.

	Multiplexer	Inputs	Output	Output Data Rate
TABLE 8–2 Digital Multiplex Equipment	M 12	4-DS 1	DS 2	6.312 Mb/s
	M 13	28-DS 1	DS 3	44.736 Mb/s
	M 23	7-DS 2	DS 3	44.738 Mb/s

carried will be disrupted. Note also the high number of fibers (8) and transmitters and receivers (8). However, the transmission data rate in the optical link (on the optical fiber) is a very nominal 1.544 Mb/s.

If the projected traffic growth does develop, it will be necessary to apply some technique to expand the link capacity. The technique shown in Figure 8-1 works, but it does not efficiently utilize the inherent bandwidth an individual fiber presents.

In Figure 8-2 we show an alternative arrangement that utilizes only two fibers. Here an M 12 multiplexer is used at each terminal end. From Table 8-1, such a unit will accept up to four DS 1 signals and consolidate them into a DS 2 signal at a 6.312 Mb/s data rate.

Note that the optical link data rate of the arrangement shown is now 6.312 Mb/s and that the system requires only two fibers and two transmitters and receivers. The M 12 multiplexers represent an increase in system cost, but this may be

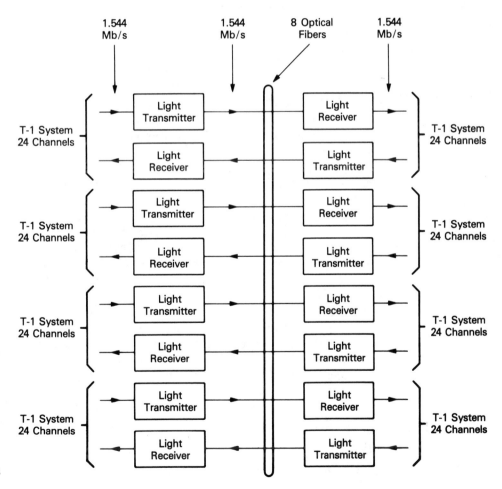

FIGURE 8-1
Multiple Fibers

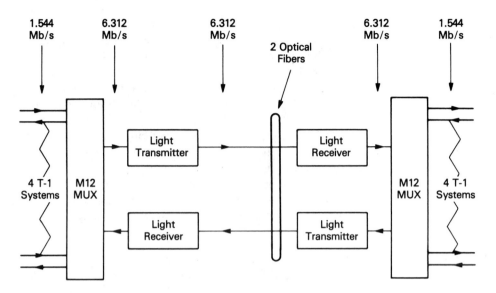

1.544 Mb/s 6.312 Mb/s 6.312 Mb/s 2 Optical Fibers 6.312 Mb/s 1.544 Mb/s

4 T-1 Systems — M12 MUX — Light Transmitter — Light Receiver — M12 MUX — 4 T-1 Systems

Light Receiver — Light Transmitter

FIGURE 8–2
Low-Level
Multiplexing

more than offset by the reduced number of fibers and terminal equipment. The failure of any fiber or any electro-optical unit, or either M 12 multiplexer, would interrupt all traffic, and this configuration cannot be gracefully expanded to handle any growth either.

Figure 8–3 shows a variation of the configuration shown in Figure 8–2. Here, two M 12 multiplexers are used at each terminal end. One DS 2 signal is applied to one fiber pair and the other to the second fiber pair, thus providing two discrete, completely independent transmission links. By dividing the traffic evenly across these two systems, a failure of any fiber, any optical terminal, or any M 12 multiplexer will only disrupt 50 percent of the transported traffic.

Note that in this configuration the optical link data rate is now 6.312 Mb/s but the system only requires four fibers, four transmitters, and four receivers. The four M 12 units are an additional expense. The system can now accommodate eight DS 1 signals (192 voice channels) with no rearrangement. This is in full compliance with the projected traffic growth requirement of 192 voice channels.

Yet another alternative is shown in Figure 8–4, where M 13 multiplexers are used with one fiber pair to provide a transmission capacity of 28 DS 1 signals (672 voice channels) without rearrangement. The optical link data rate is now 44.736 Mb/s, and this arrangement is quite vulnerable. If any fiber, any optical terminal, or either multiplexer fails, all system traffic will be disrupted. Only two fibers and two transmitters and receivers are required, however.

It should be noted that the M 13 multiplexer is more sophisticated, and thus more expensive, than the M 12 multiplexer. From the figure, it is obvious that the M 13 units provide substantially more capacity in the initial installation, with the penalty being higher data rate and sophistication and higher cost.

The commonality of equipment that is introduced by higher level digital multi-

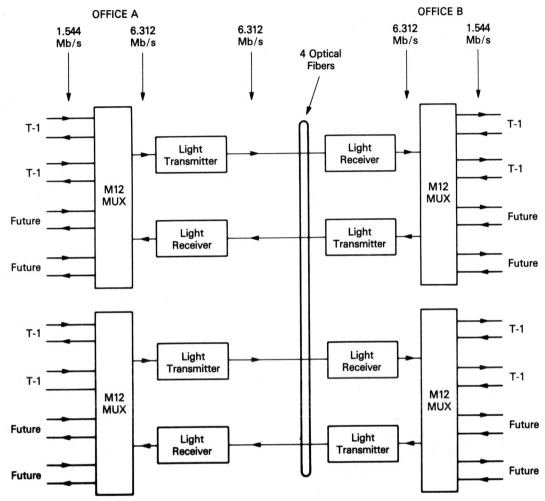

FIGURE 8-3 Independent Subsystems

plexing is clearly evident in Figure 8-4. Although the drawing shows only a single fiber pair employed to demonstrate the digital multiplexing technique, it is possible to provide and equip a second fiber pair to protect continuity of service. As shown, a failure of this single unit (the M 13) at either end of the link will introduce a 100 percent disruption of service.

EXPANDING TRANSMISSION CAPACITY _____

The design shown in Figure 8-3 can be expanded at any later date, should the need arise, by adding an M 23 multiplexer to each terminal. Figure 8-5 shows the new configuration.

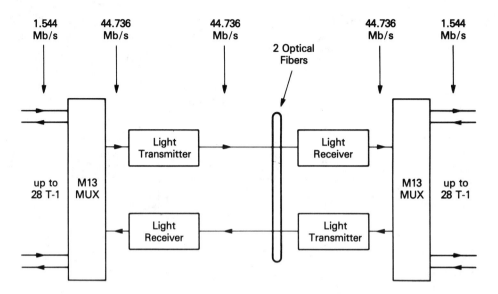

FIGURE 8–4
High-Level
Multiplexing

Note the increase in optical link data rate when the addition is made. The fibers placed initially, together with the electro-optical terminal units, would have to have been selected on the basis of this higher transmission rate even if the link was operated at a substantially lower rate initially. Note that the M 12 multiplexers shown in the figure (any number up to a total of seven M 12s can be accommodated) need not all be installed initially. These may be added incrementally as the traffic growth develops.

WAVELENGTH DIVISION MULTIPLEXING

It is possible to transmit more than one lightwave signal through an optical fiber if each signal transmitted is of a different light wavelength. This is analogous to frequency division multiplexing in cable carrier systems or multiple television channel transmission in cable television systems. By employing filtering techniques at both fiber ends, two or more lightwave carriers can be multiplexed on a single fiber. The devices required are called WDM couplers, and they are purely optical in nature, consuming no operating power. Because they are passive, the reliability will be very high.

Figure 8–6 shows a WDM coupler (filter). The separation of lightwave frequencies is indicated by the two different frequency designations. The application is assumed to be the same as in the example presented earlier in the chapter. Initial service requirement is 96 voice circuits, with expansion capability to 192 circuits at a later date.

Figure 8–7 shows a single fiber pair (two fibers) employing WDM couplers to duplicate the transmission capacity of the configuration of Figure 8–1, which was

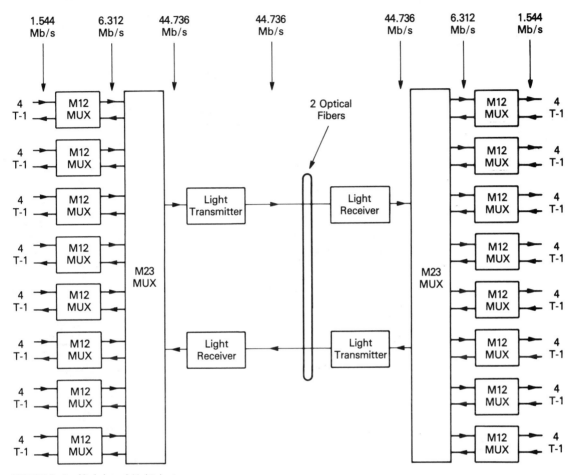

FIGURE 8-5 High-Level Multiplexing

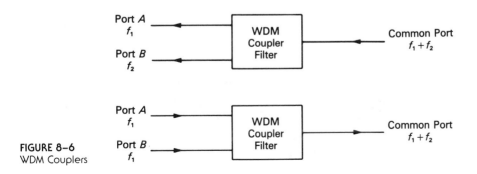

FIGURE 8-6
WDM Couplers

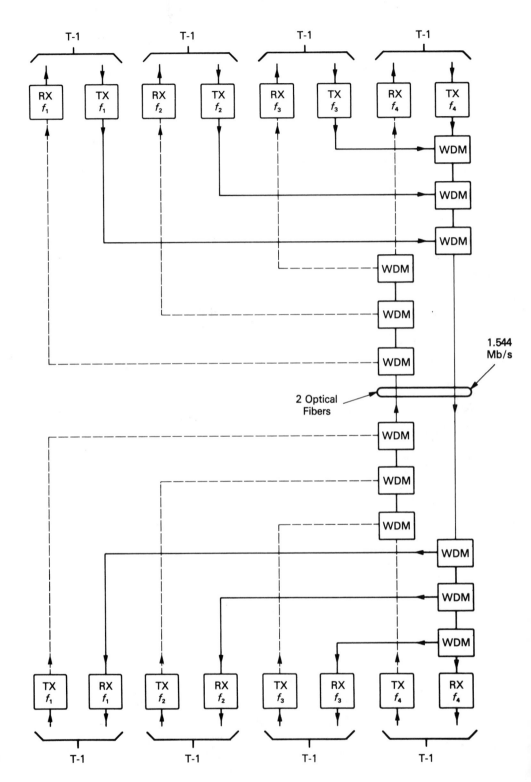

FIGURE 8–7
WDM
Multiplexing

130

four T-1 systems. Note that the data rate on the optical link is only 1.544 Mb/s and the design requires only two fibers with eight transmitters and receivers. This configuration is quite vulnerable to failure: if either fiber is disrupted, all traffic is interrupted.

This design can be improved significantly by operating bidirectionally on each fiber, as shown in Figure 8-8. Note that the data rate is unchanged and the number

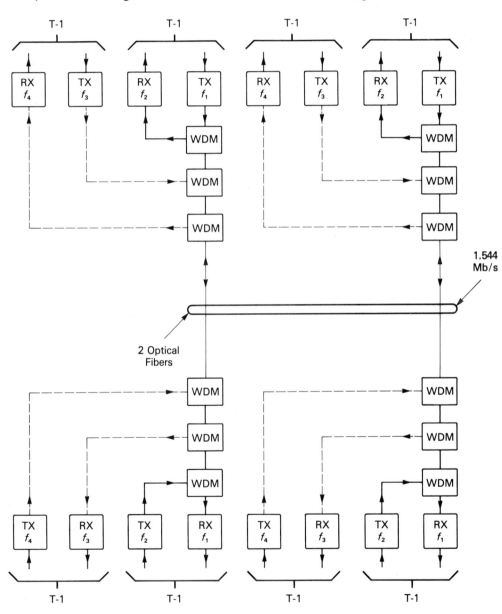

FIGURE 8-8
Bidirectional
WDM

and kind of materials and fiber count are essentially the same, but if a fiber fails in this system, only 50 percent of the traffic carried would be interrupted.

The systems in both Figures 8–7 and 8–8 are in accordance with the initial traffic capacity requirement, but if traffic growth occurs some technique to expand traffic capacity must be applied. The configuration of Figure 8–8 provides the same traffic capacity, but with a higher level of reliability, as that of Figure 8–2, which used digital multiplexing alone.

In Figure 8–9, WDM is used in conjunction with low-level (M 12) digital multiplexing to provide a capacity of eight T–1 systems on two fibers. The arrangement is bidirectional and requires only two fibers with four transmitters and receivers. A failure of any single fiber or any single multiplexer will interrupt only 50 percent of

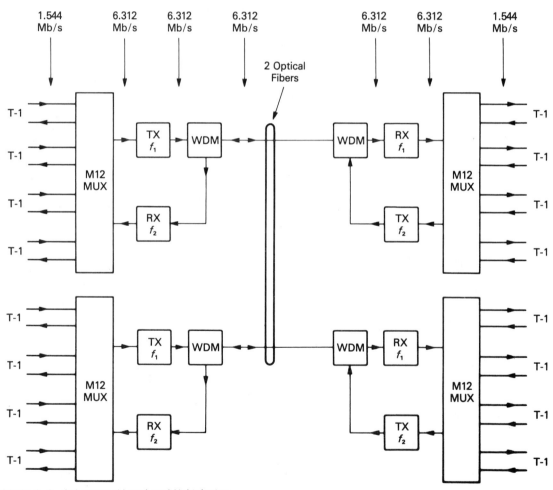

FIGURE 8–9 Bidirectional Low-Level Multiplexing

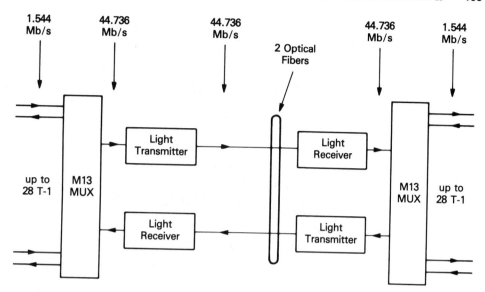

FIGURE 8–10
High-Level
Multiplexing

the traffic carried, and the configuration has a capacity of eight T–1 systems, making it fully responsive to the traffic growth prediction in the earlier example.

Figure 8–10 shows high-level digital multiplexing using M 13 units with the same transmission capacity shown earlier in Figure 8–4.

EVALUATING DESIGN ALTERNATIVES

Not all transmission systems can be characterized as to size and capacity. Neither is there one single, clearly superior design that will best fit all applications. The alternative techniques presented here all offer advantages and disadvantages. They are not presented as firm recommendations or as an endorsement of any particular approach. They may all be implemented with off-the-shelf hardware available today, and from a purely transmission point of view they are all technically acceptable.

The applications engineer is obliged to consider the initial cost of a system, the probability of growth, and several other factors such as system maintainability and reliability of service in the decision-making process. No tutorial such as this can relieve the designer of that basic responsibility, nor should it. In a separate chapter we shall address, in greater depth, the basic question of protection of service and protective design configurations.

When the application is clearly defined as a trunking interconnection between two discrete service terminals and the traffic density is substantial, the design problem itself is well defined, and the options are somewhat limited. As lightwave technology intrudes into lower levels of the network plant, in limited-capacity trunking to remote switching terminals, or in other applications in the distribution plant, the options in system design are more numerous. It is axiomatic that these same applications must be supported by a lower level revenue base.

The responsibility for developing system designs that are cost effective and that efficiently utilize the inherent capabilities of the optical fiber placed is emphasized in these lower level situations. It is in just such situations that the designer's grasp of the technology, its capabilities, and its limitations will have the greatest impact. Compared to other transmission media, such as multiple-pair or coaxial cables, even the least sophisticated optical fibers represent a substantial increase in bandwidth available for use, and thus in link capacity. The efficient utilization of optical fibers placed will depend to a very large degree on how clearly this concept is understood and how effectively this understanding is applied in the design effort.

The applications engineer should think in terms of transmission capacity (bandwidth) rather than in terms of discrete transmission facilities (such as pair or fiber counts). Every application addressed should be reviewed from the point of view of the inherent transmission capacity the placed facility will provide along its entire length, and even perhaps beyond the initially identified terminal location.

The engineer should not limit this thinking to artificial bounds imposed by stereotyped definitions of service, either. From a strictly transmission point of view, there is little distinction to be made between digital signals carrying trunking information and digital signals carrying subscriber loop or distribution plant signals. Indeed, an optical facility extended into or through a geographical area may very well serve both purposes simultaneously without compromising either level of service.

MULTIPLE SERVICE POINTS

Suppose an application is presented for design that requires a total of four T–1 systems (96 equivalent voice channels) between offices *A* and *B*. The distance (cable route length) between these two offices is 12 km, and the traffic study projects possible future growth to eight T–1 systems (192 equivalent voice channels). A design shown as Figure 8–11 is developed and proposed, with future expansion to be handled by upgrading to M 23 digital multiplexing at a later date, as shown in Figure 8–12. Initially, the traffic will be distributed across the two independent systems shown in Figure 8–11, with no protection switching provided at all.

For initial transmission design purposes, the optical link data rate is taken to be 44.736 Mb/s, and the optical fibers and optical terminal units are selected to function satisfactorily at that rate. Initially, they will be operating at the DS 2 rate (6.312 Mb/s), as shown in Figure 8–11.

In submitting this proposed design, the applications engineer has been completely responsive to the requirements he or she was given. Both initial and projected traffic can be accommodated, and transmission engineering of the proposed system can proceed. This scenario is more than academic: it is typical of the applications engineering performed in many cases today, and the engineer would be considered fully competent in such a scenario.

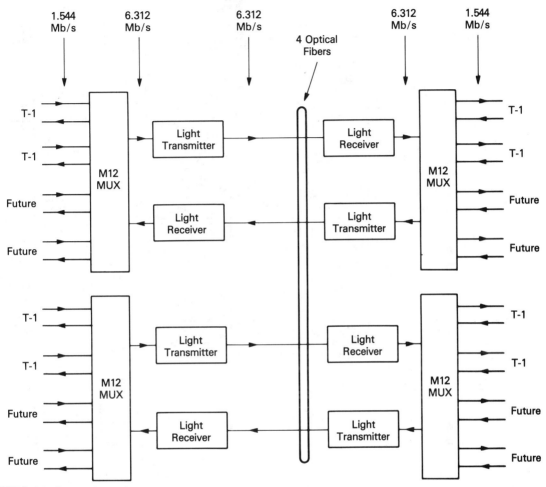

FIGURE 8–11 Sample Initial Design

However, a more enlightened approach might be for the engineer to ask a few questions before proceeding with the link transmission engineering work:

1. What transmission capacity does the facility present over and above the stated requirements?

2. What can be done at this point to preserve any such excess capacity as a usable future option?

3. What are the penalties of such preservation?

Since the initial design resolved the data rate in the optical link to be 44.736 Mb/s, the fiber was selected to operate satisfactorily at this rate over this distance. The fiber then presents the capability of carrying several lightwave signals (different

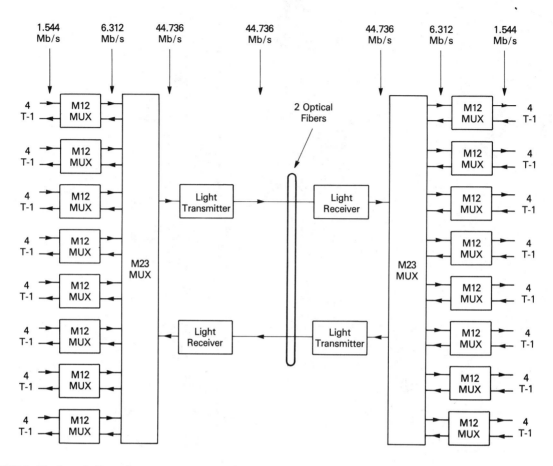

FIGURE 8-12 Sample Upgrade

wavelengths) at this same data rate or lower. The points where such additional services might be delivered need not even be determined. Indeed, a point may even be serviceable beyond the point *B* terminal, if the fiber is later extended. Note that the option just presented cannot be accomplished by digital multiplexing up to a higher level, since that process is limited to the established terminal points themselves, i.e., points *A* and *B*. However, it is possible, should the need arise, to add a second lightwave carrier and duplicate the 44.736 Mb/s capacity between points *A* and *B*.

What penalties, both technical and economic, might be incurred in preserving this option? From a technical point of view, all that is necessary is to include the insertion loss of any potential WDM couplers in the calculations of link loss performed in the initial transmission engineering process. From an economic point of view, the cost of the WDM couplers would be in the range of hundreds of dollars, an insignificant amount in a project that is costing in the tens of thousands of dollars. And even this nominal economic penalty need not be accepted until the service requirement actually evolves. Note that whether the new service is trunk or local is irrelevant as far as transmission performance is concerned.

It would seem a reasonable obligation of the applications engineer not only to consider such alternatives in every case, but to articulate them to management also, just as is required in proposing a final design or designs.

SUMMARY

Applications engineering is the process of developing system configurations that satisfy the technical requirements of the application. All configurations that are developed are evaluated as to cost, reliability, and future expandability or flexibility of use.

Lightwave facilities present substantially more transmission capacity than metallic conductor facilities, and several techniques are available for expanding the transmission capacity of lightwave facilities. The three techniques presently in use are digital multiplexing, wavelength division multiplexing, and increasing optical fiber count. Higher level digital multiplexing requires greater transmission system bandwidth: wavelength division multiplexing does not.

When high-level digital multiplexing is employed, some commonality of equipment is introduced into the system, and redundant, protection-switched terminal equipment is often installed because of this.

It is possible, using WDM techniques, to provide service at any point along a lightwave facility cable route.

WDM couplers do not require or consume operating power, and they are less expensive than digital multiplexers.

For higher traffic densities between two discrete terminals, digital multiplexing is most generally employed. For lower traffic densities, or to provide multiple points of service along a lightwave facility route, WDM techniques are more efficient.

REVIEW QUESTIONS

True or False?

T F 1. The characteristics of the terminal equipment alone determine the traffic carrying capacity of a lightwave facility.

T F 2. Adding optical fibers to a lightwave facility is one way to add transmission capacity to that facility.

T F 3. Digital multiplexing is a process whereby more than one digital signal can be consolidated into a single electrical signal with a higher transmission data rate.

T F 4. When digital multiplexing is employed to increase a system's capacity, it requires that the system provide greater transmission bandwidth.

T   F   5. A T–1 line code (a DS 1 signal) is a data rate of 6.312 Mb/s.

T   F   6. An M 12 digital multiplexer can only be used to combine two DS 1 signals into one DS 2 signal.

T   F   7. A 45 Mb/s signal can accommodate 672 equivalent voice circuits or channels.

T   F   8. The digital signal most commonly employed on twisted pair cables is T–1 (DS 1).

T   F   9. In paired cable applications, a T–1 system usually requires four conductors (two pairs).

T   F   10. An M 12 multiplexer is substantially more complex than an M 13 unit, and costs more.

T   F   11. It is possible to operate a digital multiplexer with a lower traffic load than its maximum capacity.

T   F   12. More than one lightwave signal can be carried on a single optical fiber at the same time.

T   F   13. Different wavelength light signals can be isolated from each other using filters.

T   F   14. Wavelength division multiplexing couplers require the application of local operating power to function.

T   F   15. Only unidirectional operation can be provided using WDM on a single fiber.

T   F   16. A fiber capable of supporting 90 Mb/s data transmission at one wavelength will support that same transmission data rate on any other wavelength.

T   F   17. WDM can increase the traffic carrying capacity of a fiber, but cannot provide service at more than one point along the length of the fiber.

T   F   18. WDM couplers are more complex, and generally cost more, than digital multiplexers.

T   F   19. Bidirectional WDM couplers must provide more isolation between wavelengths than unidirectional couplers.

T   F   20. The transmission capacity of any lightwave facility can be increased by providing additional optical fibers.

CHAPTER 9

Transmission Engineering
Lightwave Systems

Lightwave systems are systems that utilize lightwave energy transmission through optical fibers to transport information between two or more points. The task of examining the links in a lightwave system to determine whether the composite system formed out of any combination of particular elements will produce acceptable transmission quality and capacity is called *transmission engineering*.

If the physical length of an optical fiber is excessive, the attenuation introduced by the fiber may present too low an optical input signal to the photodetector. The detector may then be unable to produce an electrical output signal that faithfully replicates the system input signal. The result will be a noisy electrical output signal in the case of an analog system, or an unacceptably high rate of error in a digital system. Such systems are *attenuation limited*.

Even if the optical signal level input to the photodetector is of sufficient amplitude for the detector to operate properly, the optical signal may become distorted by other propagation characteristics of the interconnecting fiber. The result will again be noisy analog output signals or a high BER digital output. In a digital system, the pulse dispersion will be more pronounced if multimode interconnecting fiber is employed and will be more of a problem at higher transmission data rates also, even if single mode fiber is employed. Such systems are *dispersion limited*.

When the physical length of the interconnecting fiber is such that the optical receiver output is unsatisfactory, or when pulse distortion excessively degrades the recovered information, it may be necessary to break the system down into shorter transmission links by inserting a regenerative repeater. In such cases, each individual transmission link must be addressed as a separate link design effort.

In every instance, the design effort must examine both transmission attenuation and distortion. We can logically expect that one or the other limitation will prevail in a particular application; that is, the application will become attenuation limited before distortion becomes unacceptable, or vice versa.

SYSTEM BANDWIDTH AND ATTENUATION

The information bandwidth of any transmission system defines the spectrum of frequencies that the system presents for use in transmission. This information bandwidth (usable bandwidth) may be clearly defined regardless of the modulation technique employed or the nature of the information the system is expected to transport.

The information bandwidth is generally defined as the range of frequencies that may be transmitted through the system and recovered at the distant end with a loss of signal amplitude no greater than 3 dB below that frequency that is passed through the system with the least attenuation. Since –3 dB is a power ratio of 1 to 2, it represents a loss of signal level no greater than 50 percent of the highest amplitude signal level. This level is often referred to as the half-power point.

If the information bandwidth presented by a system is not sufficiently broad, distortion will be introduced into information passed through the system, be it analog or digital. In an analog system the distortion will be evidenced as a loss of higher frequency information. In a digital system the distortion will be evidenced in a broadening or misshaping of the input pulses, which may result in incorrect detection of pulse presence or absence at the receiving terminal.

The information bandwidth presented for use by any system is affected by the various elements comprising the system. In a lightwave system these elements are the terminal equipment and the optical fiber itself. It may be helpful to draw some analogies between optical fibers and metallic conductors, but there are some basic differences that must be clearly understood.

In conventional paired telephone cables, the transmission losses are greater for higher frequencies transmitted over the pair than for lower frequencies. These unequal losses can be compensated for by periodically equalizing the cable loss so that the end-to-end transmission is relatively flat across a range of frequencies. This rather simple treatment permits the use of the pair for long cable lengths across a relatively wide range of frequencies.

At higher transmission frequencies, say 1 MHz or so, we might still equalize the pair, but the treatment would require much more frequent equalization and probably frequent reamplification as well. Since remedial measures are readily available and generally understood, we do not usually specify the transmission bandwidth of a pair as a function of the length of the facility.

In coaxial cable systems, transmission losses are also greater at higher transmission frequencies. These differences are, however, readily compensated for by cable equalization also, and a long length of coaxial cable can, for all practical purposes, present the same usable bandwidth that a shorter length presents. Thus, in metallic conductor systems, the bandwidth of a system is considered to be independent of the physical length of the facility, because correction is easily accomplished at the terminal ends, or periodically along the facility length, or both.

This is not the case in optical fiber facilities. The factor that limits usable bandwidth in a digital system is the dispersion (pulse broadening) of the lightwave signal that is introduced as the signal passes through the fiber. At this time, we have no practical technique for compensating for this dispersion at lightwave frequencies.

A longer length of fiber will introduce more dispersion than a shorter length, so that the information bandwidth of a specific fiber is a function of the physical length of the facility in which it is employed. For this reason, optical fiber specifications must call out the usable bandwidth presented by some stated unit of fiber length. The unit employed in multimode fiber systems, for example, is bandwidth (in MHz or GHz) per kilometer of fiber length. No such reference is required in metallic conductor specifications, and this is a significant difference between optical fibers and metallic conductors.

Optical fiber bandwidth may be specified in frequency (MHz × km or GHz × km), or it may be calculated from the rise time per unit of fiber length (ns/km for multimode fibers). Attenuation, on the other hand, is simply stated as units of loss (dB) per unit of fiber length, usually dB/km.

Pulse dispersion can be directly related to the usable bandwidth that a fiber presents for transmission. In a digital application, the bandwidth required to satisfactorily transport a digital signal is generally taken to be the same as the transmission bit rate itself. For example, a 45 Mb/s transmission bit rate requires 45 MHz of fiber bandwidth; a 90 Mb/s data rate requires 90 MHz of bandwidth.

In an analog application, such as a CATV system carrying a number of television signals, the bandwidth required will be determined by the bandwidth of the information to be transported. Thus, if a single TV channel requires, say, 12 MHz (as it might if it employed frequency modulation), and we are attempting to carry five such signals, then the lightwave system bandwidth must be at least 60 MHz end to end (12 × 5 = 60).

The fiber manufacturer will specify the bandwidth/length characteristic of a particular fiber, or we can calculate the bandwidth presented for use by a particular fiber of any given length from the time domain performance specifications published by the fiber manufacturer. Thus, for any given system length, we can predict the end-to-end bandwidth that the fiber will provide. Consequently, we can easily validate the usable length of any particular fiber for any particular transmission bit rate or information bandwidth (as in the case of the television signals).

In digital applications, the pulse distortion characteristics of multimode fibers are different from those of single mode fibers. We shall make calculations for each type, but expect a gross difference in the usable length of the two types of fiber.

SINGLE MODE BANDWIDTH LIMITATIONS

Typical dispersion characteristics for single mode fibers available today are on the order of 3.5 ps/nm-km in the 1,250 to 1,300 nm wavelength range. Thus, using a light source spectral width of 3 nm, this corresponds to a bandwidth length factor of about 34 GHz-km or so. We arrive at this conclusion by first calculating the end-to-end dispersion through a 1 km length of fiber as follows. We have

$$T = \frac{t \times \text{FWHM} \times L}{1,000}$$

where

$$T = \text{End-to-end system dispersion (ns)}$$

$$t = \text{Fiber dispersion (ps/nm-km)}$$

$$\text{FWHM} = \text{Laser line width (nm)}$$

$$L = \text{Link length (km)}$$

then

$$T = \frac{3.5 \times 3 \times 1}{1,000}$$

$$= 0.0105 \text{ ns or } 10.5 \text{ ps}$$

System dispersion can be related to system bandwidth by the formula

$$\text{BW} = \frac{440}{T \text{ (ns)}}$$

or

$$\text{BW} = \frac{440,000}{T \text{ (ps)}}$$

where

$$\text{BW} = \text{System bandwidth (MHz)}$$

$$T = \text{System dispersion (ns or ps)}$$

then

$$\text{BW} = \frac{440}{0.0105} = 41,904 \text{ MHz, or } 41.9 \text{ GHz}$$

The fiber just discussed would be specified by the cable supplier to have a bandwidth/length characteristic of 42,000 MHz-km or 42 GHz-km. We can dispense with the dispersion calculation entirely, of course, if the manufacturer specifies the fiber bandwidth.

To find out how long a link of such a fiber could be before it became dispersion limited in a particular application, we have to assume or assign some transmission data rate for that application. Let us arbitrarily assume a 275 Mb/s application is involved. Then we require 275 MHz of bandwidth from the fiber end to end, however long it may be. We use one of the following equivalent equations:

$$BW_{fiber} \times (MHz\text{-}km) = BW_{sys} (MHz) \times Length (km)$$

$$BW_{sys} (MHz) = \frac{BW_{fiber} (MHz\text{-}km)}{Length (km)}$$

$$Length (km) = \frac{BW_{fiber} (MHz\text{-}km)}{BW_{sys} (MHz)}$$

Thus, given a 275 MHz requirement and a bandwidth/length factor of 42 GHz, we calculate the usable length as follows:

$$L (km) = \frac{BW (MHz\text{-}km)}{BW (MHz)} = \frac{42,000}{275} = 152.1 \text{ km (94.9 mi.)}$$

It is self-evident that a fiber of the quality of 3.5 ps/nm-km would be attenuation limited long before it was extended out to a length of 152 km. If we assume a fiber loss/length characteristic of 0.6 dB/km, 152 km of fiber would introduce 91 dB of loss even discounting all splicing losses. We would require a very high-amplitude (optical) input signal and a very sensitive photodetector indeed to accommodate such a long link. This is not possible with today's equipment.

In any event, the development, though instructional, is somewhat academic. In practice, we are usually trying to determine what bandwidth/length characteristic is required for the length and data rate of the application under design.

Let us assume a 40 kilometer link length. We know the system bandwidth requirement to be 275 MHz from a data rate of 275 Mb/s. How do we determine the required fiber characteristic for the 40 km length of fiber? We have

$$BW_{fiber} (MHz\text{-}km) = BW_{sys} (MHz) \times L(km)$$

$$= 275 \times 40$$

$$= 11,000 \text{ MHz-km, or 11 GHz-km}$$

The fiber we examined earlier had a bandwidth/length characteristic of 42 GHz-km, well in excess of this requirement. We could thus use this fiber in a 40 km long link and produce the required end-to-end bandwidth.

What if the link length were 60 km? Given the same 275 Mb/s data rate, the calculation is then

$$BW \text{ (MHz-km)} = 275 \times 60 = 16,500 \text{ MHz-km, or } 16.5 \text{ GHz-km}$$

Again, the 42 GHz-km fiber evaluated earlier would be adequate. In a single mode fiber, 16.5 GHz/km is not a particularly severe specification, but the lower requirement of 16.5 GHz-km might be a less expensive fiber specification than the 42 GHz-km.

In our example, we required an optical bandwidth of 275 MHz. If the fiber length is shorter, we can relax the fiber bandwidth/length specification. For example, if the cable route length is only 30 km instead of 60 km, the fiber characteristic required is only 8,250 MHz-km (8.25 GHz-km) calculated from

$$BW \text{ (MHz-km)} = 275 \times 30 = 8,250 \text{ MHz-km, or } 8.25 \text{ GHz-km}$$

Operations at transmission bit rates lower than 275 Mb/s present a less severe requirement, so we can relax the fiber specification accordingly. For example, a 30 km fiber operating at 45 Mb/s (45 MHz of required bandwidth) need only have a 1,350 MHz-km characteristic; that is,

$$BW \text{ (MHz-km)} = 45 \times 30 = 1,350 \text{ MHz-km, or } 1.35 \text{ GHz-km}$$

TIME DOMAIN DEVELOPMENT

System engineering based upon time domain calculations is rarely encountered any more. It is general practice simply to relate all functions to bandwidth, instead. Nevertheless, we could make the development in time domain factors alone, and we should be at least familiar with this process.

Earlier we calculated the end-to-end dispersion of the fiber under examination to be .0105 ns/km of length, including the factor of laser line width, as well as fiber dispersion. If we now assume a fiber length of 60 km, then the total fiber rise time contribution would be 0.63 ns (from 60×0.0105). In Table 5–1, we found that a transmission bandwidth of 275 MHz requires an overall rise time of 1.28 ns or less. All that is necessary now is for the designer to select a transmitter and receiver whose combined rise times, together with the fiber rise time, do not exceed this allowable limit.

We combine system element rise times to develop overall rise time as follows:

$$RT_{sys} = 1.1 \sqrt{RT_{tx}^2 + RT_{rx}^2 + RT_{fiber}^2}$$

or

$$RT_{sys} = 1.1 \left(RT_{tx}^2 + RT_{rx}^2 + RT_{fiber}^2 \right)^{0.5}$$

where

$$RT_{sys} = \text{Total system rise time (ns)}$$

$$RT_{tx} = \text{Transmitter rise time (ns)}$$

$$RT_{rx} = \text{Receiver rise time (ns)}$$

$$RT_{fiber} = \text{Fiber rise time (ns)}$$

This formula can be solved mathematically for any of the unknown variables, thus, a complete system solution can be developed, including individual specification of discrete system elements.

EXAMPLES OF SYSTEM ANALYSIS

If we make some simplifying assumptions in some of the formulas we have introduced, we can develop an effective bandwidth calculation methodology that produces a reasonably accurate basis for rapid evaluation of single mode fiber system alternatives. From before we have

$$BW = \frac{440,000}{FD \times L \times FWHM}$$

where

$$BW = \text{Effective system bandwidth (MHz)}$$

$$FD = \text{Fiber dispersion (ps/nm-km)}$$

$$L = \text{Fiber length (km)}$$

$$FWHM = \text{Laser spectral line width (nm)}$$

Solving for length L, we obtain

$$L = \frac{440,000}{FD \times BW \times FWHM}$$

If we now assume that the transmission bit rate is equal to the optical bandwidth, we can easily calculate the effect on system length of various changes in other system elements. For example, if the system requires 90 MHz of bandwidth (a data rate of 90 Mb/s), and we use a single mode fiber with 4.0 ps/nm-km dispersion and a laser with a spectral width of 5 nm, the link can be 244 km long before being bandwidth restricted

$$L = \frac{440,000}{4 \times 90 \times 5} = 244 \text{ km, or 151 miles}$$

A link of this length would obviously become attenuation limited well before it became bandwidth limited.

The same fiber, using the same laser, but applied in a 560 MHz (560 Mb/s) system, could be 39 km long

$$L = \frac{440,000}{4 \times 560 \times 5} = 39 \text{ km, or 24 miles}$$

If a lower-grade fiber were employed that had an 18 ps/nm-km dispersion characteristic, and the same laser was used in a 560 MHz system, the length limitation would be 8.7 km:

$$L = \frac{440,000}{18 \times 560 \times 5} = 8.7 \text{ km, or 5.4 miles}$$

If we now substitute a different laser whose spectral width is only 0.8 nm, but use the 18 ps/nm-km fiber in a 560 Mb/s system, the allowable length would be 54.5 km:

$$L = \frac{440,000}{18 \times 560 \times 0.8} = 54.5 \text{ km, or 33.9 miles}$$

If we use a dispersion-shifted fiber, we could get 1.5 ps/nm-km dispersion, and a late-generation laser might provide a line width of perhaps 0.8 nm. In a 560 MHz system, this equates to a bandwidth length limitation of 654 km:

$$L = \frac{440,000}{1.5 \times 560 \times 0.8} = 654 \text{ km, or 406 miles}$$

Such performance is curently available in these items. It is obvious that in using single mode fibers, system bandwidth will become less and less limiting as a design factor, at least until transmission data rates are raised high enough to approach the theoretical limits.

MULTIMODE BANDWIDTH LIMITATIONS

Typical multimode fibers available today provide a usable bandwidth on the order of 600 to perhaps 2,000 MHz-km. This variation is attributable to the composition of the fiber itself and may be reflected in a difference in fiber or cable cost. To show the impact on system designs, we can do some calculations using both fiber bandwidths. A 2,000 MHz-km fiber would represent superior performance in that it would provide more usable system bandwidth, and thus could support higher transmission bit rates.

As discussed in Chapter 5, we can apply a length-dependence factor that reflects the fact that higher order modes of propagation will incur more attenuation in a multimode fiber than lower order modes. Thus, a fiber of given length will exhibit greater usable bandwidth than its bandwidth length characteristics might suggest. For example, a 10 km multimode fiber presenting 600 MHz-km of bandwidth for use might be expected to provide 60 MHz of usable bandwidth end to end:

$$\text{BW}_{sys} \text{ (MHz)} = \frac{\text{BW}_{fiber} \text{ (MHz-km)}}{L \text{ (km)}} = \frac{600}{10} = 60 \text{ MHz}$$

The preceding calculation is appropriate for single mode fiber, but is in error for multimode fibers because it does not reflect the length-dependence factor. A generally accepted dependence factor would raise the physical fiber length to a power of 0.8 ($L^{0.8}$). Taking into account this factor, the calculation becomes

$$L^{0.8} = 10^{0.8} = 6.3 \text{ km effective length}$$

Then the physical fiber length of 10 km becomes an effective fiber length of 6.3 km. Since the fiber was specified to provide a bandwidth of 600 MHz-km, the end-to-end bandwidth (through a 10 km length) will be 95 Mhz:

$$\text{BW}_{sys} \text{ (MHz)} = \frac{\text{BW}_{fiber} \text{ (MHz-km)}}{L \text{ (km)}} = \frac{600}{6.3} = 95 \text{ MHz}$$

Note that this is substantially more bandwidth than the earlier calculation produced when the length-dependence factor was not applied. Remember that the length-dependence factor only applies to multimode fibers.

The length-dependence factor reduces a given physical fiber length to the effective fiber length by raising the given length to the 0.8 power. Conversely, raising the effective length to the 1.25 power will produce the physical length.

For a 90 MB/s data rate, we require 90 MHz of bandwidth end to end. We can calculate the maximum physical length of a fiber that will still provide this bandwidth if the fiber has a bandwidth length characteristic of 600 MHz-km, as follows:

$$L \text{ (km)} = \left[\frac{\text{BW}_{fiber} \text{ (MHz-km)}}{\text{BW}_{sys} \text{ (MHz)}} \right]^{1.25} = \left[\frac{600}{90} \right]^{1.25} = 10.7 \text{ km}$$

We can validate this procedure by calculating the effective length of such a fiber as follows:

$$L \text{ (km)} = \frac{BW_{fiber} \text{ (MHz-km)}}{BW_{sys} \text{ (MHz)}} = \frac{600}{90} = 6.66 \text{ km (effective length)}$$

Applying the length factor 0.8 to the effective length, we obtain

$$6.66^{\frac{1}{0.8}} = 6.66^{1.25} = 10.7 \text{ km (physical length)}$$

A multimode fiber with a usable length of 6.6 km would actually be 10.5 km long. Then, using a multimode fiber that is specified to provide 600 MHz/km, a 90 Mb/s system would become length limited by fiber bandwidth (by pulse distortion if in a digital application) at a length of 10.5 km (6.5 mi.). So up to this distance, we can ignore bandwidth and address the system design as a problem of transmission level and fiber attenuation alone.

We can perform the same calculation for a better grade of fiber. At 90 Mb/s we require 90 MHz of bandwidth. If the fiber has a bandwidth-length characteristic of 2,000 MHz-km, then dividing 2,000 by the required end-to-end bandwidth, 90 MHz, gives a usable fiber length of 22.2 km. The length-dependence factor will increase this distance to 48 km:

$$L \text{ (km)} = \left[\frac{BW_{fiber} \text{ (MHz-km)}}{BW_{sys} \text{ (MHz)}} \right]^{1.25} = \left[\frac{2000}{90} \right]^{1.25} = 48.2 \text{ km}$$

Then, using a multimode fiber that is specified to provide 2,000 MHz-km, a 90 Mb/s system would become length limited by fiber bandwidth at a length of 48 km (29.8 mi.). Up to this distance, we can ignore bandwidth and address the system design as a problem of transmission levels and attenuation alone.

It is instructive to calculate the effect of lower transmission data rates. Let us assume a 45 Mb/s application where we require only 45 MHz of end-to-end bandwidth. We can use the same two grades of multimode fiber we used earlier and see clearly the impact of lower data rates on fiber specifications and system length:

$$L_{600} = \left[\frac{600}{45} \right]^{1.25} = 25.4 \text{ km}$$

$$L_{2,000} = \left[\frac{2000}{45} \right]^{1.25} = 114.7 \text{ km}$$

Thus, if the fiber has a bandwidth length characteristic of 600 MHz-km, it may have a physical length of 25.4 km, and if it has a bandwidth-length characteristic of 2,000 MHz-km fiber, it may have a physical length of 114.7 km. Clearly, the better grade fiber (2,000 MHz-km) will permit a longer system, but if we only needed a link 25 km long, the lower-grade fiber would be quite satisfactory and might be less expensive.

SUMMARIZING THE FIBER BANDWIDTH FACTOR _____

We have demonstrated each of the following results by calculation:

1. At a transmission data rate of 45 Mb/s, using a lower grade of multimode fiber (600 MHz-km), we are not bandwidth limited until the fiber length exceeds 25 km. By using a better grade fiber (2,000 MHz-km), we can extend this limitation to 114 km.

2. At a transmission data rate of 90 Mb/s, using a lower grade of multimode fiber (600 MHz-km), we are not bandwidth limited until the fiber length exceeds 10.7 km. By using a better grade fiber (2,000 MHz-km), we can extend this limitation to 48.2 km.

3. At a transmission data rate of 90 Mb/s, using a typical grade of single mode fiber, we may not be bandwidth limited until the fiber length exceeds 244 km, depending upon the laser light source employed.

We could extend the preceding series of calculations to higher data rates, but the methodology is clear. For systems design efforts within these established limitations, using fibers of these qualities, we can confidently dispense with any further consideration or calculation of fiber rise time or bandwidth. This could significantly reduce the design effort by restricting it to the much simpler considerations of transmission levels and attenuation alone.

FIBER TRANSMISSION LOSS _____

Through the mechanisms discussed earlier, lightwave energy is attenuated as it propagates through optical fibers. Different types and grades of fibers will introduce different degrees of attenuation, but all fibers will introduce some transmission loss. Of course, these losses may be different for different wavelengths passing through the same fiber.

The fiber or cable manufacturer will specify the transmission loss at any particular wavelength. The units generally employed for both single mode and multimode fibers are decibels of loss per kilometer of fiber length (dB/km). Calculating the end-to-end loss of any particular fiber length requires only that we multiply the length unit loss (dB/km) by the overall cable length. For example, a fiber specified to introduce 0.6 dB of loss per kilometer will introduce 6 dB of loss end to end through a 10 kilometer length ($10 \times 0.6 = 6$).

FIBER SPLICING LOSS _____

Optical fiber cables can be neither manufactured nor placed in a single continuous length that would be adequate for most applications. Thus, some splicing of fibers is usually required, and every such splice will introduce some transmission loss. In designing links, we must estimate and allow for the losses that will be introduced by splicing.

Optical cables are physical plant, and consequently, are vulnerable to mechanical damage in either aerial or underground installations. In restoring fiber continuity in such cases, additional fiber splices may be required. It is prudent design practice to allow for some additional splicing loss to cover any repairs or restoration work. Assuming a conservative 0.3 dB per individual fiber splice, and supposing that one splice would be required per kilometer of cable length, this should be adequate to handle any subsequent splicing that might be required in the event of mechanical damage to the cable at a later date.

The loss figure of 0.3 dB per splice is entirely arbitrary. Experience or individual judgment might develop a different allowance, and such a figure could be technically defensible. We merely point out the necessity for including an estimate or allowance rather than proposing a precise value.

Given a fiber loss of 0.6 dB per kilometer and a cable length of 10 km, the fiber loss would be 6 dB as we calculated earlier, and the splicing allowance (using 0.3 dB per km) would be 3 dB. The physical plant loss would be 9 dB total in this case, for system design purposes.

In some installations it may be desirable to provide disconnect-reconnect capability between the opto-electronic terminal equipment and the optical fibers themselves. In multiple-fiber installations, this can provide for optical fiber patching between several fibers or terminal units. In such cases, an optical connector is used instead of a "hard" fiber splice. Due to the dimensions of the optical fibers and the close tolerances required of the connectors, a prudent but, again, arbitrary loss figure of 1.0 dB per connector is suggested. If a connector is employed at both terminals, then 2 dB of additional link loss must be assumed.

We are obliged to connect the terminal units (transmitters and receivers) to the interconnecting fibers. Most generally, this will involve splicing a fiber pigtail that is supplied as part of the terminal equipment, either directly to the interconnecting fibers, or to the optical connectors if they are provided. It is then necessary to define the losses incurred in these connections. For example, the transmitter manufacturer will specify the light output of the product. We must identify where this energy level is specified. If it is specified to be at the end of a factory-provided fiber pigtail, then the system designer must consider the transmission losses incurred in connecting this pigtail to the system, and this is equally true at the receiver input end of the system.

Chapter 7 dealt with a variety of optical devices that might be inserted into lightwave links, such as couplers and splitters. All such devices will introduce transmission loss, and the system designer is obliged to make allowance for these losses. In some cases, the design may be for an installation that does not include such devices initially but anticipates their addition at some later date. Adequate provision must then be made during the initial design phase, or the insertion later may compromise the performance of the link at that time.

As in any transmission system design, sound engineering practice must be employed to provide some overall system margin to accommodate such variables as equipment aging and thermal variations of the interconnecting fibers. A prudent but entirely arbitrary figure of 3 dB of additional loss is often applied in lightwave system designs.

OPTICAL TRANSMISSION LEVELS

The transmitter manufacturer defines the output level of the product in dBm or milliwatts, often both. It is necessary to determine exactly where that signal level is available, as mentioned earlier, but beyond that there is nothing else to be considered. The choice of a transmitter will be based upon the output level as required by the cumulative link losses and the receiver sensitivity.

Lightwave receivers are specified by the manufacturer to provide some stated quality of electrical output signal for a specified optical input signal level. This is sometimes called the threshold of the receiver, that is, the threshold above which the stated electrical output signal quality will be assured. In analog systems, the quality of electrical output will be stated as a signal-to-noise ratio (SNR). In digital systems, it is generally related to the bit error rate (BER) of the electrical output signal.

For example, a digital lightwave receiver may be specified to maintain a BER of 10^{-9} when the optical input signal level is –39 dBm or higher. Such a unit should consistently deliver an output signal quality of 10^{-9} or better, as long as the optical input does not fall below –39 dBm. At input levels below this, we expect the BER to be poorer.

There is also a condition of receiver overdrive that must be considered. At optical input levels that are too high, the receiver will not produce a satisfactory electrical output signal. Manufacturers generally define this point of overdrive by specifying the dynamic range of the receiver. This simply identifies that input level, above the previously specified threshold, at which the receiver electrical output quality will be degraded.

The dynamic range of a receiver is a function of the receiver design, the type of photodetector employed, and the automatic gain control (AGC) range that the unit evidences. For example, the receiver just mentioned was specified to produce a BER of 10^{-9} at an optical input level of –39 dBm or higher. This same unit might be specified by the manufacturer to present a dynamic range of 30 dB. This simply means that at optical input levels anywhere between –39 dBm and –9 dBm (which is 30 dB higher), the unit should maintain an output (electrical) signal quality of 10^{-9} BER or better. Figure 9-1 shows this dynamic range.

Appendix A provides an in-depth discussion of ratios, decibels, dBm, logarithms, etc. for readers who may not be familiar with these terms.

Transmission Level Design Considerations

It is the responsibility of the transmission engineer to select a particular fiber and the transmitting and receiving equipment so that the link will provide the transmis-

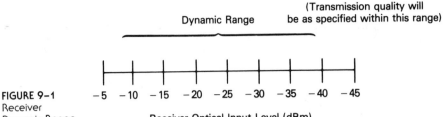

FIGURE 9–1
Receiver
Dynamic Range

sion quality and capacity that are required. Within this broad statement of responsibility, there is wide latitude for judgment and choice.

From an examination of Figure 9–1, the designer probably will have little difficulty in finding several combinations of fiber and terminal equipment that will assure optical input levels to the receiver which are above the unit's threshold. At input levels below this point, transmission quality will be unacceptable.

The problem will be to examine several alternatives quickly and easily and to select not only a technically satisfactory combination, but an economically advantageous one as well. We shall discuss this in greater depth in the next chapter, but before leaving the question of transmission level, it would be helpful if we could develop a methodology for examining the system from the point of view of transmission losses alone.

A useful practice is to prepare a link loss budget for each link designed. This technique itimizes all losses and permits a systematic examination of transmission levels, assuring that the fiber and equipment selected will actually produce satisfactory overall link performance. The budget also includes margins and allowances for additional or future losses that might be introduced at a later date, thus providing some insurance against inadvertently omitting any such factors. Figure 9–2 presents a typical link loss budget.

The loss budget addresses transmission loss and signal levels only; it does not relieve the designer of the obligation to separately address the questions of rise time and bandwidth.

SUMMARY

The transmission engineering process is not complex, but in order to assure a satisfactory system design, it must methodically address transmission levels, signal attenuation, and the various factors that contribute to transmission distortion.

In shorter systems and in applications that do not require high transmission data rates, we need only examine distortion sufficiently to establish that the link is actually attenuation limited. In such cases, transmission losses and signal levels are the only considerations that must be addressed further.

Due to the significant dispersion that is introduced in multimode fibers, their use in systems is somewhat restricted to lower data rates and shorter fiber lengths.

Link Loss Budget

1. Transmitter output power (dBm)   . . . .
2. Receiver input threshold (dBm) (for BER of 1×10^{-9})   . . . .
                                                                     ----
3. System gain (dB) (line 1 − line 2)   . . . .
4. Transmitter coupling loss (dB)   . . . .
5. Receiver coupling loss (dB)   . . . .
6. Optical connector losses (dB) (dB per connector × number of connectors)   . . . .
7. Splice losses (dB) (dB per splice × number of splices)   . . . .
8. Allowance for future splices (dB) (dB per splice × number of splices)   . . . .
9. Other optical device losses (dB) (WDM couplers, splitters, etc.) . . . .
10. System margin (dB) (equipment aging, etc.)   . . . .
11. Total system losses (dB)   . . . .
12. System gain (dB)   . . . .
13. Total system losses (dB)   . . . .
                                                                     ----
14. Total acceptable fiber loss (dB) (line 12 − line 13)   . . . .

FIGURE 9-2 Link Loss Budget (0.6 dB/km Fiber)

Single mode fibers, on the other hand, reduce dispersion to such an extent that relatively long links, even at data rates of 90 Mb/s and higher, can often be addressed simply as transmission-level problems alone.

System bandwidth and rise time can be directly related, and transmission performance over a lightwave facility can be calculated and predicted, by using either frequency or time domain data.

The development of a link loss budget is a useful technique in designing lightwave systems and reduces the possibility of inadvertently omitting any discrete transmission loss item.

REVIEW QUESTIONS

True or False?

   T   F   1. An intermediate regenerative repeater is only necessary in lightwave transmission systems that are operating at transmission data rates of 565 Mb/s or higher.

   T   F   2. The information bandwidth of a transmission system is only a factor in systems that are operating in an analog mode.

   T   F   3. In metallic conductor facilities, transmission losses are the same at all frequencies transmitted.

T   F   4. An optical fiber of longer length will always introduce more pulse dispersion than a shorter length of the same fiber.

T   F   5. The term MHz-km denotes the bandwidth/length characteristics of a multimode fiber.

T   F   6. Attenuation in optical fibers is denoted as dB/km for both multimode and single mode fibers.

T   F   7. The transmission data rate of an optical fiber system is taken to be the same as the bandwidth required of the system.

T   F   8. In an analog fiber system carrying more than one television signal, the required bandwidth is always 12 MHz regardless of the number of signals carried.

T   F   9. In digital applications, a multimode fiber will introduce more pulse dispersion than a single mode fiber of the same length.

T   F   10. Dispersion in single mode fibers is denoted in units of ps/nm-km.

T   F   11. If system dispersion is known, system bandwidth can be calculated by $BW = T/440$, where T = dispersion in ns.

T   F   12. If the length of a fiber is reduced, the bandwidth/length specification can be relaxed and still support the same transmission data rate.

T   F   13. If the transmission data rate is reduced, the fiber bandwidth/length specification can be relaxed for the same cable route distance.

T   F   14. We can combine all system element rise times and calculate the system RMS rise time.

T   F   15. If a fiber with less dispersion is introduced, a system link may be lengthened and still support the same transmission data rate.

T   F   16. A laser with a narrower spectral width will provide a narrower system bandwidth.

T   F   17. Using a dispersion shifted fiber can increase system bandwidth.

T   F   18. A multimode fiber of a given length will provide more usable bandwidth than the fiber bandwidth/length specification indicates.

T   F   19. At transmission data rates of 90 Mb/s or lower, a single mode fiber link will be attenuation limited.

T   F   20. A link loss budget will help define the transmission bandwidth that an optical fiber facility will present for use.

CHAPTER 10

Simplifying Lightwave System Design

It is possible to significantly reduce the complexity of designing lightwave transmission systems by focusing on transmission losses and operating transmission levels alone. Not that we are relieved of the obligation to consider system distortion, but the extensive use of single mode fibers with their much lower dispersion contributions greatly increases the number of applications that are strictly attenuation limited.

We can, by calculation, determine at what transmission data rates and cable route lengths designs will be attenuation limited. And then we need not address distortion at all in such cases, and might even develop system design aids. Nomographs, for example, could facilitate the design process or make it easier to examine a larger number of alternatives.

The system design effort, when confined to transmission levels alone, is not particularly sophisticated. It is by no means essential to use nomographs or other mechanical design aids in this process. Nevertheless, such aids may serve a useful purpose, and the development of appropriate nomographs may, in itself, reinforce a general understanding of the design process.

FIBER BANDWIDTH LIMITATIONS

Using the procedures presented in Chapter 9, we can establish the length of multimode fiber that will be bandwidth limiting for a given transmission data rate. Of course, we shall have to consider the bandwidth/length characteristics of the fiber under consideration.

Accordingly, consider a multimode fiber with a bandwidth specification of 600 MHz-km. At a data rate of 90 Mb/s, such a fiber would produce less than the required 90 MHz of usable bandwidth at lengths beyond 10.7 km:

$$L \text{ (km)} = \left[\frac{\text{BW}_{\text{fiber}} \text{ (MHz-km)}}{\text{BW}_{\text{sys}} \text{ (MHz)}} \right]^{1.25}$$

$$= \left[\frac{600}{90} \right]^{1.25} = 10.7 \text{ km}$$

Now consider the same application, but using a fiber with a bandwidth specification of 2,000 MHz-km. Such an installation would be bandwidth limited only beyond a fiber length of 48.2 km:

$$L \text{ (km)} = \left[\frac{\text{BW}_{\text{fiber}} \text{ (MHz-km)}}{\text{BW}_{\text{sys}} \text{ (MHz)}} \right]^{1.25}$$

$$= \left[\frac{2,000}{90} \right]^{1.25} = 48.2 \text{ km}$$

For instructional purposes, let us calculate the usable length of the higher grade fiber at a lower transmission data rate, say, 45 Mb/s:

$$L \text{ (km)} = \left[\frac{\text{BW}_{\text{fiber}} \text{ (MHz-km)}}{\text{BW}_{\text{sys}} \text{ (MHz)}} \right]^{1.25}$$

$$= \left[\frac{2,000}{45} \right]^{1.25} = 114.7 \text{ km}$$

As expected, at the lower data rate the system can be much longer before it becomes bandwidth limited.

From the calculations, then, a multimode fiber with a bandwidth/length characteristic of 2,000 MHz-km, can handle 90 Mb/s or lower rate signals over cable route lengths of up to 48 km (29.8 mi.) without regard to fiber dispersion. Hence, we can address designs up to this length, using this grade of fiber, on the basis of transmission level and system attenuation alone, providing that the data rate does not exceed 90 Mb/s.

Consider, then, a single mode fiber with a dispersion specification of 4 ps/nm-km, not a particularly stringent specification in today's marketplace. We shall arbitrarily use a source with a spectral output of 5 nm, which is typical of most lasers on the market today. At a data rate of 90 Mb/s, such a fiber would produce less than the required 90 MHz of usable bandwidth at lengths beyond 244 km:

$$L = \frac{440,000}{FD \times BW \times FWHM}$$

$$= \frac{440,000}{4 \times 90 \times 5} = 244.4 \text{ km}$$

Even if the transmission data rate were increased to 135 Mb/s, using the same fiber and light source, the system could be 162.9 km long before we need to address dispersion in the system design effort.

Consequently, for system designs within these established limitations, we can confidently dispense with any further consideration or calculation of fiber rise time or bandwidth. Also, we can, if we wish, develop nomographs that will facilitate the design of any link we require, assuming that the link in question falls within the established limitations.

Although doing so might be considered merely an academic exercise, it is likely that more and more designs will be based upon the use of single mode fiber alone, and that the use of lightwave transmission will be introduced more frequently into lower levels of the telecommunications network, that is, that more applications will develop for lower data-rate, shorter length applications in the distribution or exchange areas of the network. In that case, nomographs may very well serve a practical purpose as time passes. If fiber losses and single mode fiber dispersion characteristics become standardized on one or two values, slide rules or nomographs could become universally applicable.

FIBER ATTENUATION

For graphic purposes, such as use in design nomographs, we can construct scales that relate fiber loss to fiber length. It would be useful if such scales could include some arbitrary but prudent allowance for splicing losses also—both the initial, planned splices and the estimated number of splices that service restorations might require over a period of plant service life. In any event, we have to make some allowance or estimate for such losses. The link loss budget (Figure 9–2) shows how these estimates are handled and applied.

If several different fibers are considered for use, with each such fiber having a different attenuation characteristic, then we shall also have to construct a scale for each fiber, with each scale incorporating the same splicing allowances.

Consider a fiber whose attenuation/length characteristic is specified to be 0.6 dB/km. This loss might be associated with any operational carrier frequency, short or long wavelength. All that is necessary is that the loss characteristic relate to the proposed operating wavelength of the system under design. Also, it is irrelevant whether a single mode or multimode fiber is referenced, as long as the stated losses per unit of length are accurate for the type of fiber under consideration.

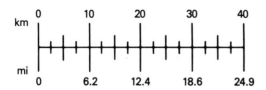

FIGURE 10-1
0.9 dB/km
Fiber
Attenuation

Before constructing a scale, we must establish a splicing loss figure. We might assume a conservative 0.3 dB per splice and assign one such splice for every kilometer of plant placed. This would appear to be quite adequate to cover any subsequent splicing that might be necessary in the event of mechanical damage to the plant at some later date.

The preceding figures are arbitrary. Experience or individual judgment might develop a different allowance, and such a figure could be technically defensible. In such cases, the reference scale that is constructed would simply reflect this different allowance.

Figure 10-1 presents a scale that reflects a fiber loss of 0.6 dB/km of fiber length and a splicing allowance of 0.3 dB/km. The scale is calibrated in both kilometers and miles to facilitate use. It actually indicates a total interconnecting fiber attenuation of 0.9 dB/km, but the calibration is given not as loss, but as overall optical cable length. Other similar scales could easily be constructed reflecting different fiber loss characteristics, different splicing allowances, and different types of fiber.

TRANSMITTED OPTICAL SIGNAL LEVELS

Regardless of which type of light source is employed in a transmitter, only a finite amount of light energy will be available for transmission. We shall define this amount as the level of optical signal actually coupled into the interconnecting optical fiber. The manner in which this energy is coupled into the fiber may vary in different installations. For example, optical transmitters may be equipped with fiber pigtails that can be directly (hard) spliced to the interconnecting fiber itself. In such cases, the specified optical signal level output from the transmitter would be considered to be the level actually coupled into the interconnecting fiber.

In some installations, for any of several reasons, it may be desirable to provide a disconnect-reconnect capability between the transmitter output and the interconnecting fiber. This can, for example, facilitate patching between several transmitters and several interconnecting fibers. In such instances, an optical connector (a mechanical mechanism rather than a hard splice) would be utilized. If so, additional attenuation to the light energy signals would be introduced; a prudent but arbitrary estimate of transmission loss through such a connector would be 1 dB per connector placed.

In some lightwave systems, WDM may be applied in configuring the link itself. This entails the insertion of various types of optical devices into the interconnecting

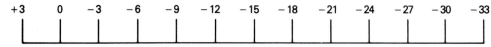

FIGURE 10-2
Transmitted
Optical Signal
Level (dBm)

optical fibers. Each such device will introduce some transmission loss, and there will be one or more such devices required at each terminal end of the interconnecting fiber. A conservative estimate for device losses might be 1 dB per device installed. The devices may or may not be installed initially, but even if the actual installation is postponed to a later date, the system design must make provision for the additional transmission loss that is anticipated.

As in any transmission system, sound engineering practice provides some overall system margin to accommodate such variables as equipment aging and thermal variations in the interconnecting fibers. A prudent but arbitrary margin is 3 dB per link. We can easily construct a scale that permits incorporation of any or all of these factors.

In Figure 10-2, the range of signal levels shown on this scale is +3 dBm to –33 dBm. This permits us to enter the scale directly at the output signal level specified by a manufacturer for a particular lightwave transmitter, and then to adjust this point of entry to a lower level to reflect any or all of the variable design factors we have discussed.

Suppose, for example, we were considering the use of a lightwave transmitter that is specified by the manufacturer to produce a –3 dBm optical signal level at the wavelength we intend to use. (In fact, this would be representative of many transmitters that are available today that use laser-type light sources.) Then we would initially enter the scale at –3 dBm. If we propose to install optical connectors at each terminal end of this link, we would simply adjust the –3 dBm initial scale entry point downward to –5 dBm to reflect the connector losses (one at each end) which will be introduced.

If we also wish to consider the use of WDM in the system, we would downgrade the entry point an additional 2 dB (1 for each terminal end device installed) from –5 dBm to –7 dBm. Finally, we would apply a system margin of 3 dB by downgrading the transmitted signal level an additional 3 dB from –7 dBm to –10 dBm. Note that this adjustment applies to every design effort, but the other corrections might or might not apply, depending upon the actual system configuration being developed.

RECEIVED OPTICAL SIGNAL LEVELS

All lightwave receivers are specified by the manufacturer to produce a stated quality of electrical output signal for a specific optical input signal level, called the *receiver threshold*. The receiver threshold is the optical threshold for the stated quality of electrical output signal.

In analog systems, the quality of the electrical output is stated as S/N, and in digital systems as BER. For example, a digital receiver might be specified to produce a BER of 10^{-9} when the optical input signal level is –39 dBm or higher. Such a receiver should consistently deliver a BER of 10^{-9} at input levels at or above –39 dBm. At optical input levels below –39 dBm, we would expect the BER of the electrical output signal to be poorer.

There is also a condition of high optical receiver input that can overdrive the detector, and degrade the quality of the output electrical signal below a BER of 10^{-9}. The manufacturer will generally specify this point of overdrive by delineating the receiver's dynamic range. This simply states that at some known optical input level above the receiver threshold, the receiver electrical output signal quality will be degraded. The dynamic range is a function of the receiver design, the type of detector employed, and the range of AGC the receiver has. For example, a receiver might be specified to have a dynamic range of 30 dB and a threshold of –39 dBm to produce a BER of 10^{-9}. Then at optical input signal levels between –39 dBm and 30 dB higher, which is –9 dBm, this unit should maintain a BER of 10^{-9} in its electrical output signal.

In developing a system design-aid nomograph, it will be useful to construct a scale reflecting all the received optical signal levels we anticipate having to consider. Such a scale is shown in Figure 10–3. The scale covers received optical signal levels between 0 dBm and –60 dBm, which should accommodate all lightwave receiving equipment we expect to encounter.

FIBER ATTENUATION SCALES

The scale shown in Figure 10–1 will be adequate for any fiber whose loss is 0.6 dB/km regardless of the operating wavelength, provided that at that wavelength the fiber does actually evidence 0.6 dB/km of loss. It is equally applicable to single mode or multimode fibers, again with the qualification that the fiber have the given loss characteristic. For fibers that have a different length/loss characteristic, or for applications where we choose to apply a different splicing loss allowance, we must construct additional scales. Figure 10–4 shows four different scales exhibiting different loss characteristics and, in the case of scale D, a different splicing loss allowance as well.

Later in this chapter, we shall demonstrate a methodology for constructing fiber attenuation scales for fibers that have loss characteristics other than those arbitrarily chosen for demonstration in the figure.

FIGURE 10–3
Received
Optical Signal
Level (dBm)

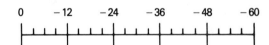

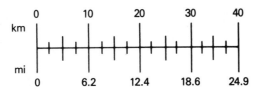

(a) 0.6 dB/km fiber loss plus 0.3 dB/km splice loss.

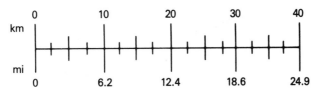

(b) 0.9 dB/km fiber loss plus 0.3 dB/km splice loss.

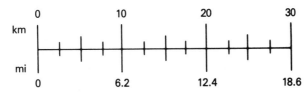

(c) 1.2 dB/km fiber loss plus 0.3 dB/km splice loss.

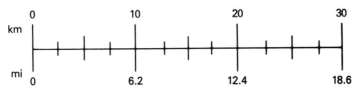

FIGURE 10–4
Fiber
Attenuation
Scales

(d) 1.4 dB/km fiber loss plus 0.4 dB/km splice loss.

TRANSMITTED OPTICAL SIGNAL LEVEL SCALE _____

Figure 10–5 shows the operational output ranges of laser and LED transmitters that are available today superimposed on the scale of Figure 10–2.

Since we desire to utilize the nomograph to evaluate alternative link designs, it is helpful to recognize the relationship and output ranges of the two different types of light source. Different light sources have different output and operating characteristics, and different basic costs as well.

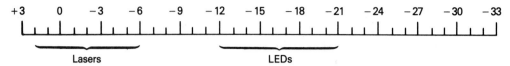

FIGURE 10-5 Lightwave Transmitters

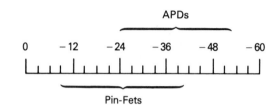

FIGURE 10-6
Lightwave
Detectors

RECEIVED OPTICAL SIGNAL LEVEL SCALE

Figure 10–6 shows the operational dynamic ranges of two basic types of lightwave receivers: an avalanche photodiode and a PIN-FET type detector. As with transmitters, these receivers have different levels of performance, sophistication, and cost.

A SYSTEM DESIGN-AID NOMOGRAPH

As discussed earlier, many lightwave applications do not require us to address rise time, or bandwidth, at all, since it will not be the limiting factor in these applications. In such cases, the design effort need consider only three factors: transmitter optical output level, receiver optical input level, and interconnecting fiber loss. If we can correctly relate the three separate scales that delineate these factors in a single graphic representation, we can construct a nomograph that might be useful in many design efforts.

Figure 10–7 shows this completed nomograph for a composite fiber loss characteristic of 0.9 dB/km (0.6 dB/km fiber loss plus 0.3 dB/km splicing loss). A straight line cutting all three coplanar curves (the three scales) will intersect the related values of each of the three variables. If any two variables are known, the corresponding value of the third variable can be quickly and easily determined.

Figures 10–8, 10–9, and 10–10 are other nomographs for other composite fiber loss figures. The fiber loss used is separately identified for each of these other figures and is arbitrary. A scale of the same kind as these can be constructed for any fiber loss characteristic and any splicing allowance.

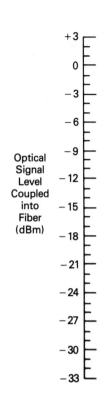

Optical
Signal
Level
Coupled
into
Fiber
(dBm)

+3
0
−3
−6
−9
−12
−15
−18
−21
−24
−27
−30
−33

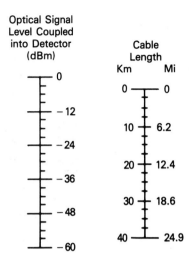

Optical Signal
Level Coupled
into Detector
(dBm)

0
−12
−24
−36
−48
−60

Cable
Length
Km   Mi

0 ——— 0
10 —— 6.2
20 —— 12.4
30 —— 18.6
40 —— 24.9

FIGURE 10–7
Chart for 0.9
dB/km Fiber

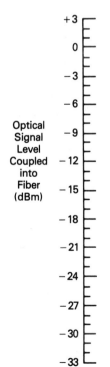

Optical
Signal
Level
Coupled
into
Fiber
(dBm)

+3
0
−3
−6
−9
−12
−15
−18
−21
−24
−27
−30
−33

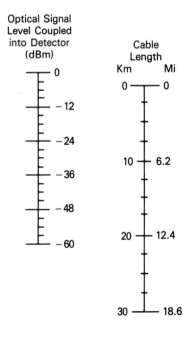

Optical Signal
Level Coupled
into Detector
(dBm)

0
−12
−24
−36
−48
−60

Cable
Length
Km   Mi

0 ——— 0
10 —— 6.2
20 —— 12.4
30 —— 18.6

FIGURE 10–8
Chart for 1.8
dB/km Fiber

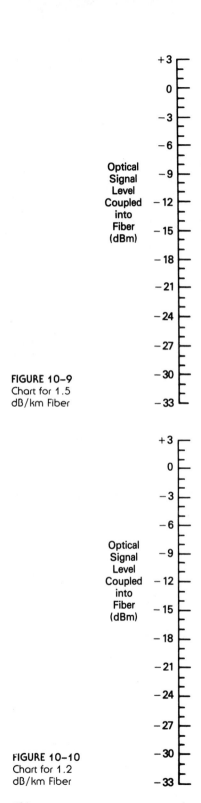

Optical Signal Level Coupled into Fiber (dBm)

+3
0
−3
−6
−9
−12
−15
−18
−21
−24
−27
−30
−33

FIGURE 10–9
Chart for 1.5 dB/km Fiber

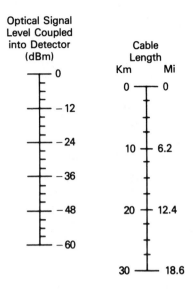

Optical Signal Level Coupled into Detector (dBm)

0
−12
−24
−36
−48
−60

Cable Length

Km   Mi

0 ——— 0

10 ——— 6.2

20 ——— 12.4

30 ——— 18.6

Optical Signal Level Coupled into Fiber (dBm)

+3
0
−3
−6
−9
−12
−15
−18
−21
−24
−27
−30
−33

FIGURE 10–10
Chart for 1.2 dB/km Fiber

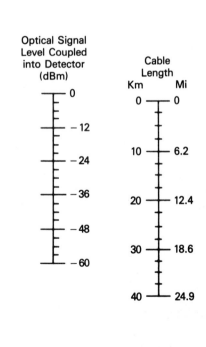

Optical Signal Level Coupled into Detector (dBm)

0
−12
−24
−36
−48
−60

Cable Length

Km   Mi

0 ——— 0

10 ——— 6.2

20 ——— 12.4

30 ——— 18.6

40 ——— 24.9

VALIDATING THE NOMOGRAPH'S ACCURACY _____

Before attempting to utilize the nomographs, we must thoroughly validate their accuracy. This may be accomplished by making a series of calculations and comparing the calculated results with the solutions provided for the same problems by the nomograph.

Nomograph results are easily obtained by placing a straightedge between the values of any two known variables and reading the value of the third variable directly on the appropriate scale. For example, to establish the optical input level to the receiver when the fiber is 20 km long and the optical input to the fiber is –6 dBm requires placing a ruler such that it connects –6 dBm on the left-hand scale of Figure 10–7 with 20 km on the right-hand scale. The answer, –24 dBm, can be read directly from the center scale at the point where the ruler intersects that scale.

A calculation of these same values is made as follows:

1. Given a 0.9 dB/km composite fiber loss (0.6 dB/km fiber loss plus 0.3 dB/km splicing allowance), a 20 km length of fiber would introduce a total of 18 dB of attenuation (20 × 0.9 = 18).

2. If the input level to the fiber is –6 dBm, and the total attenuation through the fiber is 18 dB, then the receiver optical input level is –24 dBm (–6 + (–18) = –24). This result correlates nicely with the result obtained using the nomograph in Figure 10–7.

Further comparisons of calculated and graphical results at several different input signal levels and lengths of optical cable will validate the nomograph. The reader is encouraged to perform these comparisons on his or her own to reinforce confidence in the use of the nomograph. Properly applied, the nomograph greatly facilitates the evaluation of several technical alternatives in developing any particular system design. A few examples will bear this remark out.

Example 1 (Initial Design Approach)

Suppose we have an application that requires an interconnecting cable 20 km (12.4 miles) long. The transmission data rate required has been determined to be 45 Mb/s.

The initial design approach is to utilize a laser light source (transmitter) capable of coupling 0 dBm into the interconnecting fiber. We propose to utilize optical connectors at each system terminal, allowing a total of 2 dB transmission loss for these devices. We do not intend to apply WDM either initially or subsequently, and we do require a 3 dB system margin. The fiber under consideration is specified to introduce 0.6 dB/km at the proposed operating wavelength.

The nomograph in Figure 10–11 presupposes a fiber loss of 0.6 dB/km with an additional 0.3 dB/km allowance for splicing losses. We enter the left-hand scale at 0 dBm (point A) reflecting the proposed laser output level. We next degrade this level 2 dB to –2 dBm (point B) to reflect the optical fiber connector losses. We then

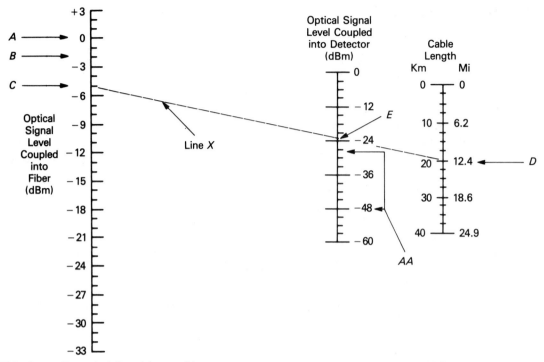

FIGURE 10-11 Example 1 (Initial Approach)

degrade this value an additional 3 dB to –5 dBm (point C) to provide the required 3 dB system margin.

Point C represents the light level we anticipate coupling into the interconnecting fiber. A straight line (X) is drawn connecting point C with the 20-km point (point D) on the right-hand scale. Point D reflects the required cable length of 20 km.

The point E where line X intersects the middle scale is read off to show –22.5 dBm as the optical signal input level to the receiver.

Let us now suppose that we are considering a receiver that is specified to produce a BER of 10^{-9} in the electrical output signal level with an optical input signal level of –48 dBm. Suppose further that this unit is specified to have a dynamic range of 20 dB. Then on the nomograph, we can bracket the range of optical input level we can present to this receiver and still be certain of maintaining a BER of 10^{-9}. This is shown as AA in Figure 10–11.

Examining the figure, we see that the optical input to the receiver in this design is –22.5 dBm, shown as point E. However, this signal level falls outside the range AA; that is, the input level is too high in amplitude to ensure the required quality of output electrical signal. Thus, if we were to build the system using this particular combination of fiber and terminal units, we would overdrive the receiver and could not expect to produce a BER of 10^{-9}. Accordingly, we shall have to modify the design in some manner.

Example 1 (Alternative 1)

We could consider using an LED transmitter rather than the laser originally utilized. An LED light source will have a lower level light output, which may resolve the design problem satisfactorily, and would be less expensive and possibly have a longer service life as well.

Consider, then, an LED transmitter that has an output of –18 dBm. We can easily evaluate the performance of this alternative on the nomograph. In Figure 10–12 we enter the left hand scale at –18 dBm (point A), the LED source output level. We then degrade this 2 dB, as before, to reflect optical connector losses, and an additional 3 dB for system margin. These corrections are shown in the figure as points B and C. Note that C, the optical signal we expect to couple into the fiber, is –23 dBm.

A straight line, X, is then constructed connecting points C (–23 dBm) and D, the 20-km point on the right hand scale. The optical input signal level to the receiver can then be read off on the middle scale as –41 dBm (point E in the figure).

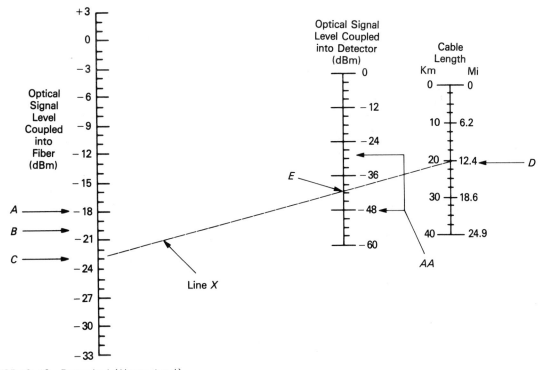

FIGURE 10–12 Example 1 (Alternative 1)

Since we have not changed the receiver from what it was in our original design approach, it would still have the same threshold and dynamic range as before, and we see this as AA in the figure. An operating input level of –41 dBm (point E) falls comfortably within the dynamic range of the receiver.

Example 1 (Alternative 2)

Alternative 1 is not the only solution to the design problem. We could instead consider the use of a higher loss interconnecting fiber, which may, incidentally, be less expensive and possibly reduce system cost also. Reverting back to the initial design approach, we have a laser-type transmitter with an optical output of 0 dBm. Again we degrade this to –5 dBm to provide for 2 dB of fiber connector losses and a 3 dB system margin.

In Figure 10–13 we use a nomograph that presupposes a composite fiber loss of 1.5 dB/km, which includes a fiber loss of 1.2 dB/km and a splicing allowance of 0.3 dB/km. Connecting the –5 dBm fiber input level (point C) with a 20-km cable length (point D), we find the optical input to the receiver to be –29 dBm (point E on the middle scale). This falls nicely within the receiver dynamic range shown by AA in the figure.

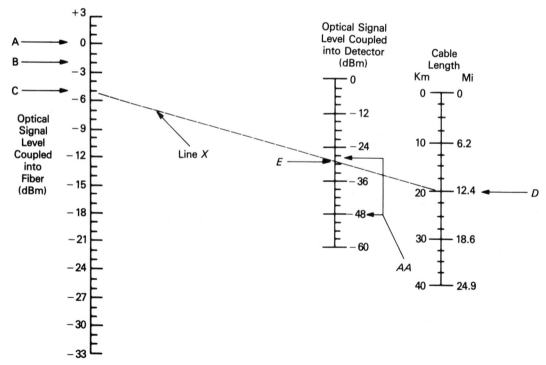

FIGURE 10–13 Example 1 (Alternative 2)

This solution is technically acceptable, and since optical cable cost is a significant portion of the total system cost, the approach may even be cost advantageous, if a higher loss fiber can be purchased at a lower cost.

Summary of Design Example 1

Using the nomographs, we were able to quickly examine several alternatives from a technical point of view. Marginally acceptable operation was clearly indicated, and the effects of corrective actions were immediately available for examination.

Optical fiber costs, as well as lightwave terminal equipment costs, are presently in a state of evolution where they change very rapidly. Hence, it is difficult to present cost data which will not become outdated quickly. It is quite evident that if current cost data were available in a convenient tabulated format, it would complement the nomographs and permit us easily to compare the cost differentials of various technically acceptable design alternatives.

A suggested tabulation of lightwave terminal equipment is shown in Table 10–1. The specifications are representative, and not necessarily items actually available on the market today. Note that the cost of each unit has been left blank. The intent is to suggest a convenient format for collecting current cost data for comparison purposes. Such cost information would have to be periodically updated, of course.

A similar tabulation of optical fiber cables, including transmission loss and bandwidth characteristics, fiber count, and costs, would be equally useful and would require the same frequent updating.

Example 2

To simulate an actual system design problem more realistically, we shall present an example in which the design effort is broadened to include alternative system configurations as well as alternative transmission designs. We shall assume that the

TABLE 10–1
Equipment
Comparison

Light Source	Lightwave Transmitters			
	LED	Laser	Laser	Laser
For use with	MM	MM	SM	SM
Output (dBm)	−18	0	−3	0
Wavelength (nm)	1300	1300	1300	850
Cost ($)	. . . .	. . . .	. . . .	. . . .

MM = Multimode fiber. SM = Single mode fiber.

	Lightwave Receivers			
Sensitivity (dBm)	−43	−39	−46	−37
Dynamic Range (dB)	30	30	20	20
Detector (Type)	PIN-FET	PIN-FET	APD	APD
Cost ($)	. . . .	. . . .	. . . .	. . . .

Specifications provide a BER of 10^{-9}

requirement is to provide 144 equivalent voice circuits between two points, such as might be required for trunking between two telephone switching centers. Traffic studies indicate that growth may require an additional 48 circuits within a five year period, but we shall suppose that the system is to be installed in a resort area, where the additional requirement is uncertain and may be substantially higher. The cable route length between the two terminal locations has been determined to be 22.5 km.

Example 2 (Initial Design Approach)

It is proposed to install a four fiber optical cable, dedicating one fiber pair to traffic usage and the second fiber pair as a service protection facility. To be able to handle the predicted traffic growth, we shall design the link to operate at a 45 Mb/s transmission rate, which can accommodate DS 3 level digital signals. We shall equip both terminal locations with M 13 digital multiplexers, providing an initial transmission capability of 28 T–1 systems (672 voice circuits). We shall only equip six T–1 systems (144 voice circuits) initially, however, since that is all the immediate requirement calls for.

Figure 10–14 shows a block diagram of this initial design approach. Note that since the M 13 multiplex equipment is common to all traffic carried, we propose to support this unit, and the interconnecting lightwave facilities as well, on a 1 × 1 protection switching basis.

Since we propose to multiplex up to a DS 3 digital level, the transmission data rate is established to be 45 Mb/s (actually 44.763 Mb/s). The optical cable route length was given previously to be 22.5 km.

Earlier, we determined that even if we utilize a lower grade of multimode fiber that provides only 600 MHz-km of transmission bandwidth, in a 45 Mb/s application we need not address fiber bandwidth at all if the interconnecting fiber length is 25 km or less. Accordingly, for this application we may utilize a transmission level nomograph design aid to facilitate the design process.

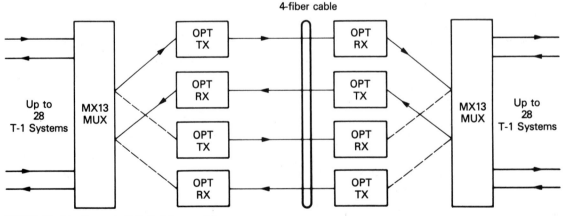

FIGURE 10–14 Example 2 (Initial Approach)

We shall first consider using a laser-type transmitter that presents a 0 dBm optical output signal level. Since we intend to provide a spare fiber pair as a service protection facility, there seems little need or justification for any optical connectors, and we shall not include any connector losses in our initial design effort. Also, since the traffic requirement is initially only 144 voice circuits, and the projected growth is only 48 additional circuits, there seems little reason to reserve any transmission loss for WDM at a future time. The 28 T-1 capacity we are providing seems more than sufficient to handle any growth that may develop.

We must allow some system margin, however, and we shall arbitrarily assign 3 dB of transmission loss to cover this. We shall attempt to utilize a fairly low-loss fiber initially, and this is reflected in the nomograph shown in Figure 10–15.

Since the laser output was 0 dBm, we enter the nomograph at 0 dBm (point A) on the left-hand scale in the figure. We then degrade this to –3 dBm to provide the system margin (point B). Next, connecting point B with the 22.5-km point on the right-hand scale (point C), we find that the interconnecting straight line (X) intersects the middle scale at point D indicating a receiver input of –23 dBm. Consequently, we can consult Table 10–1 and select any of the optical receivers that are compatible with the received input level of –23 dBm. An APD type of detector, for

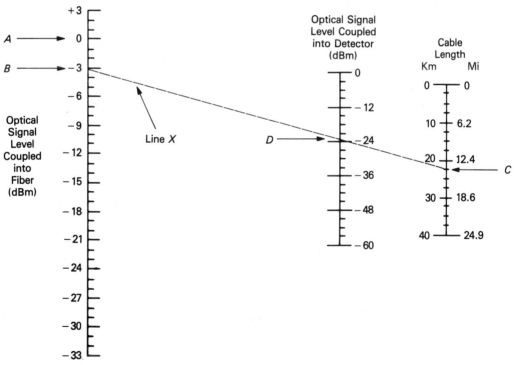

FIGURE 10–15 Example 2 (Initial Approach)

example, with a specified dynamic range of 20 dB and sensitivity of –37 dBm will produce a BER of 10^{-9} across an input signal level range of –37 dBm to –17 dBm. A PIN-FET type of detector with a specified dynamic range of 30 dB and sensitivity of –39 dBm will present the same quality output signal across an input signal range of –39 dBm to –9 dBm.

Either of these receivers should perform quite adequately at the initial design input level of –23 dBm, and we could make a judgment and decision based upon receiver cost alone. However, there are other design alternatives to be considered.

Example 2 (Alternative 1)

Although we have not presented any fiber cost data in this example, from past experience we know that lower grades of fiber, that is, fibers with higher transmission loss characteristics, may be significantly less expensive. It might, then, be worth while to examine the technical practicality of using a laser-type transmitter and a higher loss fiber in designing this system.

Suppose we use a laser type-transmitter with a 0 dBm output and degrade this value for a 3 dB system margin. We enter the nomograph of Figure 10–16 on the left-hand scale at point B. Connecting this point with point C on the right-hand

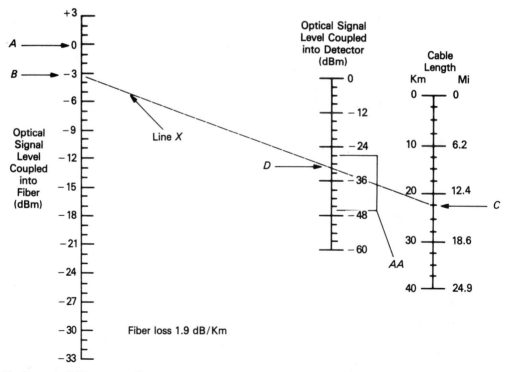

FIGURE 10–16 Example 2 (Alternative 1)

scale, 22.5 km, we establish line X, which intersects the middle scale at –30 dBm. Note that this nomograph is for a composite fiber loss of 1.2 dB/km.

Table 10-1 presents an APD type receiver that has a sensitivity, or threshold, of –46 dBm with a dynamic range of 20 dB. This unit should produce a BER of 10^{-9} at any input level within the range of –46 dBm and –26 dBm.

Our design input level of –30 dBm falls comfortably within this range, as shown by AA in Figure 10-16. If there is actually a cost advantage in using this higher loss fiber—and remember that there are 22.5 km of fiber involved, it should be quite acceptable from a transmission point of view.

Example 2 (Alternative 2)

Our initial traffic requirement, even projected out for five years growth, was only 8 T-1 systems. Although we did have some concern with possible growth beyond this level, it is possible to satisfy this requirement with an entirely different system configuration.

Figure 10-17 shows a configuration with digital multiplexing to a DS 2 signal level, which can accommodate 4 T-1 systems. What we have designed is two completely independent subsystems: there is no individual unit of equipment that is common to all traffic carried. Even if an individual fiber was disrupted, the impact on traffic would be restricted to only 50 percent of the derived circuits.

Since we anticipate a gross failure of this type to exist only a very small percentage of the time, it may be reasonable to eliminate any redundant facilities or equipment entirely. This would eliminate all protection switching and sensing sophistication also, and would produce a significant cost reduction.

We must still consider the possibility of traffic growth beyond the projected 48 additional circuits, however. In the figure, this is provided for by designing WDM capability into the system for future use. In this manner, by simply duplicating the initially provided terminal equipment on another operating wavelength, the system capacity is easily expanded from eight to sixteen T-1 systems at any time.

The system could be expanded more gracefully by simply adding four T-1 systems on only one of the potential WDM links. In this manner, capital expenditures for additional traffic capacity can be made incrementally, that is, from an initial capacity of eight T-1 systems to 12, and then subsequently to 16 T-1 systems. Since 16 T-1 systems can provide 384 voice circuits, this certainly represents an adequate capability for unexpected traffic growth.

It is readily apparent in the figure that the failure of any single lightwave terminal unit (transmitter or receiver), the failure of any single digital multiplexer, or an interruption of continuity through any single optical fiber would affect only 50 percent of the traffic being carried. This same level of protection, through independent subsystems rather than through duplicate facilities or equipment, would prevail to a very respectable degree if both WDM links were fully implemented and operational.

Although the WDM couplers are shown as subsequent additions, there may be some merit in installing these units initially. Being entirely optical, and consum-

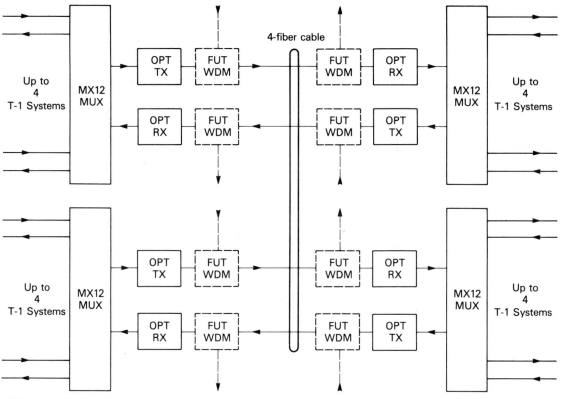

FIGURE 10-17 Example 2 (Alternative 2)

ing no operating power, they are passive in nature. The reliability of such devices should be very high, and installing them initially would eliminate the necessity for an interruption of service at a later date when, or if, the WDM option is exercised.

If, when we install the initial lightwave terminal equipment, we select different wavelengths for the two independent transmission links, then either terminal could be patched over to operate on the other fiber pair through the WDM couplers in the event of a fiber failure. In effect, a patching spare transmission facility is derived using the initially installed four fibers alone. Restoration of service would be expedited, and repairs to damaged fibers could be effected during normal working hours rather than on the basis of an emergency call.

Figure 10-18 shows the design of the link employing WDM, using the same nomograph as was used in Figure 10-16. We enter the nomograph at point A as before, again degrade 3 dB for system margin, and then degrade an additional 2 dB for the insertion loss due to the WDM coupling devices. The light energy level actually coupled into the interconnecting fiber is shown to be –5 dBm (point C) in the figure.

The reader should now be sufficiently familiar with the use of nomographs to

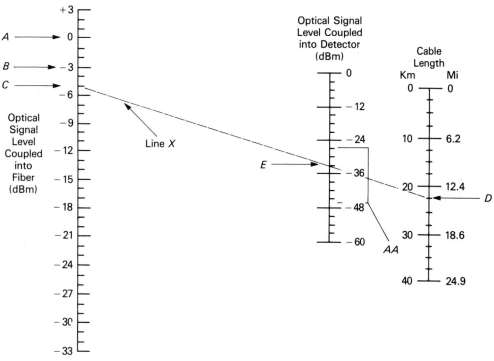

FIGURE 10-18   Example 2 (Alternative 2)

recognize that the additional transmission loss incurred by the insertion of the WDM devices does not significantly alter the overall link design results developed earlier. Note that the receiver input level still falls quite comfortably within the receiver's dynamic range as shown by AA.

Although it reduces the transmission data rate from 45 Mb/s to the DS 2 level of 6.312 Mb/s, this new design may not really present any significant economy in the optical fibers themselves. We know that at data rates below 45 Mb/s the link will always be attenuation limited, and reducing fiber bandwidth specifications at lower data rates may have no beneficial impact at all on fiber cost.

Summary of Design Example 2

There may be some real merit in considering less sophisticated, lower transmission data-rate equipment, sometimes referred to as *thin route system* equipment. Such equipment does present lower transmission capacity, of course, but we have provided adequate capacity by other means, mainly by the system configuration. The thin route type of equipment may be significantly less expensive, but this is really an equipment selection problem, and will have little if any impact on the transmission design effort.

If there are no compelling cost advantages, in either fiber or equipment, to be gained by dropping down to a lower transmission data rate, it is prudent to consider any possible disadvantages such an action might present. Although we shall not discuss them, there are some, though they might not be considered highly persuasive.

If the fiber bandwidth and terminal equipment were selected for a 45 Mb/s operating data rate, the entire system, including both WDM links, could be easily upgraded to a DS 3 digital signal level at any time. This would be possible without actually discarding the initially installed M 12 multiplexing units; instead, we can simply add the necessary new multiplexing capability. Even though we may not anticipate such traffic growth, if no cost penalty were incurred—perhaps even if a nominal penalty were involved—preserving such expansion capability for future operations might be a wise choice.

If we opt for the 45 Mb/s data rate, the transmission design problem is identical with that in our initial approach in this example. The alternatives we examined earlier for that approach would be equally applicable here, and just as before, we must include the transmission losses introduced by the WDM couplers. It should be noted that although 45 Mb/s capable optical terminal equipment might be installed initially, these can operate quite satisfactorily with lower data rate electronic terminals, of course.

DEVELOPING ADDITIONAL NOMOGRAPHS

In Figures 10–7, 10–8, 10–9, and 10–10 we presented nomographs for four different fiber loss characteristics. All of these nomographs are equally applicable at any operating wavelength, provided that the fiber attenuation figure used is correct for the wavelength being considered. The nomographs are useful for both single mode and multimode fibers, with the same qualification that the fiber loss characteristic agrees with the fiber under examination.

In developing these introductory charts, we applied a 0.3 dB splicing loss allowance per kilometer of fiber length. This is an arbitrary figure, and other values might be quite appropriate if they were based upon other experiences with fiber systems.

We might also have to consider fibers with different loss characteristics than the ones presented. Accordingly, it is helpful to consider constructing nomographs to cover any variations in the factors we used.

Toward that end, Figure 10–19 is identical with the other charts presented earlier, except that the right-hand scale has not yet been calibrated. For instructional purposes, we shall calibrate this scale using a splice allowance figure of 0.2 dB/km and a fiber loss of 1.7 dB/km, simply because none of the earlier charts have used these figures. However, the methodology we shall present will be applicable to any fiber loss figure and any splicing allowance factor.

In our new nomograph (Figure 10–19), the title of the figure shows the total fiber loss per kilometer of fiber length to be 1.9 dB. We can easily calculate the total

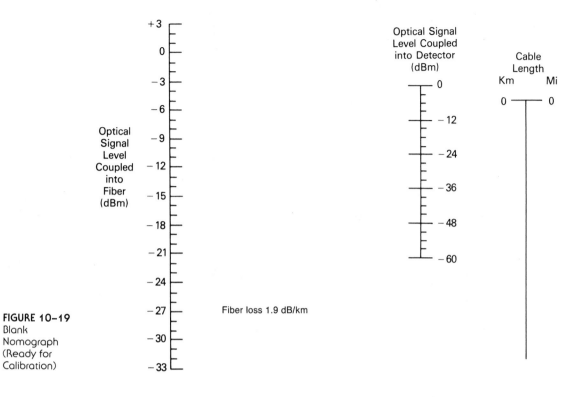

FIGURE 10-19
Blank
Nomograph
(Ready for
Calibration)

loss for any given fiber length by multiplying this figure, 1.9 dB/km, by the fiber length. A series of such calculations will produce the data shown in Table 10-2.

To ensure the accuracy of the nomograph we are calibrating, we shall calibrate the right-hand scale of the chart at one optical input signal level shown on the left-hand scale of the chart, and then verify the precision by using two other input signal levels. We will use 0 dBm, -15 dBm, and -30 dBm.

If we input 0 dBm into a fiber 5 km long (9.5 dB total fiber loss), the optical input to the receiver is -9.5 dBm. At 10 km of fiber length, the receiver input is -19.0 dBm. A series of similar calculations will produce the data shown in Table 10-3.

TABLE 10-2
Fiber
Loss Data

Fiber Loss (dB/km)	×	Fiber Length (km)	=	Total Fiber Loss (dB)
1.9		5		9.5
1.9		10		19.0
1.9		15		28.5
1.9		20		38.0
1.9		25		47.5
1.9		30		57.0

TABLE 10-3
Receiver
Input
Level (dBm)

Fiber Length (km)	at 0 dBm Fiber Input	at −15 dBm Fiber Input	at −30 dBm Fiber Input
5	−9.0	−24.5	−39.5
10	−19.0	−34.0	−49.0
15	−28.5	−43.5	−58.5
20	−38.0	−53.0	−68.0
25	−47.5	−62.5	−77.5
30	−57.0	−72.0	−87.0

Obviously, some of these receiver input levels will be below the sensitivity (detection threshold) of the receivers that are available for use, and thus will not fall within the usable range of our nomograph. Since the nomograph was intended for use in applications that were completely attenuation limited, we shall simply discard any data that do not fall into the nomograph's working range.

We can apply the data in the table directly to the nomograph itself. At 0 dBm fiber input, when the fiber length is 5 km, the receiver input level will be –9.5 dBm. In Figure 10–20 we construct line X that connects 0 dBm, point A on the left-hand scale, with 9.5 dBm on the center scale. The point where extended line X intersects the right-hand scale represents 5 km of fiber length.

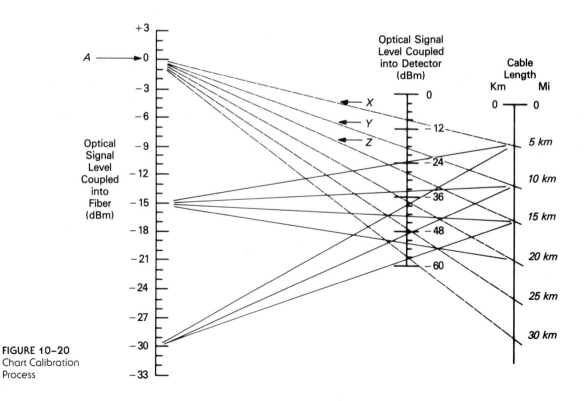

FIGURE 10-20
Chart Calibration
Process

Similarly, a line constructed to connect point A with –19.0 dBm on the center scale will intersect the right-hand scale at a point representing 10 km of fiber length; and a line Z connecting point A with –28.5 dBm on the center scale will indicate 15 km of fiber length on the right-hand scale. Without further discussion, we have constructed, in Figure 10–20, similar lines to identify fiber length out to 30 kilometers, in 5 kilometer increments.

The right-hand scale is now tentatively calibrated. The process required some interpolation of points on the chart's scales, but the degree of accuracy achieved should be quite adequate for system link design purposes.

All the lines constructed so far have been rendered as dashed lines. We shall now validate the calibration points on the right-hand scale by constructing a series of new, solid lines from the other two fiber input signal levels on the left-hand scale (–15 dBm and –30 dBm). We simply use Table 10–3 and connect the input signal level points with their associated receiver input points and verify that the fiber length indications in fact conform closely to the tentative length calibrations of the right-hand scale.

After drawing in the lines, we see that, within the limitations of the accuracy of the scales themselves, all calibration points on the right-hand scale do agree with each test we have applied. We have presented this explanation in some detail to avoid confusion, but it is not a complicated or lengthy process, and it would be quite easy to generate a new chart for any combination of fiber loss and splicing allowance figures.

In using the nomographs, no other adjustments or corrections need to be applied other than those made in entering the left-hand scale. The process permits a variety of system configurations to be quickly and easily demonstrated and tested.

Figure 10–21 shows the new nomograph with the right-hand scale just as we calculated, plotted, and verified it.

LIMITATIONS OF NOMOGRAPH DESIGN AIDS

The simplicity and flexibility of nomograph aids in designing systems is self-evident, but we must keep in mind that however useful such a technique may be, it does suffer some limitations, and we must be aware of these at all times.

At transmission data rates above 135 Mb/s or so, system bandwidth and pulse dispersion may become design limitations before straight signal attenuation does. It will not always be easy to identify these instances because the line of demarcation may not be positively defined. When any doubt exists on this point, it is sound judgment to apply conventional design procedures and develop an analysis of system rise time or bandwidth.

In most instances where a design must be examined in greater detail, the application will be for higher traffic densities, and the capital investments involved, plus the revenue potentials, of course, will easily justify a more sophisticated engineering effort.

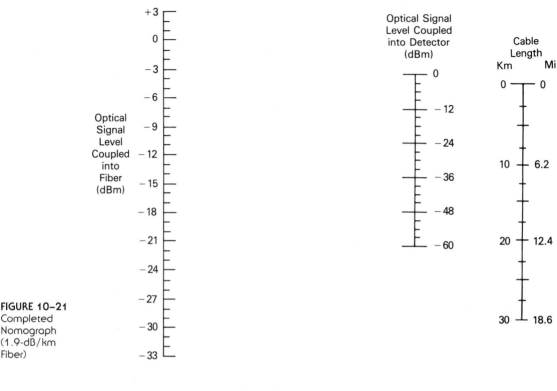

FIGURE 10–21
Completed
Nomograph
(1.9-dB/km
Fiber)

SUMMARY

The design technique presented here, and the development and usage of nomograph design aids, will be most applicable in shorter length or lower data-rate applications. In such situations it may be applied with a high level of confidence and should help produce more cost-effective installations.

Although the technique permits rapid examination of several technical alternatives, its major value may be more in developing cost comparisons of such alternatives, particularly if it is complemented and supported by relevant cost data. Individuals applying the nomographs in this manner would be well advised to review and update all cost data periodically.

In the earlier evolutionary period of lightwave technology, the variety of fibers employed—particularly the extensive use of multimode fibers with their greater dispersion contributions—made it necessary to pursue system designs in greater detail. This situation has changed considerably, however, and we now find much smaller variations in fiber performance, particularly among different single mode fibers. Optical cables now often exhibit the same dispersion characteristics across a much wider range of cable costs.

When all or most fibers available present the same dispersion performance, it is useful for fiber manufacturers to present fiber data in a format similar to the nomographs we have discussed here. Certainly it would not be economically prohibitive for them to do so.

The greatly simplified approach to system design that nomographs make possible could stimulate the use of optical fibers in lower density traffic applications. Using the technique presented, or a similar approach, we might be able to employ a wider range of operating company personnel, perhaps even craftspeople, in examining possible applications of lightwave transmission.

It is in lower traffic density applications that the simplified technique is most useful, and it is exactly these applications that can least support heavy engineering costs for system design and evaluation. At some point in time, lightwave transmission must evolve from a sophisticated scientific oddity to a practical technology. The use of nomographs, as discussed here, should facilitate this transition.

REVIEW QUESTIONS

True or False?

T   F   1. At a lower transmission data rate, a lightwave link could be physically longer before it became bandwidth limited, than a link operating at a higher transmission data rate.

T   F   2. A typical single mode fiber carrying a 90 Mb/s digital signal will be bandwidth limited at a fiber length of approximately 40 km.

T   F   3. A multimode fiber with a bandwidth/length characteristic of 600 MHz-km cannot be used in any lightwave system operating at a transmission data rate of 45 Mb/s or higher, regardless of the physical length of the required interconnecting cable.

T   F   4. WDM is an effective technique to increase the transmission capacity of an in-place lightwave facility that may have been bandwidth limited in the initial system design.

T   F   5. We can establish the usable bandwidth of a proposed design by dividing the fiber bandwidth/length characteristic (BW fiber, MHz-km) by the required interconnecting fiber length (L, km).

T   F   6. Any system design employing a single mode fiber will provide more usable bandwidth than that same design employing a multimode fiber.

T   F   7. An optical fiber connector will usually introduce more attenuation than a hard fiber splice.

T   F   8. The optical signal level presented as an input to a lightwave receiver will determine the transmission data rate that the link can support.

T   F   9. It is possible to present too low a level of optical signal input to a lightwave receiver.

T   F   10. It is not possible to present too high a level of optical signal input to a lightwave receiver.

T   F   11. If a lightwave receiver is presented with too low an input signal level, the receiver rise time will be altered.

T   F   12. In an attenuation limited lightwave application, the system designer is obliged to correlate and select transmitter output level, fiber bandwidth, and receiver input level, so as to ensure satisfactory transmission performance.

T   F   13. It is sometimes necessary to insert optical attenuation into lightwave links to avoid overdriving the lightwave receiver.

T   F   14. If a system design provides an optical input signal to a receiver that is comfortably below the receiver overdrive input level, there is no compelling need to insert optical attenuation into the link.

T   F   15. Often, more than one combination of receiver, transmitter, and optical fiber will be technically acceptable in a system design effort.

T   F   16. A higher transmission data rate requires a higher level input to the lightwave receiver.

T   F   17. Laser type lightwave transmitters are mandatory at transmission data rates above 45 Mb/s.

T   F   18. In lightwave system designs, system margins are typically on the order of 10 dB or so.

T   F   19. When system length and data rate permit, some economy in system cost may be possible by employing LED type transmitters rather than laser type transmitters.

T   F   20. APD detectors are more sensitive—that is, they have a lower threshold of detection—than PIN-FET detectors.

CHAPTER 11

Service Protection in Lightwave Systems

The disruption of service in any transmission facility is obviously undesirable, and protecting a facility or system against such disruption is a primary responsibility of the design engineer. Protection should be designed as a result of a logical analysis of the exposure of the facility and the probability of failure of each component of the system. A fact easily overlooked is that service protection of any kind increases system costs, but does not increase the revenue produced by the system. Such measures certainly may improve the reliability of the service and, based upon a considered judgment, may also be necessary or desirable. However, they generally add neither usable circuits nor transmission capacity to the system—only duplicate circuits. Thus, they do not generate additional revenue nor do they improve the grade of service, e.g., the number of times a user may experience an ''all trunks busy'' delay.

A lightwave transmission system can be logically divided into three basic elements or zones, shown in Figure 11–1:

1. The interconnecting optical fiber cables.
2. The opto-electronic terminal equipment, such as the transmitters and receivers.
3. The electronic terminal equipment, such as the digital multiplexers.

These three elements represent distinct levels of sophistication, and there is no reason to believe that all three elements will evidence the same probability of failure, since they obviously are not subject to the same degree of exposure. One might logically conclude that the three elements should be protected to different degrees.

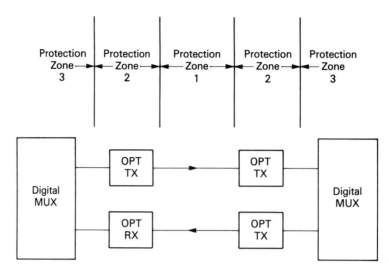

FIGURE 11–1
Protection Zones

OPTICAL FIBERS AND CABLES

The physical facilities—the interconnecting fibers themselves—would seem to present the highest level of exposure to disruption due to their sheer physical length and the environment in which they are placed. The type of plant construction utilized, that is, aerial or underground, certainly must be a factor in when, or to what extent, redundant fibers are provided.

Also important is the nature of the optical fiber cables. The fibers themselves are very small, and even a cable that includes many such fibers, 100 or more perhaps, is a relatively delicate structure. Consequently, optical fiber cables are generally less robust than cables of equivalent capacity employing metallic conductors. For example, when a multiple pair cable providing, say, 300 pairs, is placed between two service points, there is some logic to utilizing redundant pairs within the same cable sheath to protect a specific service. A cable of this size might suffer substantial mechanical damage, and some of the pairs could still remain serviceable.

An optical cable, on the other hand, is much more fragile. It is highly probable that such a cable, sustaining any kind of mechanical damage, would suffer disruption of all fibers, not just some. It follows that protective configurations that depend upon redundant, automatically switched fibers that are under the same cable sheath may not realistically provide a significantly higher level of service reliability at all. Of course, redundant fibers within the same sheath do provide some protection against an individual fiber splice or fiber connector failure, but it is incumbent on the designer who employs such a technique to justify the relatively small improvement in reliability against the relatively high cost penalty it imposes. Such justification might be a statistical history of high failure rates for in-service fiber splices or connectors.

There is a tendency on the part of many system designers to indiscriminately add fibers to cable fiber counts without any justification at all. Thus, even when a redundant fiber pair is provided and equipped as a protection facility for another, traffic carrying fiber pair, some designers specify additional, unequipped fibers without offerring any rationale. This trend is somewhat disturbing. The cost penalty is not insignificant, and since the designer has already evidenced a predisposition to four fibers in a fully protected configuration as the minimum acceptable facility, it cannot be seriously argued that two additional fibers really provide any additional traffic handling capacity.

And one would have to question the indiscriminate addition of two fibers as offering effective protection for a service that already was equipped with two fully operational fibers as protection. In any event, no submission has been made to support a high probability of fiber, splice, or connector failure.

It is not at all farfetched to conclude that the action was simply a reflex suggested by previous experience with multiple pair cables, where additional pairs did translate directly to additional cable capacity. This correlation is no longer strictly true with optical fibers.

In any case, when it is deemed necessary to reinforce or protect an individual optical fiber, the only practical alternative is to add another fiber.

OPTO-ELECTRONIC EQUIPMENT

The basic nature of opto-electronic equipment, which includes lasers and photodetectors, suggests that this type of equipment might be the system element most susceptible to failure. This is somewhat reinforced by the fact that some manufacturers include laser aging alarms as a standard function in their products. Presumably, they anticipate a higher failure rate in this kind of equipment than that expected in purely electronic units.

When it is deemed necessary to protect an individual opto-electronic unit, the only alternative available is to add a second unit.

ELECTRONIC EQUIPMENT

Although digital multiplexing to higher levels, say 45 Mb/s or higher, seems electronically sophisticated, the circuit boards and chips that make this possible have become quite conventional in design and are not intrinsically highly unreliable. However, in considering protection against the failure of such units, their commonality to all traffic carried argues persuasively for some level of redundancy. Usually several options are available.

Many higher level digital multiplexers have an architecture similar to that shown for an M 13 multiplexer in Figure 11-2. An M 13 unit can accept up to 28 DS-1 signal inputs (28 T-1 systems) and consolidate these into a single digital output signal at a DS-3 level (44.736 Mb/s). Note that the units identified in Figure 11-2

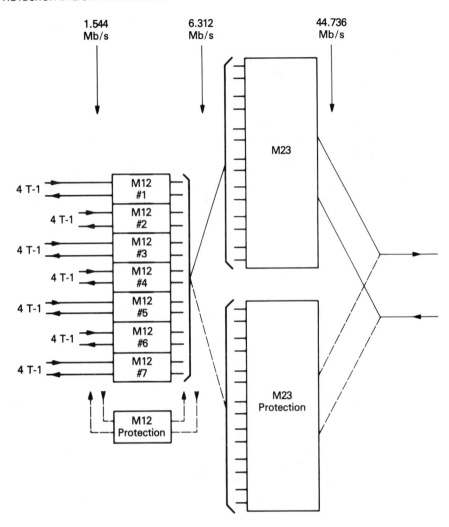

FIGURE 11–2
High-Level
Multiplexer

as M 12 multiplexers (numbered 1 through 7) may not actually be so identified in some products, but they are constructed to individually input and output four standard DS–1 signals as shown.

The output of each of the M 12 units shown is 6.312 Mb/s, and the seven outputs at this rate (if seven units are actually equipped) are consolidated within the unit identified as an M 23 multiplexer in the figure. Thus, the unit shown can consolidate 28 T–1 carrier systems (672 equivalent voice channels) into a single digital input/output of 44.736 Mb/s (a DS–3 signal).

Two levels of protection are usually provided. The higher level multiplexer (the M 23 unit) is protected with an identical unit on a one-for-one basis. As the drawing shows, a failure of the in-service M 23 unit would be sensed by circuitry and equipment (not shown for clarity), and the standby or protective M 23 would be automatically switched into service. Both the high-level (44.736 Mb/s) signals and all seven of the low level (6.312 Mb/s) signals would be switched.

The second level of protection provides a standby or protective M 12 multiplexer (identified as "M 12 Protection" in the drawing) as a switch-in replacement for any one of the seven in-service M 12 units. This level of protection, then, is one for seven, and if more than one M 12 unit fails, no more than one unit can be restored by the protective multiplexer.

Figure 11–2 shows only one terminal end of a link, of course, and a high-bit-rate signal would be extended to and from the opto-electronic equipment (the light transmitter and receiver) at this same terminal.

Different switching control functions, which require and include signal quality sampling, are provided by different manufacturers. A significant level of sophistication is introduced into the terminals, and in many cases some "handshake" transfer of information is essential with the distant terminal of the link also. Note that the configuration shown does not conflict with the protection zones shown in Figure 11–1; it simply presents one approach at supplying the requirements of protection zone 3.

A more common equipment configuration for high-bit-rate link protection is shown in Figure 11–3. This approach effectively preempts any discriminate choice of protection within a system or link.

Since the fibers, the opto-electronic terminals, and the digital multiplexer are all interconnected as an integral terminal package, any graded protection such as was suggested by Figure 11–1 will not be possible. The multiplexer functions as before, with a 1×1 backup level for the M 23 multiplexer and a 1×7 backup for the lower level M 12 units. A fiber pair, and its associated opto-electronic units (TX and RX), are dedicated to service with a particular M 23 unit. Thus, switching the multiplexer in this manner automatically switches the interconnecting lightwave transmission facility in its entirety.

As mentioned, there is no flexible discrimination in such a setup. If a failure of the in-service lightwave terminal unit (TX or RX) was concurrent with a failure of the protection M 23 multiplexer, all service would be interrupted, even though a perfectly serviceable lightwave link, and a perfectly serviceable M 23 multiplexer, may have survived elsewhere in the system.

Many manufacturers design and sell lightwave terminal equipment as complete systems, rather than as components of systems. However rational a philosophy of discriminatory, graded service protection such as that discussed earlier and depicted in Figure 11–1 may be, such an approach may be precluded by the configurations of equipment that are available on the market.

We have limited our discussion to the DS–3 signal level, but multiplexing above this level is certainly possible, and is frequently employed in high-density systems. The architecture of the multiplexer as we have shown it is simply expanded in higher level units, and the protection ratios are the same also—that is, high-level signals are protected on a one-for-one basis.

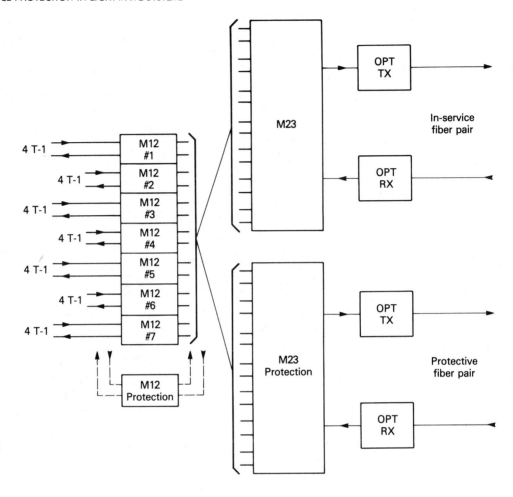

FIGURE 11–3
High-Level
Multiplexing

HIGH-TRAFFIC-DENSITY APPLICATIONS

A lightwave link connecting two major traffic centers is expected to carry substantial traffic. There is every expectation that it will produce significant revenue also. It may or may not be true that at one time circuit or service protection was basically motivated by a "sense of public service," but it is certainly true today that protection of service is primarily a matter of revenue protection. If a particular network or system is unreliable or produces a poor quality of transmission, then traffic will migrate to some competing network, and so will the service revenues. And those revenues may be very substantial.

The greater the fiber utilization, that is, the more service carried on a fiber, the more persuasive the argument to protect continuity of service becomes. This is reinforced by the fact that the more circuits or services a fiber supports, the lower the cost of service protection becomes per circuit (or service). Applications of this nature, with heavy traffic and high revenues, clearly justify, and perhaps demand, high levels of service protection, and of course they produce revenue on a scale that can easily support the costs involved. The question of whether or not to design protection into these installations becomes largely academic, but some judgment is necessary as to what constitutes a high volume of traffic or a high-priority installation.

To date in the telecommunications industry the majority of lightwave installations have been high-traffic-density applications. However, the basic technology has much to recommend it for lower levels of the network as well—in distribution or subscriber loop plant, for example. Here, the demographics and economics are dramatically different, however, and the conventional trunking philosophy of protection may not be appropriate in all cases. Indeed, in some instances, it will surely not be affordable.

LOW-TRAFFIC-DENSITY APPLICATIONS

There is no generally recognized line of demarcation between a high-density and low-density application. The question is subjective. There is, however, an abrupt demarcation in the North American digital hierarchy between a DS–2 signal (96 equivalent voice channels) and a DS–3 signal (672 equivalent voice channels). For purposes of this discussion, since standard terminal equipment is readily available for both levels (DS–2 is 6.312 Mb/s and DS–3 is 44.736 Mb/s), we shall establish an arbitrary demarcation that traffic requirements up to and including 96 voice channels (DS–2) are low-density applications.

Then the highest level digital multiplexing necessary to accommodate all low-density applications (by our qualification) would be an M 12 unit, which accepts four DS–1 signals and outputs one DS–2 signal. This is a relatively unsophisticated level of equipment, and it is interesting to note that at least one major manufacturer does not even offer this low-level a product in a redundant (1 × 1 protection) configuration. The implication is that the operational history of this level of equipment does not warrant redundancy because the failure rate is very low.

If a low-density system, as we have defined it, requires a transmission data rate no higher than 6.312 Mb/s, it would appear that some economy in opto-electronic terminal equipment, and in the optical fibers themselves, is possible. A study of fiber and terminal equipment costs, however, indicates that such economies may be largely illusory.

The nature of the light sources and detectors, and of the fibers themselves, is such that a 45 Mb/s rate, and even a 90 Mb/s rate, is almost as easy to achieve as 6 Mb/s. Even the higher 90 Mb/s may not impose any significant cost penalty on fiber quality unless the cable route length is excessive.

Thus, there appears to be little cost advantage in specifying and purchasing such items as terminal units and fiber for a transmission rate lower than 45 Mb/s.

At the nominal cable route lengths usually encountered in low-traffic-density applications, there are rarely any intermediate repeaters required, and consequently there is usually little or no economic penalty if transmission engineering assumes the minimum transmission bit rate to be 45 Mb/s. Accordingly, fiber and opto-electronic equipment should probably be specified and purchased for operation at 45 Mb/s even if the system is actually placed in service at some lower transmission bit rate. Units that are capable of operating satisfactorily at 45 Mb/s can certainly function when carrying lower data rate signals.

These recommendations should be verified with fiber and equipment suppliers before actually making a decision to purchase. However, since the conditions that hold today may not prevail indefinitely, some installations could actually incur a cost penalty.

LOW-LEVEL MULTIPLEXER PROTECTION

The system designer has a great deal more latitude in how to protect service in lower level systems, and the simplest approach is distribution of the traffic. Consider an application requiring four T–1 systems, which involves four DS–1 level signals providing 96 voice channel capacity.

As shown in Figure 11–4, one approach is to utilize one fiber pair and an M 12 multiplexer at each link terminal. Traffic capacity is four T–1s, as required, but no growth can be accommodated without modifying and adding to the initial installation. This approach is quite vulnerable to interruption, since a failure of any single fiber, of any optical unit, or of either M 12 multiplexer will disrupt 100 per-

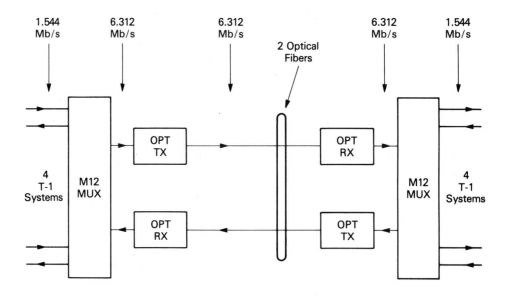

FIGURE 11–4
Low-Level
Multiplexing

cent of the traffic carried. We can improve on the preceding design by utilizing WDM as shown in Figure 11–5.

Here, the initially installed two fibers are utilized to provide two completely independent transmission systems, each using a different light carrier wavelength. Two M 12 multiplexers are required at each system terminal.

A respectable level of service protection can be provided simply by applying half the traffic to one independent system and the other half to the other system. A failure of either fiber will disrupt all traffic, but a failure of a multiplexer or any single opto-electronic unit can only affect 50 percent of the traffic.

In Figure 11–6 we have improved the situation even more by utilizing the two fibers as bidirectional links. Now a failure of either fiber can only disrupt half the traffic carried.

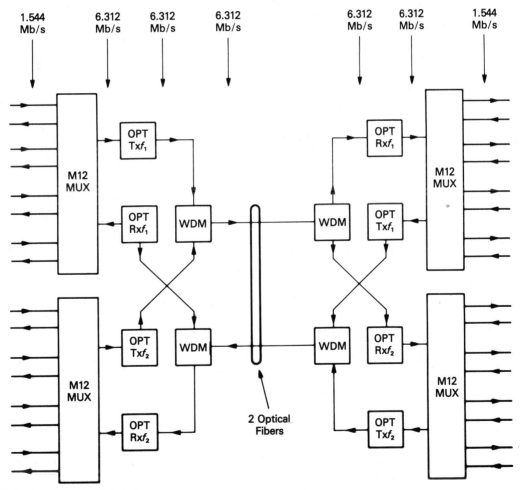

FIGURE 11–5 Low-Level Multiplexing

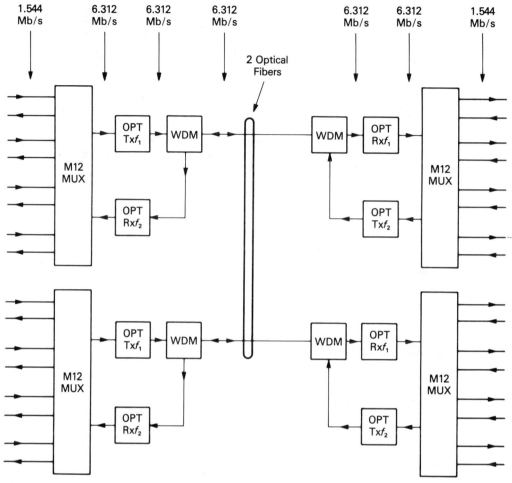

FIGURE 11-6 Bidirectional Operation

One basic advantage of the designs shown in both Figures 11-5 and 11-6 is the additional transmission capacity that each provides. Both configurations can handle eight DS-1 signals (eight T-1 systems) without any later modification at all. Additional circuit cards must be added to realize this additional capacity, but both designs exceed the original requirement for a capacity of four T-1 systems.

OPTO-ELECTRONIC EQUIPMENT PROTECTION

If it is determined that a higher probability of failure actually does lie in the interconnecting fibers or in the opto-electronic terminal equipment, a level of protection such as that shown in Figure 11-7 might be provided.

Here, a single M 12 multiplexer is used, which satisfies the original service

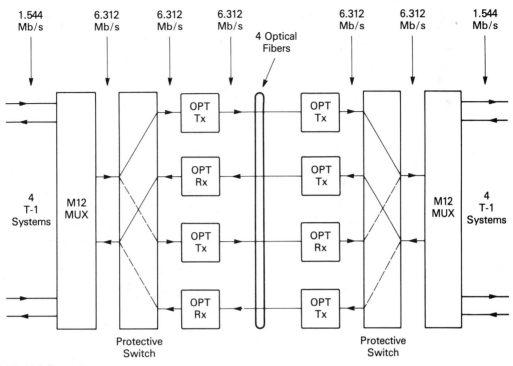

FIGURE 11-7   Link Protection

requirement for a capacity of four T-1 systems, but the lightwave terminal units and the interconnecting fibers are both fully and automatically protected. The configuration may not be widely available as a standard product, but it is certainly technically possible. Note that a failure of either multiplexer would introduce a total interruption of service, but the approach assumes that the M 12 equipment has already been qualified as having a low and acceptable failure rate.

INTERCONNECTING FIBER PROTECTION

The configuration shown in Figure 11-8 will protect against a fiber failure only, but may not be available as a standard product. Optical switches are available today that would permit such a configuration, but special sensing and switching control arrangements might have to be developed.

SUMMARY

There are a variety of protective configurations possible and a number of levels of protection that can be provided for lightwave transmission systems. The difficulty is in resolving when, and to what extent, protection is either justified or required.

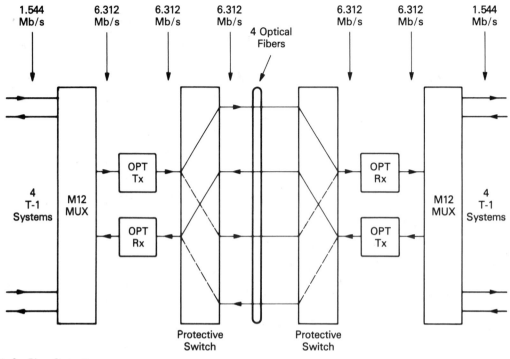

FIGURE 11-8 Fiber Protection

It is incumbent on the design engineer to adopt a logical approach to protecting any facility or service, and not simply to implement a standard response for all cases. The latter would be engineering by rote, not by reasoning.

The basic elements of a lightwave transmission facility are identifiable and quite different in nature. It is reasonable to expect that the probability of failure of these different elements will not be the same. It follows that they may not require the same level of protection or redundancy.

Higher level digital multiplexing within a system can be a very cost-effective technique for increasing the system transmission capacity, but it does introduce a high degree of equipment commonality. When all of the traffic carried on any facility is passed through a common unit of equipment, the arguments for backing up that common equipment in redundant configurations can be persuasive.

The expenditure of capital for protecting continuity of service is often unavoidable, but it imposes an obligation for cost justification. The system designer must consider such factors as supporting revenues as well as the more obvious technical advantages. Any and all alternatives should be considered and evaluated before a final decision is arrived at.

Protecting a transmission facility against a complete disruption of all traffic carried is, in itself, an identifiable objective in system engineering. If the continuity of a significant percentage of transmission capability can be assured, even if the total transmission capacity cannot, the protection so provided is substantial and respectable.

Complete isolation of a terminal within a system is catastrophic. A partial loss of transmission capacity to that terminal is not the same level of system disruption, and may be an acceptable penalty, if the periods of loss are not excessive, and the economic benefits are substantial.

REVIEW QUESTIONS

True or False?

T   F   1. The design engineer bears responsibility only for the technical performance of a system, and not for the economic efficiency of the system design.

T   F   2. Service protection must, in every case, provide alternative transmission capability for every system element, regardless of the economics of the application involved.

T   F   3. There are options available in protecting system service continuity other than providing a complete backup for each individual system element.

T   F   4. Several optical fibers, all under the same cable sheath, represent a significantly higher level of fiber reliability than a single pair, or two pairs, of fibers under the same cable sheath.

T   F   5. It is possible to provide a degree of service protection without increasing the optical fiber count by providing independent transmission capacity using WDM.

T   F   6. In lower traffic density applications, there are more options available for protecting service continuity, than there usually are in higher traffic density applications.

T   F   7. The choice, when such a choice is possible, between higher level or lower level digital transmission rates will involve economic judgments as well as transmission engineering considerations.

T   F   8. Protection-switching configurations in higher level digital multiplexers may provide different levels of protection within the multiplexer itself. Lower levels of such units, which often involve several identical subassemblies, may not be protected on a one-for-one basis at all.

T   F   9. When a facility is designed and equipped to protect service continuity, a higher grade of service is provided if grade of service is defined as an "all circuits busy" condition.

T   F   10. Engineering responsibility is confined entirely to technical matters and imposes no obligations for cost effectiveness or economy.

T   F   11. There may be merit in a protection configuration that significantly reduces the probability of total isolation of a transmission system terminal, even if such a configuration cannot provide total protection for all system transmission capacity.

T   F   12. Protection-switched facilities or equipment must transfer transmission capacity in a manner that does not introduce interruptions in other sections of the network. For example, switching transient times must be short enough to avoid circuits being disconnected by the normal functions of other elements within the system.

T   F   13. It is possible to configure systems to protect only those system elements that are more vulnerable than other elements.

T   F   14. The provision of independent, parallel transmission paths through system configuration is a viable alternative to conventional protection-switching configurations in some applications.

T   F   15. In addition to offering an effective method for increasing system transmission capacity, WDM can be employed in alternative service continuity configurations as well.

T   F   16. There is substantial evidence in operating history data to indicate that optical fiber splices are a major cause of lightwave system service interruptions.

T   F   17. As a general observation, lightwave transmission systems have been found to provide a lower level of reliability of service than other transmission disciplines.

T   F   18. Fully protection-switched terminal configurations are the most frequently encountered arrangements in lightwave systems constructed to date.

T   F   19. All technically acceptable lightwave transmission facilities must provide, at a minimum, four discrete optical fibers.

T   F   20. Alternate cable routings can be incorporated into various service continuity protection configurations, but such configurations may impose substantial economic penalties.

CHAPTER 12

Optical Fibers and Cables

Much of the material published on lightwave transmission includes detailed explanations of the processes by which optical fibers are fabricated. But since our objective is to present the technology from a practical applications perspective, and introducing material on fabrication will not contribute significantly to that end, we will address fiber and cable strictly from the point of view of understanding their structure and characteristics, and not pursue the manufacturing or fabrication processes at all.

The cost of optical fibers and cables fabricated with such fibers is changing rapidly. Although we have included some current cost information, we caution the reader against accepting the data as authoritative. Some of the cost differentials mentioned may continue to be applicable, even though basic cable costs may increase or decrease, as the case may be.

OPTICAL FIBERS

Optical fibers are lightguides through which light rays will propagate. One way of classifying optical fibers is by the material they are constructed from. There are plastic fibers; glass core, plastic-clad fibers; and a variety of types of glass in all-glass fibers. Most of the optical fibers that have been employed in telecommunications networks to date have been made of silica glass.

The physical dimensions of optical fibers are very small, with outside diameters of 125 μm or so being typical. Although glass as a material may have greater strength than a metallic wire of comparable size, the inherent strength of glass depends on the absence of stress-producing chemical or physical defects in the glass itself.

Flaws in an optical fiber that may be introduced during fabrication, handling, placement, or even exposure to the atmosphere all represent points of potential fiber failure. A fiber once flawed may be further degraded by stress corrosion, and the defect may expand or grow. Moisture, mechanical stress, and chemical attack can all accelerate stress corrosion in an optical fiber.

Because of their small physical size and the mechanisms of light propagation through them, optical fibers are inherently more vulnerable to defects than metallic conductors. A metallic conductor may experience minor surface abrasion during handling, but the impact on transmission through that conductor will be negligible compared to the effect of that same defect in an optical fiber.

OPTICAL CABLE DESIGN CONSIDERATIONS

In the design of optical fiber cables containing one or more fibers, a major objective is isolating each glass element from any and all stresses, insofar as is possible. This includes stress due to moisture as well as the mechanical stress that cable placement and normal thermal expansion and contraction of other cable materials may introduce.

There are unique vulnerabilities in lightguides that necessitate meticulous attention to stress relief in fibers. For example, the phenomena of microbending and macrobending are not commonly encountered in metallic cable technology.

Microbending may be defined as small abrupt changes in the optical fiber core mechanical structure, or sharp irregularities at the interface between the fiber cladding and the core materials, as shown in Figure 12–1. Such irregularities may be minute, but they may still introduce significant additional attenuation to light transmission through the fiber. Microbending may be introduced in the manufacturing processes.

Macrobending occurs when a fiber bend radius decreases to the point that light rays within the core start to escape into the cladding material, as shown in Figure 12–2. The radius at which macrobending occurs depends upon the fiber type and

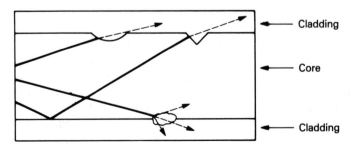

FIGURE 12–1
Microbending

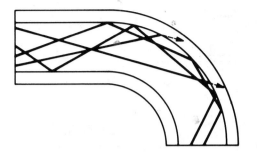

FIGURE 12–2
Macrobending

size and the operating light wavelength. Longer wavelengths are more susceptible to losses due to macrobending.

In the design of multiple pair metallic conductor cables, the size, strength, and number of conductors involved can impart substantial strength and antibuckling qualities to the cable structure. This is not the case in fiber cables. There, the design problems are complicated by the necessity to isolate each fiber from the cable structure to some degree. For example, to avoid transferring stress to optical fibers, excess fiber length must be provided per unit of cable length to accommodate fiber bends and thermal excursions. In effect, the fibers must be free to move within the confines of the cable structure at a different rate, and perhaps even in a different direction, than the rest of the cable material moves.

Before addressing the complexities of fiber cabling, some level of protection for each individual fiber alone must be provided. Most fibers are individually coated with one or two polymeric coatings, typically 65 to 185 μm in thickness, to preserve the intrinsic strength of the fiber, to provide some protection against macrobending, and to permit fiber handling. Generally, it is necessary to remove these coatings either mechanically or chemically before any fiber splicing or connecting is begun.

FIBER BUFFERING PROTECTION

Over and above the coating process, in many cable structures it is advisable to provide additional fiber protection against stress, moisture, or chemicals. This can be done by encasing each individual fiber within an annular *buffer tube*.

Figure 12–3 shows one technique of buffering fibers using what is generally referred to as a *loose tube*. Here, the protective jacket or tube is substantially oversized compared to the outside diameter of the coated fiber. This allows considerable freedom of fiber movement within the tube, as shown in the three parts of the figure. It also allows excess fiber length to be accommodated.

The tube material itself provides some lubrication for the independent fiber movement, and the tube is often flooded with an elastic compound or gel to inhibit entry of moisture.

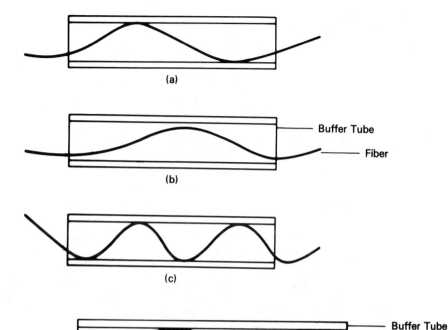

(a)

Buffer Tube

Fiber

(b)

FIGURE 12–3
Loose-Tube
Buffering

(c)

FIGURE 12–4
Tight-Tube
Buffering

Buffer Tube

Fiber

A different buffering technique is shown in Figure 12–4. In *tight tube buffering*, the inner diameter of the tube is only slightly larger than the coated fiber outside diameter. In feel and appearance, this configuration might resemble a plastic insulated metallic conductor in a conventional multiple-pair cable.

The earliest tight-tube buffer techniques used buffer materials that were not closely compatible thermally with the optical fiber itself. Consequently, when the buffer tube contracted due to thermal change, some stress was transferred to the fiber. This problem has been largely overcome in later cables by selecting buffer material that is more thermally stable and compatible with the fiber.

Note in Figure 12–4 that the tight buffer tube cannot accommodate much excess fiber length, although it does actually allow some independent movement of the fiber within the tube. To offset this, in cables employing tight buffered fibers, the lay or helical serving of the fibers throughout the length of the cable is closer or tighter to provide extra fiber length within any discrete cable length.

CABLE DESIGN OBJECTIVES

As in metallic conductor cable design, the fiber cable structure is intended to protect the relatively fragile fibers from mechanical damage and exposure to moisture and chemicals, and to provide sufficient mechanical strength and rigidity for the fibers to survive normal cable handling and placement and long term exposure to the environment. Many of the materials and manufacturing techniques that have evolved for metallic conductor cables find direct application in fiber cables. The major difference between the design problems for each is the increased necessity in fiber cables of isolating the fibers themselves from any stresses that may develop within the cable structure.

Essentially, today's fiber cables incorporate only two basic structural designs. The first type of structure provides a protected core void or tube throughout the cable by means of cable architecture. This space or core in the center of the cable is not greatly different in function than a separate cable conduit would be, although the optical fibers are placed within the core as a part of the cable fabrication process, not pulled into the "conduit" at a later time as a separate work operation. The cable architecture relies entirely on the outer wall structure of the cable to provide mechanical strength for pulling and placement, and for rigidity to avoid cable buckling. This type of cable might be characterized as being of thick-walled construction.

The second basic cable structure more closely resembles conventional multi-pair cable designs in that individual fibers retain their identity and individuality within the overall structure, just as metallic pair conductors do, even though they may be stranded together in bundles. Typically, cables of this type include a so called *strength member,* which may be made of any of several materials and which may occupy the center of the cable structure. The optical fibers are grouped around this element. Various levels of protection can be designed into the outer wall of such a cable, including metallic moisture barriers, several extruded jackets, and armor tapes.

RIBBON CABLE STRUCTURE

Figure 12–5 shows a conduit type of cable structure which has become known as *ribbon cable.*

Note that all fibers are positioned within the core space provided in the center of the cable. Individual fibers are assembled in a parallel relationship and then encapsulated into a flat "ribbon" type of construction that becomes an integral assembly. A ribbon of this type can include as many as twelve individual fibers, and several or many ribbons can be stacked as a single mechanical assembly. The ribbon assembly is spiraled or helically served through the length of the cable to provide extra fiber length to avoid stress, and the core void enclosing the ribbons is often flooded with a moisture-excluding compound or gel.

One unique aspect of these high-fiber-count cables is a companion method for splicing single ribbons in a single work operation, rather than splicing individual fibers separately. Figure 12–6 shows a mechanical splicing technique used with ribbon cables of this type.

In general, ribbon cable is expensive to fabricate, and to date the configuration has largely been limited to high-traffic-density cable routes connecting major population centers where a large number of fibers would be required.

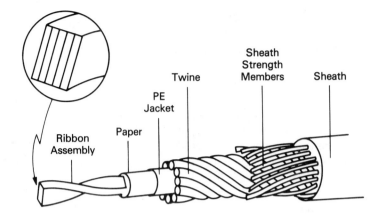

FIGURE 12-5
Ribbon Cable
Structure

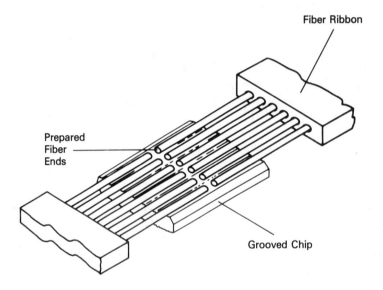

FIGURE 12-6
Ribbon Cable
Splicing

STRENGTH MEMBER STRUCTURE

Figure 12-7 shows a cable that uses a strength member type of construction. The so called strength member, which may be made of any of several materials, occupies the center of the cable structure, and the optical fibers are grouped around this element. Various levels of protection can be designed into the outer wall of such a cable, including a metallic moisture barrier, several extruded jackets, and armor tape.

Although the central member is often called a strength member, it usually does not provide all or even most of the cable mechanical strength, although it does serve to inhibit fiber or cable buckling when the cable is bent or flexed.

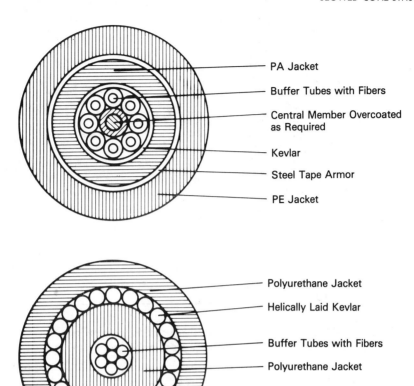

FIGURE 12-7
Strength
Member
Structure

PA Jacket

Buffer Tubes with Fibers

Central Member Overcoated
as Required

Kevlar

Steel Tape Armor

PE Jacket

Polyurethane Jacket

Helically Laid Kevlar

Buffer Tubes with Fibers

Polyurethane Jacket

FIGURE 12-8
Central Cluster
Structure

Some variations on this type of construction do not provide a centrally located strength member at all, but instead cluster the coated and individually buffered fibers along the axis of the cable, as shown in Figure 12–8.

SLOTTED CORE STRUCTURE

Another technique uses a somewhat oversized core member of extruded plastic material that has slots or grooves around its outer edge. This core may or may not incorporate a separate tensile or strength member at its center.

The fibers themselves are positioned in the outside grooves or slots, as shown in Figure 12–9. Since there is sufficient room and the fibers are loosely placed, this type of cable does not usually use individual fiber buffering tubes at all. The structure is often called a *slotted core* cable.

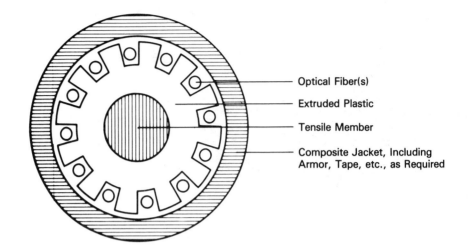

FIGURE 12-9
Slotted Core
Structure

Optical Fiber(s)

Extruded Plastic

Tensile Member

Composite Jacket, Including
Armor, Tape, etc., as Required

CABLE DESIGNS AND FIBER STRESS

In all of the preceding configurations, the fibers are helically served throughout the cable length, and flooding with elastic compound or gel is a common practice to inhibit moisture. An examination of the different cable structures shows that, in varying degree, all designs offer substantial isolation to the fibers from mechanical and other stresses that normal cable handling and environments might introduce.

When any cable is *pulled* into place, or *plowed,* a multiplicity of strains will be transferred to the cable structure, and it can be shown that eventually these stresses migrate to the cable core. This can introduce cable elongation or other distortion, but in good cable designs few of these stresses will be transferred directly to the optical fibers themselves.

The selection of thermally compatible materials, as in the case of the tight buffer tube, and due attention to potential fiber buckling and the exclusion of moisture produce fiber cables today that should give years of trouble-free service in environments that telecommunications cable can be expected to be exposed to.

Almost any of the cable structures, with the possible exception of the ribbon cables, can be procured with a wide variety of fiber counts, i.e., the number of optical fibers provided. In addition, most manufacturers can and will provide a small-gauge metallic pair or two for maintenance communications purposes.

METALLIC ELEMENTS IN FIBER CABLES

The advent of glass fibers as a transmission medium presents an entirely new option in physical transmission facility design and construction. In multiple-pair metallic cable or in coaxial cable systems metallic continuity throughout the transmission system was unavoidable since the metallic elements themselves were the transmission medium. Optical fibers, however, are made of glass, and in themselves are inherently dielectric. Theoretically, it is possible to install a complete lightwave facility that introduces no metallic elements at all and thus presents no electrical continuity along its length.

This would eliminate exposure to lightning damage, sheath currents, and other possibly disruptive electrical disturbance. Such a facility would certainly be technically attractive, although it may not be entirely practical in many installations.

One of the advantages of a metallic element in any cable structure is its effectiveness as a barrier to moisture. An all-dielectric cable would sacrifice this advantage, but perhaps complete flooding of the cable interior or development of an alternative moisture barrier could offset the loss.

A second advantage of metallic elements is structural strength, but with the multiplicity of materials available today, it would appear that mechanical strength could be provided using other cable designs and materials.

The third argument is the mechanical protection that metallic elements can offer. An example might be the use of armor tapes to limit or discourage damage due to rodents. By using alternative construction techniques, such as large diameter plastic conduits or even split conduit that inserts the cable during the plowing operation, it is at least possible to provide adequate or even equivalent mechanical protection by other means.

The placement of all-dielectric cables introduces other problems, such as locating buried cables at some future date. Moreover, the alternatives to metallic element cables may be costly. It remains to be demonstrated that the advantages of a nonconductive facility offset the difficulties or costs.

Perhaps the all-dielectric plant will find initial application in areas of high lightning incidence, even if it does impose cost penalties. Perhaps an all-dielectric self-supporting cable would find applications in low-density, rural aerial plant, where lightning exposure might be a problem. Certainly the option of nonconductive transmission facilities is a unique characteristic of lightwave transmission, and a study on how, if, when, or at what cost, such facilities might actually be constructed merits serious consideration.

OPTICAL CABLE COSTS

Although fiber and cable costs change rapidly in today's market, the cost differential between fiber types and cable structures can fairly confidently be expected to remain relatively constant. The cost data we offer here are only representative, but they do provide a reasonable basis for evaluating alternative system designs and configurations.

Fiber costs do not reflect changes in fiber characteristics as much as one might expect them to. In multimode fibers there may be drastic variations in fiber bandwidth, but these variations may not be evidenced proportionally in fiber costs. Thus fiber with a length/bandwidth characteristic of 600 MHz-km would present substantially less bandwidth for use than a fiber with a performance specification of 2,000 MHz-km, but the cost of such fibers might be nearly the same.

In single mode fibers, unless we are considering the more sophisticated fibers such as dispersion-shifted or dispersion-flattened structures, there is little difference in the bandwidth/length characteristics of the fibers available on the market today, and thus little cost differential is evident. If an application required 90 Mb/s capability, and a single mode fiber were under consideration, the cost of such a fiber would not change significantly if the data rate were reduced to 45 Mb/s or increased to 135 Mb/s. Indeed, it may not change at all.

This condition does not exist regarding fiber loss, however. Here, we will find some cost differential between a lower loss fiber and a higher loss fiber. This differential is, to some extent, the result of selection of production output, rather than variations in fiber design or material composition.

It is common practice today for cable manufacturers to quote cable costs in units of cable length. Most generally used is the term "cable meter," which relates a unit cost to a meter length of finished cable. Table 12–1 lists some cable cost differentials for four basic cable structures. Note that the cost data are representative of and applicable to single mode fiber cables only.

From the table, the cost penalty for an additional two fibers is approximately $.50 to $.60 per cable meter. For optical cables, the fibers represent about 60 percent of the total cable cost. For multimode fiber cable, costs are not as predictable, but cable can be supplied in any of the configurations listed and several others as well. Cable costs will be quoted by most cable manufacturers on request.

Cable reels holding 2-km lengths of cable are standard today, and longer reel lengths can be supplied upon request.

SUMMARY

The fabrication of rugged, highly reliable cables has been amply demonstrated in many applications, and there are no compelling reasons that exclude the placement of optical fiber cables in any environment. Indeed, optical submarine cables have been placed in intercontinental lightwave links.

TABLE 12–1
Cable Cost
Comparison

	Four-Fiber Cable (per meter)	Six-Fiber Cable (per meter)
Armored cable with nonmetallic center strength member	$2.21	$2.80
Armored cable with steel center strength member	$2.01	$2.60
Unarmored cable with nonmetallic center strength member	$1.88	$2.47
Unarmored cable with steel center strength member	$1.70	$2.29

Although optical fibers are immune to electromagnetic interference, such as induced currents and electrical noise, the fibers are more vulnerable to minor mechanical flaws than are metallic conductors of comparable size.

A variety of mechanical cable structures are available to suit almost any environment, and many of the materials employed and coatings or shields that may be included are similar to those used in metallic conductor cable structures.

The major consideration in cable design is to isolate the optical fibers from all mechanical stress that may be applied to, or developed within, the composite cable structure.

A unique characteristic of optical cables is that, from a transmission point of view alone, there is no necessity to provide any metallic conductivity end to end throughout a transmission link. For the first time, it is possible to construct completely dielectric interconnecting links in physical transmission plant.

REVIEW QUESTIONS

True or False?

T   F   1. Optical fibers may be fabricated from glass, plastic, or a combination of both.

T   F   2. Optical fibers are not inherently stronger than metallic conductors of comparable size.

T   F   3. Optical fibers are inherently more vulnerable to minute defects in structure than metallic conductors of comparable size.

T   F   4. In optical fiber cables, the fibers themselves contribute substantially to the antibuckling properties of the cable.

T   F   5. Optical cable designs are intended to restrict the freedom of movement of the optical fibers within the cable structure.

T   F   6. Individual fiber coatings are not required on fibers that are individually protected in buffer tubes.

T   F   7. In tight-tube cable construction, the helical serving or "lay" of the optical fibers is tighter than in loose-tube cable construction.

T   F   8. The strength member provided in some optical cable structures is designed primarily to provide antibuckling properties in the cable structure.

T   F   9. In ribbon-type cables, fiber splicing can splice all fibers within a single ribbon in a single work operation.

T   F   10. In slotted core cables, no more than one fiber is positioned within each individual "slot."

T   F   11. In loose buffer tube cables, no more than one fiber is positioned within each individual buffer tube.

T   F   12. All-dielectric optical cables are defined as cables that do not include any metallic pairs for use in transmission.

T   F   13. Metallic elements are often introduced into optical cable structures to provide moisture barriers.

T   F   14. One available construction technique inserts optical cable into a flexible conduit structure during the actual cable-plowing operation itself.

T   F   15. All-dielectric optical cables cannot be fabricated as a self-supporting structure for aerial plant construction.

T   F   16. Multimode fibers generally present a wider selection of bandwidth/length characteristics than do single mode fibers.

T   F   17. A lower loss fiber will usually cost significantly more than a higher loss fiber.

T   F   18. Cable costs are usually presented as units of cost per meter of finished cable length.

T   F   19. Optical fiber cables are not suitable for aerial plant construction.

T   F   20. A primary concern in designing optical cable is to isolate the optical fibers from the mechanical stresses that will be imposed on the cable structure.

Lightwave Plant Construction and Splicing

Although construction methods for lightwave plant do not differ radically from more conventional metallic conductor installations, the two methods are distinct, and we need to be familiar with the differences. Some of the differences have been influenced or made possible by the basic nature of optical fiber cables. For the transmission capacity they present, optical fiber cables are very light, very small, and somewhat more fragile than metallic conductor structures.

The nature of the fibers themselves, both in the material used in their fabrication, and in the extremely small physical dimensions involved, makes fiber splicing a bit more difficult than splicing metallic conductors. The splicing sophistication required is a persuasive argument for reducing the number of splices to be made, insofar as this is possible.

In one early lightwave experience that the author is familiar with, the construction program was designed specifically to test several conventional construction techniques. The project involved aerial plant on new messenger strand, some aerial plant overlashed to existing telephone cable, and a significant amount of direct burial plant also. Telephone company craftspeople were employed exclusively, although none of these people had any previous experience with optical fiber cable placement. Even fiber splicing was performed by the telephone personnel after only minimal training.

The resultant system performed satisfactorily, but the most important observation is that the telephone craftspeople had no problems at all with this alien technology. Not only did they place the plant using conventional methods, but they had a very respectable production output as well. This experience has been duplicated many times since those early days, and there is no reason or evidence to indicate that exotic construction methods are ever required.

These earlier systems did not make any particular effort to reduce the number of fiber splices required, but given the apparent ease with which the optical fiber cables could be handled due to their small size and light weight, it was a logical extension to address this possibility next. When it became evident that fiber cables could be fabricated and supplied in greater lengths, the basic construction methods were reviewed to determine whether these lengths could be utilized during construction to reduce the amount of splicing required. The factor of greatest concern was the relatively low tensile strength that most optical fiber cables offer. If longer lengths were to be employed, then cable pulls would have to be restricted in length or other construction techniques used.

UNDERGROUND CABLE PLACEMENT

In underground conduit systems, manhole spacings reflect both the lengths of metallic cable that can be mechanically handled, and the strength that such cables must inherently have to avoid damage from the pulling operation. The first efforts at longer pulls of optical cable in such systems was for personnel to manually assist the cable pull at intermediately located manholes. This was reasonably successful, but the pulling tensions that were actually being transferred to the cable were difficult to determine or control with any precision, and there was a limit to how much manual assistance could actually be rendered.

The second technique to come into use was to make a series of shorter cable pulls possible, as shown in Figure 13–1. Here, the entire length of the cable is pulled off the reel and through the first conduit section.

This necessitates storing the excess cable in some way, until the second cable pull can be made, and a technique of coiling the cable up in a figure eight was developed. After completing the first conduit pull, the pulling winch was moved to the third location, as shown in the figure, and a second pull was made, drawing the stored cable out of the figure-eight temporary storage configuration. In this manner, no single conduit pull was excessively long or imparted significant tensile stress to the cable. The handling and storage of the cable at intermediate points did present some exposure to possible damage to the cable, and such operations required quite a bit of manpower and excellent communication between work locations.

A refinement on this technique is to utilize powered capstans or other powered traction-applying machines at intermediary pull locations, as shown in Figure 13–2. With several turns of cable around such a capstan, which must have an acceptable bend radius, when tension is taken up on the cable past the work location, the cable is tightened around the capstan drum.

The capstan then automatically imparts pulling power to the cable already in place ahead of that work location. In effect, a number of short pulls can be made as one work sequence, and since the intermediary capstans remove pulling power

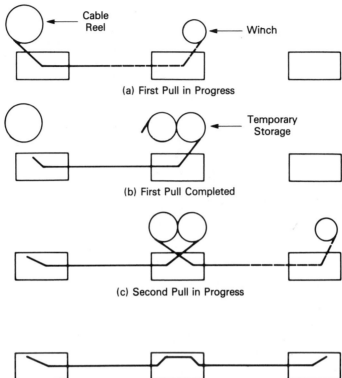

(a) First Pull in Progress

(b) First Pull Completed

(c) Second Pull in Progress

FIGURE 13-1
Temporary
Cable Storage

(d) Second Pull Completed

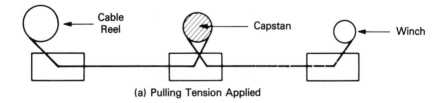

(a) Pulling Tension Applied

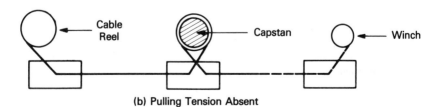

FIGURE 13-2
Power-Assisted
Pull

(b) Pulling Tension Absent

when the tension is removed from the next pulling section, each capstan tends to apply pulling power only when necessary.

This technique works very well, is controllable, and permits very long pulls to be easily subdivided into several shorter pulls. The technique is unique to optical fiber cables and to small-diameter metallic conductor cables, since it depends to a large extent on the allowable bending radius of the cable to be placed and the cable's inherent flexibility.

The method can be used for long pulls in placing aerial cable also. It is not applicable to cable plowing operations, however, but in such operations the tension placed on the cable is not a function of the cable length and can be handled quite adequately in other ways. Although capstans are shown in Figure 13–2, other types of equipment are available.

In underground placement operations, cable lubricants are usually applied to further reduce the tensile load on the optical cable.

DIRECT BURIAL CABLE PLACEMENT

As used here, cable placement means cable plowing as a single-placement work operation, not placing cable into open trenches. Cable plow operations are not new by any means, and the same basic equipment and techniques developed for conventional cables are directly applicable to optical cables.

The placement plow shoe may be of a larger radius to assure acceptable bending of the optical cable, and some plowing equipment has been equipped with cable tension measuring devices to further assure gentle cable handling.

Several alternatives for laying cable exist. For example, sometimes an all-dielectric cable is placed inside a protective, semi-rigid plastic material conduit. One technique plows the conduit alone first, and then pulls the cable into the conduit in the conventional underground installation method. A variation of this is used in underground conduit systems, where an inner liner conduit —perhaps even several of these—is placed within the existing conduit structures, as a separate work operation. Cable is then pulled into these conduits in the same manner as it is for the original conduit structure.

A second technique draws the fiber cable through the plastic conduit before the plowing operations commence, and then plows under the composite assembly of the conduit with the cable already inside. This technique substantially limits the stress that might be transferred to the optical cable during the placement process.

Yet another technique uses a "split" conduit of plastic material that has a longitudinal opening or slit along its entire length. The conduit is plowed continuously, and the cable is inserted into the conduit through the slit as the plowing operation proceeds.

Where plowing operations cannot be continuous, such as at road or river crossings, either a splice location must be established or some arrangement for temporarily storing excess cable length, such as the figure eight arrangement discussed earlier, must be employed.

AERIAL CABLE PLACEMENT

In placing optical or conventional cable, there is almost no difference in the construction methods used. Just as in multiple-pair cable structures, optical cables can be self-supporting, providing an integral strength or supporting member in the basic cable structure. Optical cables can also be lashed to previously placed telephone cables or to new supporting messenger strand that has been placed for that express purpose.

The techniques, hardware, and placement tools, such as lashing machines, are identical to those used in conventional plant construction. Optical cables do not usually add any substantial load to existing plant, since they are both light in weight and small in diameter.

PROVIDING CABLE SPLICING ACCESS

In most installations, whatever the nature of the plant, standard splice enclosures are utilized. In plowed cable, for example, above-grade pedestals are often used which are identical to similar units employed in telephone plant. Similarly, manhole or handhole structures are also found, again, identical to existing units. In such installations, however, particularly in underground splicing, enough slack cable is stored within the manhole to permit the actual splicing to be performed above grade, often in a mobile splicing vehicle with a clean and controlled environment. This facilitates the sophisticated splicing techniques that have prevailed to date, such as fusion splicing. These measures are not essential when mechanical fiber splicing is employed.

This same rationale has been applied in some aerial cable installations where slack cable is provided and stored on the supporting strand, so that the entire splice assembly, including the splice enclosure itself, can be brought down to ground level. Splicing operations can then be done within a vehicle or enclosure. Note that although the small size and inherent flexibility of optical cables permits this type of splicing operation, there is some exposure to possible cable damage when a substantial length of cable must be handled and repositioned in this manner.

ELECTRICAL BONDING AND GROUNDING

When optical cables include a metallic element in their structure, the question of what to do with the metallic element at splice locations or repeater installations is introduced. Should the metallic element be bonded through at such locations? Should it be grounded also?

Let us consider bonding first. *Bonding* describes the electrical interconnection of different devices or sections of a system, or even two entirely different types of plant. It is primarily intended to protect personnel and plant from differences of potential that might exist between devices or systems. Thus, telephone or cable television strand is bonded to power line grounds frequently throughout the plant to protect personnel who may inadvertently come into simultaneous contact with both plants. When locally bonded together, the two are maintained at the same electrical potential.

Grounding is designed to disperse high voltages or currents from plant or devices by providing a low-resistance conducting path to ground, often by a ground rod. Bonding and grounding are usually required in joint use outside plant, as, for example, along pole lines that carry both telephone and power facilities.

When conventional multiple-pair or coaxial cables are involved, both bonding and grounding are generally required. The fact that the transmission medium in an optical fiber cable is dielectric in nature, that is, is nonconducting glass fibers, is irrelevant. The inclusion of a metallic element requires the same protective measures that any other cable plant would require. Thus, the same practices regarding grounding and bonding should be followed.

Note that if all-dielectric cable is involved, then no grounding or bonding is mandatory at all. Even in aerial installations, if self-supporting fiber cable is employed that does not include any metallic element, and such cables are commercially available, then no particular grounding or bonding practices or specifications apply.

UNIQUE APPLICATIONS FOR OPTICAL CABLES

When power companies require communication services into and out of their facilities, e.g., switch yards or substations, unusual and difficult problems may be presented. For example, consider a simple telephone or telemetry circuit that must enter a power substation or generation plant to connect the terminal equipment or telephone instrument to the public switched telephone network, which is predominately electrical in nature and which uses metallic conductors throughout. Under certain conditions of fault in the electrical power distribution grid, the power company facilities may be subjected to high ground potential conditions, or unusually high currents may be induced to flow in the servicing telephone plant conductors or cable sheaths. These disturbances can affect the telephone plant itself, or even be extended through those facilities to disrupt the telephone switching devices.

The ideal condition would be complete electrical isolation between the two plants, even though transmission continuity must be maintained. This can be done by radio or microwave transmission links, of course, and now even by cable, providing only that the cable be all-dielectric—as indeed optical fiber cables can be. This potential application of lightwave transmission was recognized early, and many local telephone facilities have been extended into potentially hostile environments in this manner, using all-dielectric entrance cables.

There are more dramatic, and possibly more extensive applications of all-dielectric cables, although they are not strictly within the domain of the telecommunications carrier company.

POWER NETWORK APPLICATIONS

Consider the power distribution network. Within this network, many fault conditions can have far-reaching effects. For instance, a lightning strike on a high-tension transmission line can operate disconnect breakers at distant terminal switch yards that unnecessarily interrupt subscriber service in areas that may not have been directly affected by the strike. To limit such chain reaction effects, power companies interconnect switching facilities with what are known as *protection circuits* or systems. Quite briefly, these systems sense fault or trouble conditions and disconnect the affected facility from the network very quickly, hopefully before the effects of the fault condition are passed through the switching facilities to other sections of the network.

For obvious reasons, the reliability of such protection services must be very high. But basically, the information is rather conventional in nature—usually telemetry tones or relatively slow-speed digital data. The protection circuit or facility is a conventional transmission requirement. Power companies have used microwave, radio, and power line carrier transmission systems, as well as using leased telephone circuits, for such services.

The power grid consists of its own transmission facilities, including transmission towers or pole lines between all switching points. It is a logical development for power companies to consider the use of their rights-of-way—even of their existing pole line structures—to provide protection services for their own use.

In all of the preceding cases, the use of lightwave transmission on all-dielectric cables can be most attractive. But how might such plant be constructed?

BURIED POWER PROTECTION FACILITIES

Since the power company owns or controls the power line right-of-way, and since that right of way extends the distances between the network switch points, it certainly is possible to bury an optical fiber cable to interconnect the terminal points. If the buried cable has no metallic elements at all, it should introduce no technical complications.

On the other hand, it may be necessary to provide mechanical protection for such a cable with conduit, perhaps, since the cable would not enjoy the protection of metallic elements. If it is not economically inhibitive, this is certainly possible. Many high tension transmission lines traverse very rough terrain, however, and under some conditions, particularly in sparsely settled wilderness areas, buried cable plant might be prohibitively expensive to construct.

AERIAL POWER PROTECTION FACILITIES

What about the aerial power transmission facilities that are already in place and operating? Could they be used in some manner to support the optical cable plant?

In many power transmission lines, one or two non-current-carrying conductors called *static wires* are provided. These are relatively high-strength conductors, usually placed above the load conductors. They are intended to attract, collect, and divert to ground any foreign disturbance, such as lightning strikes. Static wires are frequently grounded at supporting structures along the length of the system, be they poles or transmission towers.

A great deal of interest has been evidenced on the part of the power industry in incorporating lightwave transmission systems into their existing transmission plant. One approach that has been tried is to fabricate entirely new static wires which incorporate a number of optical fibers within the core of the static conductor. The resultant conductor structure is very expensive and replaces one of the more conventional static wires that was already in service along the route. Several pilot installations of this kind have been constructed, and extensive tests have been made. It should be noted that under certain system fault conditions, the static wire can be conducting extremely high current flow—on the order of hundreds of amperes or more.

One might ask, why not simply lash an optical fiber cable to the in-place, in-service static wires, which are very similar to conventional telephone messenger strand in both size and strength? There appears to be no persuasive technical reason not to do so, and the cost would be substantially lower.

The optical cable would introduce a tolerably small weight load and an acceptable wind load. It might even be a superior technique of construction, since fibers enclosed in the core of a static wire that was carrying an excessively high current would be exposed to rather high temperatures. Overlashed cables would not be so exposed—at least, not to the same extent.

It might even be practical, instead of complete replacement of one existing static wire with a very expensive composite static wire/optical fiber cable, to over-lash an individual optical cable to *both* existing static wires. This would provide a degree of optical cable redundancy to improve overall communications reliability.

INTERCONNECTING OPTICAL FIBERS

There is a practical limitation to the lengths of optical fibers that can be drawn, as well as some restriction on the lengths of cabled fibers, or cable lengths, that can be readily placed in the plant. It is also necessary to connect the fibers to terminal equipment such as transmitters or receivers. Thus, the interconnection of fibers is unavoidable.

Two clearly identifiable categories of interconnection are those where a disconnect-reconnect facility is required or desirable, and those situations that permit a more permanent or "hard" splice to be employed. In either category, the problems of efficient transfer of energy, minimized optical reflections, and mechanical integrity must be addressed.

Disconnect-reconnect capability is available in a wide variety of optical connectors. The dimensional control and stability of such units are quite critical if they are to retain satisfactory transmission characteristics after several disconnect-reconnect operations. Such connectors are sometimes employed in lightwave systems to fabricate fiber patch panels at terminals, and many are used as optical interconnections for transmitters or receivers that are configured as plug-in type cards, or modules that insert into frames or baskets. Due to the limitations of precise dimensional control and the problems of fiber core realignment, connectors usually introduce significantly more transmission loss than hard splices, and thus the use of connectors is generally limited to terminal locations.

FIBER INTERCONNECTION LOSS MECHANISMS

Sources of optical signal loss through any fiber interconnection, be it splice or connector, are either intrinsic or extrinsic to the fiber. Intrinsic losses are due to the variations and mechanical characteristics of the two fibers being joined and cannot be compensated for by connector features or splicing quality. Fiber core size discrepancies and differences in numerical aperture between the two fibers are examples of intrinsic connection losses. In a connector-type fiber interconnection, intrinsic losses may be as great as 1.0 dB when fibers from two different manufacturing lots are joined, and could be worse if the two fibers were from different manufacturers entirely.

Extrinsic losses are not due to fiber composition or characteristics and may be reduced by a superior connector design or by better splicing methods. Some extrinsic interface problems are shown in Figure 13-3.

Fresnel reflection losses may be introduced by both the glass-to-air and air-to-glass interfaces if end separation between fibers is excessive, as shown in Figure 13-3(a). Fresnel reflection loss can be minimized by the use of index-matching oil or gel at the joint.

Figure 13-3(b) shows lateral displacement between the two fibers to be joined. Obviously, the transfer of light between the two fiber cores will be less than optimal.

In part (c) of the figure, the fiber ends are properly prepared and correctly aligned, but there is an air gap between them. The loss in this case will not be Fresnel reflection loss alone, but will also be due to the conical distribution of the light exiting the one fiber end. The greater the separation, the smaller the conical area that will be intersected by the receiving fiber. Thus, some power loss will be incurred. This same condition could be present, to some extent, if there were a difference in the numerical aperture of the two fibers to be joined.

In part (d) we show properly prepared fiber ends, but an axial displacement between the fibers themselves. This obviously prevents maximum energy transfer

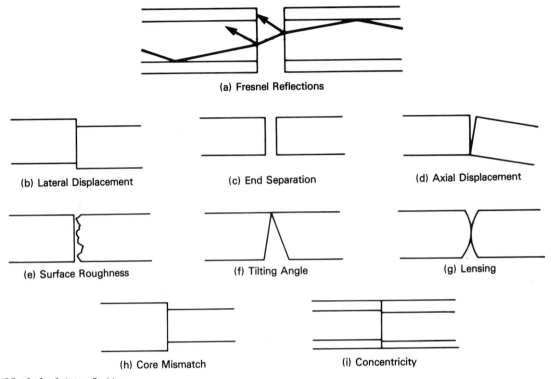

(a) Fresnel Reflections

(b) Lateral Displacement

(c) End Separation

(d) Axial Displacement

(e) Surface Roughness

(f) Tilting Angle

(g) Lensing

(h) Core Mismatch

(i) Concentricity

FIGURE 13–3 Splicing Problems

through the splice. The numerical aperture would be a factor in such a connection also, just as it was in the case of end separation.

The quality of both fiber ends has an effect on the power coupling. This is presented in diagram (e), which shows not only end separation, but also how a rough fiber end may scratch or fracture a polished fiber end.

Another loss mechanism is the lack of perpendicularity of each fiber end to each fiber axis. The amplitude of the loss is determined by the magnitude of the tilted angle, which may be different for each fiber end, as shown in sketch (f).

If fiber ends are polished improperly and produce convexities as shown in part (g) of the figure, the transfer of light energy may be less than maximal.

Diagram (h) shows core diameters of different sizes. A loss will be incurred in the transfer of energy from the larger to the smaller core, although no loss might result in the reverse direction of transmission.

Finally, in drawing (i) we show two fibers, one of whose cores is not concentric within its cladding. The result is similar to the lateral displacement shown in (b).

The complexities of interconnecting fibers demands careful attention in connector design and a high level of precision in connector fabrication. Different fiber connectors, as well as mechanical fiber splices, may require different fiber end preparation methods. Sometimes special jigs or fixtures must be used, and many such devices require that the fiber end be polished, sometimes in conjunction with the connector itself.

Cutting optical fibers requires care. A method of scratching a fiber with a diamond scribe can induce a fiber break called *cleaving,* but in some splice or connector practices polishing is still required as a subsequent work operation.

Although any of the mechanisms illustrated may be present and contribute to interconnection loss in either cleaved or polished fiber ends, one factor will usually dominate a particular connection technique. Some mitigation can be provided by the use of index-matching fluids or materials within the connector or splice itself.

Surface debris and foreign matter, such as dust or dirt on fiber end faces, can result in substantial connection power losses.

FIBER SPLICING

Direct fiber-to-fiber splicing can be accomplished using mechanical manufactured splicing devices or by fusing the glass fiber ends together by means of a flame or electric arc. In either technique, the fiber must be cleaned and all coatings (not cladding) completely removed. This may be done mechanically or chemically.

Usually a cleaving tool is used to fracture the fiber, and with proper technique, a clean and perpendicular end surface can be produced. The mechanical splices employ some type of ferrule or tube into which the fiber ends are inserted. Mechanical splicing is a relatively easy hand operation and requires a minimum of special tools or jigs. The operation can be performed in aerial or underground environments, and some types even permit reuse of the splice unit itself.

Fusion splicing requires both a fairly complicated machine and some precision in positioning the two fiber ends. Some fusion splicers include high-power microscopes, which makes them difficult to operate in some work locations. For example, with such units, fusion splicing operations in an aerial bucket are not very satisfactory, due to movement of the work platform. Also, for obvious reasons, the open flame or arc required to fuse the fibers cannot be employed in potentially explosive atmospheres such as manholes.

Both splicing techniques are used, but the fusion method predominates at this time. As mechanical splicing techniques are improved, and higher levels of confidence develop, this may change.

It is generally accepted that fusion splicing can produce individual splice losses on the order of hundredths of a dB, while mechanical splices are more conservatively rated in tenths of a dB. For all practical purposes, either method is quite adequate, and the selection of one or the other ought to reflect considerations other than technical performance alone.

Although many of the installations made to date have employed fusion splicing exclusively during the plant construction phase, presumably on the basis of technical superiority, many of these operations rely on mechanical splices entirely for restoration of service and cable repairs. This would appear to be incongruous if there were any real quality differential between the methods.

Sometimes, reviewing the evolution of earlier technical developments can be instructional, and the following may be a case in point.

Not too many years back, in toll cables that employed copper pairs, it was common practice to solder all splicing connections in pairs intended for carrier use throughout the cable plant. This was awkward to do and involved additional costs, but nevertheless, the practice was widely adopted. It was later abandoned, however, not simply to reduce costs, but because over time little evidence could be found to support any technical necessity for it.

A similar development can be found in microwave technology. In the earlier systems, extraordinary effort was taken in assembling antenna waveguide runs in the field. Waveguide connectors were used that had tunable slugs that allowed the waveguide installations to be fine tuned for minimum reflections. This required a very sophisticated test setup to maximize waveguide performance, of course.

Today these techniques are used only in the most sophisticated networks. The reason is that although the installation was theoretically superior with the level of care and attention given, it had little or no real beneficial impact on the system's performance.

When technologies are embryonic—and lightwave systems might reasonably be considered to be so at this time—there is some uncertainty as to how much sophistication is actually necessary. Systems tend to be very conservatively designed. The author has seen several installations that were so conservative that the receivers were overdriven with high-level optical input, and optical attenuation had to be subsequently inserted into the link.

Prudent engineering practice does, and should, provide margins for systems with which there has been little previous experience. However, although this is not only understandable but even commendable in early efforts, we are obliged to adjust this philosophy after experience has been gained.

In very long systems where even minor reductions in individual splice losses might eliminate an active repeater installation, extraordinary measures or exotic splicing techniques are obviously appropriate. Of course, such cases are most generally on high-fiber-count routes, where traffic density and revenues will be substantial and can easily support higher levels of sophistication.

In lower levels of plant, such as in exchange or distribution applications, the cable route lengths will be nominal, as will the supporting revenues. Thus, given a route length of, say, 20 km, and optical cable lengths of, say, 2 km, only ten splice locations are involved. It may then be rather academic if an individual splice loss is 0.05 dB or 0.08 dB in such cases.

At the present time, fusion-splicing machines cost on the order of several thousands of dollars, and the training involved to achieve very low-loss splices with fusion methods is rather extensive, and hence expensive also. Reusable mechanical splices, on the other hand, are quite inexpensive.

If we expect lightwave technology to become widely applied, then the bulk of fiber splices will eventually be in the shorter, less restrictive exchange and distribution plant applications. In such cases, not only may mechanical splicing be technically adequate, it may be economically mandatory.

Fiber Splice Testing

The emphasis on reducing splicing loss is accompanied by what might be considered excessive sophistication in testing splices. If splicing techniques can truly provide losses on the order of hundredths of a dB, then tests to that level of precision become a necessary part of the splicing process.

The practice of measuring the transmission loss of each individual fiber splice during splicing operations is widely accepted and generally applied today. This may be done by coupling lightwave energy, at a discrete, known level, into the fiber which is about to be spliced to another fiber. An optical detector coupled to a (power) meter is connected to the far end of the fiber spliced into, and the optical signal level is measured at that point.

If the attenuation of each individual fiber in each individual length of cable is known, simple addition establishes the attenuation to be expected when two individual cables are spliced together. Any variation between this predicted total attenuation and the actual measured attenuation through both cables spliced together can be attributed to the transmission loss introduced by the connecting splice itself. If this splice loss is judged to be excessive, the splice may be immediately broken and a new one made. Since no disruption is made in the test equipment connections, no new variables are introduced, and the comparative quality (as regards transmission loss) of several subsequent splice operations is plain.

Figure 13–4 shows the splicing location and test equipment setup for connecting two lengths of optical cable together. In actual installations, several interconnecting lengths of cable may be required, and there may be many splice points involved.

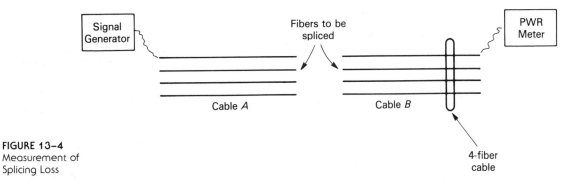

FIGURE 13–4
Measurement of
Splicing Loss

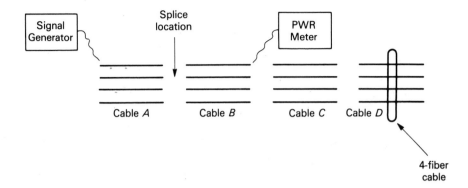

FIGURE 13–5
Multiple-Splice
Testing

Figure 13–5 shows four cable sections identified as *A, B, C,* and *D.* The hypothetical project involves fiber splicing at three intermediate locations. One approach is to connect the test signal source to the fiber under test at one end of the system. The optical signal level meter is taken out to the splice location between cables *B* and *C* and connected to the fiber under test on Cable *B,* as shown in the figure.

The splice is then made between cables *A* and *B,* and the signal level measured on the power meter. If the signal level correlates well with the predicted signal level at the end of cable *B,* the splice is considered acceptable, and both the signal source at the terminal and the meter at the end of cable *B* are disconnected and reconnected to the next fiber to be spliced through.

If the signal level does not correlate well with the predicted level at the end of cable *B,* the quality of the splice is suspect, and the splice is broken and remade. If individual cable section measurements were taken after the cable was physically placed initally, it is quite easy to identify splices with excessive transmission loss, and any required number of subsequent splices and measurements can be made to minimize actual end-resultant loss.

If individual cable section measurements were not taken after the cable was initially placed, for reasons of economy or whatever other reasons obtained, then a discrepancy between the predicted signal level and the level actually measured cannot be clearly attributed to the fiber splice itself, since it might be the result of mechanical cable damage incurred during placement.

If, after remaking the same fiber splice several times, we cannot correlate the measured signal level reasonably well with the predicted level, it could be indicative of cable damage. A logical procedure would be to abandon that particular fiber splice for the moment and attempt the entire process on a different pair of fibers. This entails connecting the test equipment at both locations to a different fiber.

If the experience with two new fibers is similar, that is, if the measured signal level does not correlate well with the predicted level, suspicion of cable damage would be reinforced. To verify or isolate the condition, we could relocate the power meter at the splicing location and measure the signal level through the first cable section alone. If it were normal, we would then make attenuation measurements through the second cable section alone.

Alternative Fiber Splice Testing

It might appear that individual cable section attenuation measurements are well worth making in every instance. This is perhaps true from a purely academic point of view, but for an installation with 10 or 12 individual cable sections the testing procedure will be both extensive and expensive. Moreover, experience with fiber installations to date does not indicate that a high probability of cable damage will be incurred, and there is a persuasive economic argument not to make such tests unnecessarily. The cost savings might be quite substantial in a large-scale project.

An alternative arrangement would be to equip the splicing location itself with a second optical signal level meter. Now the attenuation of each individual cable section can be determined as part of the splicing operation itself without a separate visit to each working location. All that is necessary is for the splicer to measure and record the optical signal level into the splice location before actually splicing the fiber. This is easily and quickly accomplished, and the technique might even justify borrowing or leasing a second meter for the relatively short period that splicing operations are being conducted.

As splicing techniques and methods improve, the repeatability of fiber splicing will also. We might seriously consider not measuring each individual fiber splice, particularly if we had extensive previous splicing experience and had found only a very small percentage of splice rejections.

Consider, in Figure 13-5, if we had simply spliced through cables *A* and *B,* and *C* and *D,* without any measurement at all. The final splicing operation would be at the junction of cables *B* and *C.* We could measure the signal level input to this location on every fiber, compare these measurements with the predicted signal levels at this point, and they might even correlate reasonably well.

The economy might not be significant in the case in Figure 13-5, with only three splice locations, but what if we had a project with 23 splice locations? Of course, we might have to troubleshoot a splice or two using this method, but the time and cost involved in this supplemental activity might be more tolerable than the time and cost of extensive testing to avoid it entirely.

Perhaps splicing methods are not yet sufficiently perfected to make this method entirely practical, but its advantages are very real, and in lower budget projects could be quite attractive.

It should be noted that it is not mandatory that splicing be performed sequentially along the cable route. The objective is to verify that acceptable transmission loss has been introduced by each fiber splice made, and that each individual optical cable section presents the attenuation it was specified to introduce. It is entirely practical to conduct splicing operations in different sections of a system simultaneously, with the final splicing being the connection of two or more sections together. This is shown in Figure 13-6.

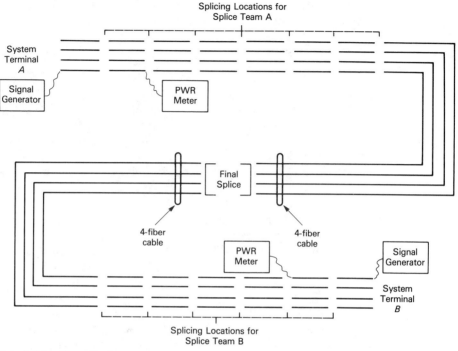

FIGURE 13-6 Multiple-Splicing Operations

Splice Testing Equipment

It is important that test equipment used in splicing operations be calibrated frequently. Since we are really only interested in verifying attenuation rather than in a precise measurement of lightwave energy levels, we might consider calibration adequate if the same meter indicates the same level from the same signal source all or nearly all of the time. We can satisfy this requirement by simply checking the meter indication against the same signal source periodically.

In an effort to reduce the manpower requirements for testing splices, a relatively new technique has been developed called *local signal injection and detection*. In this technique, by bending the fiber on each side of the splice across a mandrel, light can be injected into, and recovered out of, the fiber itself. The efficiency with which this injection and detection can be performed is not critical. All that is needed is a stable level of light energy being passed through the splice and being measured after passing through; we are not concerned with the precise level of this energy. Now the transfer of light energy between the two fiber ends can be measured, even as we reposition fiber ends or after we fuse the ends together.

The relative transmission loss of a single splice operation can be easily established, as can the relative loss of several or many subsequent splice operations. This testing technique is equally applicable to mechanical or fusion splicing operations, although most devices offered to date have been presented as integral parts of a fusion-splicing machine.

There is a tendency toward high sophistication in optical fiber-splicing operations that resembles the focus on sophistication in microwave waveguide installations discussed earlier. In the thin-traffic-route systems, which we might expect to encounter more frequently as lightwave technology intrudes into lower levels of the national telecommunications network, many links will be characterized by nominal cable route lengths. The experience with lightwave installations to date seems to indicate a higher probability of overdriving receivers in such systems rather than in presenting too low an optical signal input. Indeed, in many of these systems, optical signal levels actually had to be reduced by optical attenuation or other means. At the same time, fiber performance has been steadily improving, and long-wave, single mode installations compound the problems of possibly excessive receiver input levels.

Some manufacturers are developing lower output laser light sources. This might improve the yield from laser manufacturing processes and thus reduce transmitter costs. The use of lower laser transmitter output levels could be entirely acceptable in many shorter length rural applications. In addition, longer cable lengths are being placed, drastically reducing the number of splices required.

In view of these facts, insistence on highly sophisticated splicing techniques, which require costly apparatus and complicated testing simply to ensure minimum splice losses, might be unwarranted. Accepting testing cost penalties in the interconnecting fiber plant, and then padding down the receiver input to make the system work well, is a bit incongruous. Splice losses on the order of a tenth of a dB or so might be acceptable.

For high-traffic-density routes between major cities, minimum fiber and splice losses are very important, since they may eliminate intermediate repeaters. In typical distribution installations, these benefits are somewhat academic and may be more than offset by the practical or economic considerations we have presented.

SUMMARY

Almost all conventional plant construction techniques are suitable for use in constructing lightwave transmission systems.

The inherent small size, light weight, and flexibility of fiber optic cables makes it possible to place longer continuous lengths of cable than in conventional plant construction. Some new construction techniques have been developed to handle these longer cable lengths.

In fiber cable placement, provision for slack cable is often provided at splice locations, to allow fiber splicing operations to be conducted inside vehicles or splicing enclosures. Of course, fewer fiber splices are required when longer continuous cable lengths are utilized.

There are no transmission considerations that prohibit the use of completely dielectric optical cables. All-dielectric cables are immune to lightning, electromagnetic interference and noise, and electromagnetic pulse (EMP).

Optical fiber cables are particularly suitable for use in power grid protection systems, because they allow complete electrical isolation between sections of a system. It is possible that more power companies in the future will be placing their own optical cables, along their own rights-of-way, for their own use.

Fiber ends can be joined together by fusing the glass itself or by using mechanical splicing devices. Fiber ends may also be joined together using connectors that permit multiple disconnect and reconnect operations.

A fiber splicing technique frequently encountered today includes testing the transmission loss through each fiber splice as it is completed. As splicing operations are simplified and made more repeatable, it is likely that a much lower level of splice testing will prevail.

There are no insurmountable problems in constructing lightwave plant; indeed there are a number of new techniques possible that can significantly reduce plant construction costs. It is difficult to conceive of an environment where practical lightwave transmission facilities could not be installed.

If a certain timidity or reservation has been evident in the earliest installations, they have been monitored sufficiently now for methods and product improvements to evolve and to be adequately demonstrated.

REVIEW QUESTIONS _____

True or False?

T   F   1. Splicing optical fibers is more complex and difficult than splicing metallic conductors.

T   F   2. Most of the techniques and tools used in constructing conventional multiple-pair cable plant cannot be used in constructing lightwave plant without major modification.

T   F   3. Longer cable lengths are usually available and employed in metallic-conductor plant construction than in lightwave plant construction.

T   F   4. Powered capstans have been used extensively for years in constructing conventional multiple-pair cable plant.

T   F   5. Protective conduit can be plowed with optical cables as a single work operation.

T   F   6. Optical cable splices usually employ a different construction technique than multiple-pair cable splices.

T   F   7. Electrical bonding and grounding techniques for metallic-element optical cables are dramatically different than those used for multiple-pair metallic conductor cables.

T   F   8. All-dielectric optical cables experience transmission interference and noise introduced by electrical induction.

T   F   9. Optical cable is uniquely adapted for applications that require, or could benefit from, complete electrical isolation between sections of a system.

T   F   10. Self-supporting, all-dielectric optical cable can be placed parallel to power conductors, using power poles for support, without any special consideration.

T   F   11. Optical connectors for joining two fiber ends together can be taken apart and reconnected many times without degrading system performance.

T   F   12. Fusion splicing of optical fibers can employ an electrical arc or an open flame.

T   F   13. If fiber ends are not properly prepared before splicing, excess splice loss may be introduced.

T   F   14. One technique commonly used for cutting an optical fiber is called chopping.

T   F   15. Some optical fiber connectors require polishing the fiber ends as a separate work operation.

T   F   16. Standard fusion-splicing techniques are suitable for use in manholes without special arrangements.

T   F   17. When measuring splice losses, the precision with which the optical signal level is measured is of primary importance.

T   F   18. Transmission splice losses through fusion splices are typically on the order of 1 dB or more.

T   F   19. The only way to measure transmission loss through a fiber splice is to introduce light energy into the distant end of the fiber.

T   F   20. Transmission losses through mechanical splices are too high to permit their use in systems that are longer than 3 km.

CHAPTER 14

Cable Television Applications

Although most of this text has been directed at digital transmission applications, lightwave systems can be operated as analog transmission facilities as well. The transmission of video information is the most commonly encountered, and most compelling reason for, analog operation.

The use of lightwave links in simple, short-length, single-channel applications connecting one camera to one video monitor, as in a closed circuit surveillance system, is too fundamental to merit further discussion. Cable television systems, however, wherein a multiplicity of television signals must be transmitted, present a large potential for the construction of analog lightwave systems.

Whether we consider television signals as base-band video (unmodulated) information or as RF carriers modulated with video information, television signals are significantly more complex than most conventional digital signals. If fidelity is to be preserved through the transmission process, a substantial transmission bandwidth is required.

There is a subtle and sometimes overlooked difference in the basic characteristics of analog and digital transmission systems. When an analog signal is converted to digital information, the process is rather straightforward, and is primarily a function of the circuit or equipment design. Hence, conventional considerations such as modulation depth or index adjustment need not be addressed by operating personnel, since no adjustment is required or even possible. In this sense, a digital system is truly "user friendly."

In an analog television system, assuming modulated carriers are involved, it is possible to have all carrier levels correctly adjusted throughout the entire system, but if the modulation of each individual carrier is not properly set, the transmitted pictures can be completely unusable. This is equally true of other analog transmission systems. For example, in a high-density analog telephone carrier system, say, 300 voice channels or so, the same condition can be experienced. Even if all transmission levels for the carrier signals are exactly right, if modulation for each channel is not adjusted correctly, it is possible that no operational circuits at all would be available.

Digital transmission systems, on the other hand, are extremely easy to use, largely because of this lack of critical operating adjustment. This has contributed to the wide use of digital transmission throughout the national telecommunications network, but digital modulation techniques are available for the transmission of video signals also, and many of the inherent advantages of digital transmission would be applicable.

To understand the application of lightwave transmission in cable television systems, it is helpful to briefly review the basic principles involved in multiple-channel RF transmission of video signals over coaxial cables. If we understand the noise and interference mechanisms in such systems, then we can compare them directly to optical transmission technology.

CONVENTIONAL CATV SYSTEMS

The introduction of noise into transported television signals which pass through analog transmission facilities is well understood, and the mechanisms that are involved are the same as those involved in transporting any other analog information, such as voice channels.

Any electrical device, even a simple resistor, will generate some electrical noise due to random electron movements within the conducting material itself. This may conveniently be called the *thermal noise threshold* of the device, and if signals are introduced at the device input at the same amplitude as that of the thermal noise, then those signals would be indistinguishable from the noise. In such a case, the signal-to-noise ratio would be zero, since there would be no difference in amplitude between the noise and the desired information.

An electrical device such as a receiver or an amplifier will be specified by the manufacturer to have a *noise figure* (NF), usually stated in dB. This identifies how much noise over and above the irreducible thermal threshold noise will be generated by that particular unit. The system designer or operator must maintain desired signal levels at some higher amplitude than the thermal noise threshold of the electrical devices that are inserted into the network or system.

The difference in amplitude between a signal before modulation or after detection of a modulated carrier and the noise present in the spectrum occupied by the signal is called the *signal-to-noise (S/N) ratio,* when both measurements are made at the same point in the network or system; the difference in amplitude between a signal after modulation or before detection and the noise present in the spectrum occupied by the signal is called the *carrier-to-noise (C/N) ratio.*

After the selection of the electrical devices has been made, the designer can improve the S/N or C/N ratio of transported information only by raising the transmission levels of the carriers at the input to all devices that are inserted into the system.

The introduction of noise into transported information in analog lightwave systems follows this traditional principle, and the quality of the transmission is stated in terms of the S/N ratio, in dB. Lightwave receivers will be specified to produce a stated S/N ratio at a given optical input signal level. The system designer must calculate the receiver optical input signal that a particular network or system will experience, establish the total contribution of noise that the network will introduce, and evaluate the acceptability of that level of performance.

Other forms of distortion are introduced in conventional cable television systems, and these may be generally referred to as *intermodulation distortion products*. All manufactured electrical devices exhibit some degree of nonlinearity between their input and output signals. The more composite signal power that is passed through a broadband device, the less linear the device will tend to be.

Intermodulation distortion has many forms, including cross-modulation, second order beats, third order beats, and composite triple beats. In television signals, the unit of measure is generally given in terms of a reference to 100 percent modulation by the desired information, or a reference to the degree of visibility of the interference generated.

Analog devices used in cable television systems are usually specified in terms of the amount of intermodulation distortion introduced at some specific operating output level. Of course, the amount of distortion experienced is also a function of the number of channels passed through the device, since the more carriers present, the higher is the level of composite RF energy.

Since cable television systems using RF carriers require frequent reamplification of the signals, because of cable and other transmission losses, such systems often become distortion limited in practical applications, since each amplifier inserted will introduce some distortion. After the selection of electrical devices has been made, the designer can improve the intermodulation distortion performance of the system only by reducing either the number of carriers passed through the system or the operating output level of all the devices that are inserted into the system.

The introduction of distortion into transported information in analog lightwave transmission systems does not follow quite the same pattern as in coaxial cable RF distribution systems. The nature of cable television transmission (multiple RF carriers through coaxial cables) generally requires that a relatively large number of amplifiers be provided, all cascaded or connected serially. Since each such unit introduces some distortion, the cumulative distortion at the end of a cascade becomes a significant factor in system designs. Because this multiplicity of contributing units is usually unavoidable in coaxial systems, the system designer must manipulate amplifier operating input and output levels by limiting amplifier spacing in such a manner that the quality of the transmission is acceptable. It is the number of carriers (channels) involved plus the number of sequential analog devices required that imposes these restrictions.

Note that in such a network, passive units such as subscriber service taps or line splitters do not introduce any significant noise or intermodulation distortion products into the transported information.

In a lightwave system, generally only two devices are involved: a transmitter at one end and a receiver at the other. Thus, the number of potential contributing sources of distortion is dramatically reduced from the number present in a conventional coaxial system using many amplifiers.

CATV SYSTEM CONFIGURATIONS

In urban cable television systems, the economics can be improved by designing two subsystems which operate in tandem. A "trunk" design philosophy is employed, so that signals are arterially distributed through the service area and this level of plant is not placed along every single street where subscribers are located. By operating this subsystem very conservatively, that is, at relatively high input levels and relatively low output levels, "trunk" plant noise and distortion can be limited to better than acceptable levels.

A second level of plant, called feeder plant, is then constructed to actually provide the service connections for all subscriber locations. By intentionally limiting the physical length of such plant, we can then restrict the number of cascaded feeder amplifiers required to two or three units. Since only a few such amplifiers will be required, we can operate them at higher input and output levels. Operating the feeder plant at higher transmission levels improves the efficiency of tapping the plant for subscriber service.

At higher output levels the intermodulation distortion from each individual feeder amplifier will be higher, of course. But this can be tolerated if we intentionally limit the number of such units that are cascaded in the feeder system, and if we design the trunk for minimal noise and distortion, as mentioned earlier.

The worst-case end-of-system performance will be either at the most distant subscriber's premises or at the premises which are fed by signals that pass through the greatest number of amplifiers. The system designer calculates the noise and distortion contributions from both the trunk and the feeder subsystems individually under the proposed operating signal levels (both input and output) and then combines these figures. The final selection of equipment and operating transmission levels is made to assure that the end-of-system (worst case) transmission quality will be acceptable.

When the composite service area is large, the number of trunk amplifiers that must be connected serially may become prohibitive. The designer could then be forced to reduce operating output levels, necessitating closer amplifier spacings and increasing system construction costs. If this precaution is not taken, the noise and distortion contributed by the system may result in end-of-system performance that fails to meet the system specifications.

There are also considerations of system reliability when a great many amplifiers must be cascaded. The problems involved are compounded by the fact that most modern cable systems are required to carry more and more television signals, thus increasing the intermodulation distortion contribution from each and every system amplifier.

CABLE TELEVISION SUPER-TRUNKING

A refinement on basic system design philosophy can significantly alleviate these problems. By introducing a new level of signal transmission plant, a large service area can be subdivided into smaller areas, each requiring fewer amplifiers in cascade to reach all of its subscriber service locations. These multiple distribution centers are called *service hubs,* and a new level of high-quality (low noise and distortion) transmission plant is provided to connect each such hub with the main signal origination point, which is generally called the *system head end.*

Within each hub service area, the system design is quite conventional, but the entire approach is contingent upon the interconnecting transmission links (between the head end and each hub) introducing a low level of noise and a minimum of intermodulation distortion. The transmission requirements for all these links are quite severe in that a relatively large number of television signals must be transported over each link.

The earliest approach to this type of design was called *super-trunking,* and an example is shown in Figure 14–1. This technique applied the same CATV technology using wideband analog amplifiers, each passing many television signals, and all being interconnected with coaxial cable. To reduce the number of amplifiers required in such links, and consequently to limit the distortion contribution of each link, larger size, lower loss coaxial cable was usually employed, and often FM modulation was also used, for the noise reduction it can produce.

The transmission design of such links was extremely conservative to limit both the noise and distortion that such links introduced. The technique worked, but the reliability of the service provided within each established hub service area now depended upon the reliability of the interconnecting super-trunk plant itself and the several or many amplifiers that it required. Also, the additional amplifiers employed added to system powering and maintenance costs.

CATV MICROWAVE DISTRIBUTION

An alternative to coaxial cable super-trunking was the use of microwave transmission links between the system head end and the subsystem distibution hub locations. The earlier microwave links used FM equipment that could accommodate a single

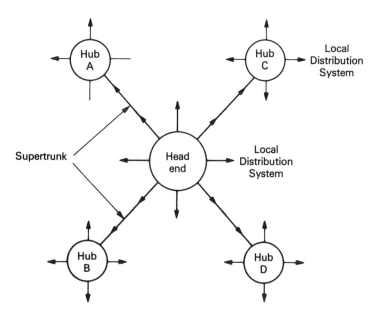

FIGURE 14-1
Coaxial Cable
Trunking

television channel for each microwave transmitter–receiver combination provided. A simple system of this type is shown in Figure 14-2.

This scheme worked reasonably well, but was expensive for transporting a large number of television signals. Of course, such systems required transmission line-of-sight, which limited site selection or imposed the penalties of towers or other antenna support structures. And microwave links, however well designed, are inherently vulnerable to atmospheric disruption.

A later development used AM techniques. This did not actually require individual transmitters in the conventional sense of the word. Instead of each channel transmitter including a modulator, the equipment took the television signals as already modulated RF carriers and simply converted each carrier to a higher, microwave frequency. Several such converted carriers were amplified in a broadband microwave frequency amplifier, and the relationship between the carriers, as well as the depth of modulation of each carrier, was not disturbed at all. Figure 14-3 shows an AM microwave system.

A significant advantage of this technique was the simplicity of the equipment at receiving locations. Transported television signals required no individualized treatment at all. The entire band of frequencies, including all the television carriers, was passed through a relatively unsophisticated converter, and all the signals were converted back down to their original RF frequencies. Since there were usually several receiving sites, this simplicity, together with the lower costs it presented, was quite attractive.

In general, the technique worked reasonably well, but passing many carriers through a single analog device such as the broadband RF amplifier and the con-

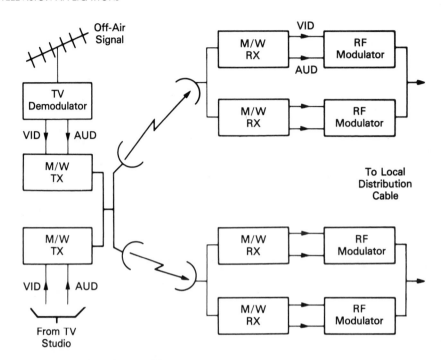

FIGURE 14–2
Frequency-
Modulated
Microwave

verter introduced a significant amount of intermodulation distortion, just as an individual coaxial cable RF amplifier did. A refinement of the technique used individual higher power RF amplifiers for each individual channel carried, but this increased the costs, of course.

Many large urban CATV systems today operate with this type of microwave equipment. Of course, these systems require FCC licensing and are somewhat restrictive as to where transmission terminals can be located.

CATV LIGHTWAVE DISTRIBUTION

As lightwave transmission technology evolved, it was a natural candidate for replacing microwave or super-trunking approaches. It offered a substantial amount of transmission bandwidth for use, which television signals require, and it was impervious to electromagnetic interference along the cable route itself. System lengths on the order of many kilometers were possible, and could be constructed without any intermediate optical repeaters at all in most cases.

This eliminated both the maintenance costs and the inherent unreliability of a multiplicity of line amplifiers, along with their noise contributions and distortion

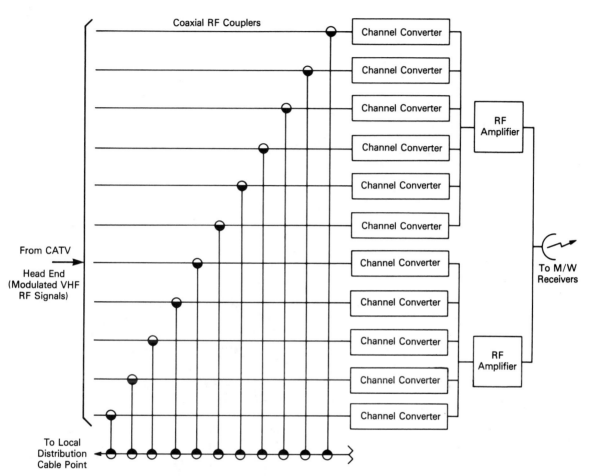

FIGURE 14-3 Amplitude-Modulated Microwave

accumulation. It also avoided the potentially disruptive atmospheric exposure that a microwave system would naturally have been exposed to.

Not all noise contribution can be avoided in lightwave systems, but existing photodetectors introduce a tolerable amount of noise power. The problems of inter-modulation distortion are, at least theoretically, still a factor, since the devices in-volved are broadband analog devices and the intention is to pass a multiplicity of carriers through them. The quality of the available devices, however, and the rela-tively low power levels at which they operate, reduce the intermodulation products to manageable levels. Careful design is necessary, but lightwave transmission has proven to be a practical and relatively noise-free alternative to the other television distribution techniques.

MULTIPLEXING AND MODULATION

If a lightwave link is to transport several television signals and still produce the high carrier-to-noise ratios that are required at the distant end, the use of vestigial sideband modulated video signals, such as are directly employed in television broadcasting and conventional CATV distribution systems, is not practical in most cases. Alternative schemes that have been demonstrated to be practical are frequency modulation of the video signals or conversion of the analog video information into digital video signals.

We can improve performance by using FM (see Figure 14–4) with a very wide deviation in the modulation of each carrier. The inherently wide transmission bandwidth of the lightwave link itself makes it possible to accept the wider bandwidth per channel required by this higher deviation.

In an FM system of this type, separation of individual television signals is achieved by frequency division multiplexing, just as individual channels are separated by frequency in commercial television broadcasting. If we use wideband FM modulation and limit the number of signals transmitted by each laser or lightwave transmitter, the system is relatively insensitive to laser nonlinearity.

Systems of this type have been constructed to carry as many as 12 television signals per laser, and many quite satisfactory systems are in day-to-day service carrying eight video signals per lightwave transmitter. The basic limitations of this technique are system bandwidth, of course, and the material dispersion experienced in the optical fiber itself. Single-frequency laser improvements should reduce the dispersion problems somewhat.

It is possible to digitalize a video signal. One line of equipment is presently available that produces a 65 Mb/s signal for each individual video channel. Such a system is shown in Figure 14–5. Using the available equipment, two such signals can be digitally multiplexed into a pulse stream at a 140 Mb/s data rate, and the resulting signal can be used to modulate the lightwave transmitter laser. Using wavelength division multiplexing, systems have been constructed that transmit two video channels per lightwave carrier, and two carriers per optical fiber, as shown in the figure. Digital processing of a video signal is particularly advantageous in longer lightwave systems because regenerative repeaters can be used that contribute minimal distortion to the transported signals.

Digital terminal devices of this type are still quite expensive, and to date, the bulk of television transporting lightwave systems have utilized FM modulation and frequency division multiplexing. Using either technique, system capacity can be further expanded by applying more than one lightwave carrier to each fiber using wavelength division multiplexing devices. The FM technique is less expensive and, at the present time, permits more television signals to be carried per optical fiber provided.

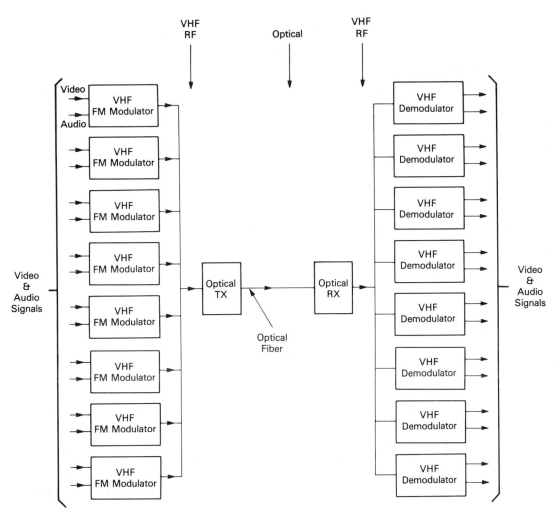

FIGURE 14-4 Analog Video Transmission

However, the digital technique might have a distinct advantage in a very long system that required multiple optical repeaters.

When a large number of television signals are present, as in a modern CATV system, distributing the signals across several optical fibers may provide an attractive redundancy of service. In the event of a failure of a single fiber, or of individual units of terminal equipment, at least some subscriber television service would persist, and this may be advantageous.

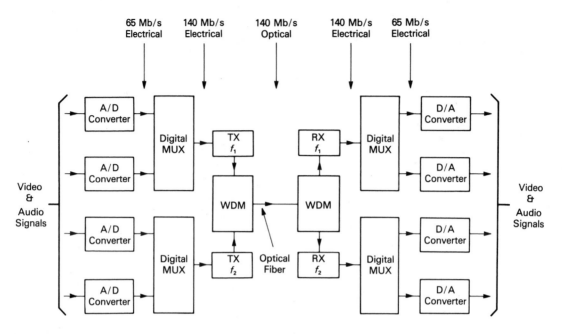

FIGURE 14–5 Digital Video Transmission

LIGHTWAVE HUB DISTRIBUTION

The advantages of lightwave transmission links for distribution of multiple television signals to CATV hubs are very persuasive. The fact that hub locations are no longer dictated by transmission considerations alone is important. In many urban situations, CATV operators have branched out into offering closed circuit television and private data link services to commercial subscribers. Interconnecting such subscribers within an urban area, and even providing interconnection to the national telecommunications network itself, may be greatly facilitated when hub locations can be strategically positioned within the area.

It is possible that a local telephone company will offer to provide such optical fiber trunking facilities for lease by a local CATV operator. Telephone companies have also shown interest in providing complete CATV facilities, including hub trunking and hub locations that coincide with existing telephone switching centers. In some instances, such companies already have optical fibers in place between these locations. The availability of lightwave transmission links as an alternative for hub trunking will have a substantial impact not only on the economics of conventional CATV services, but also on the introduction of new services. It is quite possible that a basic change in roles of both CATV operators and telephone companies may result.

It should be noted that any lightwave transmission facility like those discussed is quite capable of providing other services also, such as high-speed data links or a multiplicity of voice channels. Such services could be carried on a fiber along with one or more television signals.

LOCAL CATV DISTRIBUTION

Although the use of lightwave links in hub trunking applications has been amply demonstrated, and it seems highly probable that more of these links will be constructed, it is much less clear whether or how optical fibers may be widely employed in local CATV distribution networks. The primary question is whether or not optical fibers are economically practical as individual transmission facilities dedicated for use by a single subscriber at a single point of service. Perhaps they would be more efficiently used in a wide-band mode serving a number of subscribers, with the demarcation of individual loops being effected in the distribution plant itself.

Although CATV and telephone networks may appear to be similar in some respects, they are not identical, and the problems presented are not the same. A telephone network may be thought of as a multipoint-to-multipoint system, with interconnection between any two points being randomly provided on demand. A CATV system is more a single-point-to-multipoint network: the signals carried are all made available to all subscribers, and any discrimination or exclusion is usually effected at the subscriber's terminal.

From a transmission and design point of view, the CATV network is more analogous to the commercial distribution of electrical power. In both cases all service points require equal availability to what has been generated (TV signals or power, respectively), which originates at a single point. A power distribution grid might be thought of as a big party line, even as a wide-band type of transmission system, at least wide band in that all service points require the same bandwidth for service.

In this respect, conventional CATV distribution systems are very similar, except that they are more complex in the bandwidth that they must supply and the sophistication of the signals they carry. We cannot ignore the technical drawbacks of the current technology (multiple RF carriers on coaxial cables), which are the accumulation of transmission noise and distortion, but the technology does exist, is economically viable, and is reasonably mature and acceptably reliable.

If all service points require access to the same information, then there is little reason for all information to be transmitted over parallel individual pairs or facilities. The mere fact that a new technology is available which could be substituted (optical fibers for coaxial cables) does not, in itself, carry any conviction. There have to be compelling technical and/or economic reasons to change. At the present time at least, there do not appear to be any convincing reasons of either kind for directly replacing coaxial technology with the lightwave alternative for cable television services alone.

HYBRID LIGHTWAVE SYSTEMS _____

The arguments become much more convincing if we assume that some other services could or would be involved. Certainly, there is little question that an optical fiber facility extended from a telephone switching center to each and every point of service could accommodate both television and telecommunications services. The extended bandwidth that fibers present could even expand greatly on the variety of services that are currently available to subscribers. Data, telephone, utility metering, electronic mail, and alarm systems of various types are only some of the services that could be delivered, along with a wide selection of television signals.

Under this scenario, the provision of a dedicated fiber (or fiber pair) to every point of service may be justified. We cannot lose sight of the fact, however, that such a dedicated facility would probably underutilize the inherent transmission capacity of the fibers placed. Also, a network of this nature might require increased sophistication of the equipment or devices located at each subscriber's premises, and if so, the economic impact would be quite severe. The lowest common denominator in distribution plant cost is the subscriber's premises. If the initial and/or maintenance costs of such units increase dramatically, the sheer number of such installations may very well be prohibitive.

Notwithstanding the preceding considerations, there is every reason to anticipate some intrusion of optical fibers into the distribution plant. What remains unclear is just what form this intrusion may take. It could be a wideband facility extending out into the general distribution area, with a number of intermediate terminals of some complexity where individual subscriber loops will break out of the wideband mode and be extended on individual fibers, on twisted pairs, or even, perhaps, on coaxial cable. Or it might be that an individual optical facility will be extended from a central-switch/head-end installation to each individual point of subscriber service. There have been a number of pilot programs of both types, and many articles have been written on the issue, but without resolution as yet. Also, it would be naive to ignore those factors other than economics or technology that may influence the resolution of the issue.

For example, our society has evolved separate operating utility entities for all subscriber services. These corporate entities have been polarized and held apart by both law and custom. Thus, we have, in many communities, an economically sound electric utility owning and operating its own plant facilities. Along the same pole line routes, and serving the identical subscriber locations, we have a parallel telephone transmission system, also self sustaining and profitable as is. Yet a third layer of transmission plant, the CATV system, has been added, and it too is independently viable.

There is a great deal of duplication in these three operations. Pole line plant is common, plant and transmission engineering functions are similar, and even service staffs and billing are comparable. But regardless of how practical it might be, or seem to be, to consolidate these individual operations to some degree, it is highly improbable that the political and regulatory conditions will permit this in the foreseeable future. Accordingly, it is highly questionable whether the development of hybrid transmission systems, capable of efficiently serving a multiplicity of diverse services, will evolve, despite both the technical possibilities and the potential economic advantages.

SUMMARY

The basic transmission requirements of CATV systems can be divided into trunking requirements between hub distribution centers and the requirements of the local signal distribution plant that directly feeds subscribers within the local distribution area.

CATV trunking is analogous to the trunking between switching exchanges within a telephone network and can be accomplished using RF transmission on coaxial cable, microwave radio links, or lightwave transmission facilities. Of these, lightwave technology is the most recent development and offers many advantages, but other considerations may dictate the use of other technology. For example, microwave may be the only practical way to cross a large body of water or a river.

In using lightwave systems to connect CATV hubs, either analog (usually FM modulated) or digital transmission techniques may be applied. Despite the promise of low noise contribution that digital transmission offers, the limitation on the number of channels that can be multiplexed on a single fiber and the cost of digital equipment make digital transmission more practical only for longer transmission systems at the moment.

Although there are persuasive arguments for using optical fibers in the local CATV distribution network, it is unclear as yet just how this may evolve. If hybrid distribution systems providing a multiplicity of unrelated subscriber services were feasible, then the evolution of fiber into CATV plant might be accelerated. The question as to whether, and if so, when such consolidation of services may actually occur may be more political than technical, but significant economic advantages can be clearly identified.

REVIEW QUESTIONS

True or False?

T   F   1. Video transmission requires substantially more transmission bandwidth than does voice transmission.

T   F   2. When many video signals are transmitted through an analog system, the accumulation of noise in such signals is dramatically different than in analog voice transmission systems.

T   F   3. When many video signals are transmitted through an analog system, the distortion introduced into such signals will be the same as the distortion introduced if a single video signal were being transmitted through that same system.

T   F   4. In a typical CATV coaxial cable system, only a few intermediate amplifiers will be required.

T   F   5. In a typical analog lightwave transmission system, few, if any, intermediate repeaters will be required.

T   F   6. In a typical coaxial cable CATV system, both "trunk" and "feeder" plant operate with the same transmission signal levels.

T   F   7. The primary motivation for trunk/feeder system designs in CATV systems is to improve the efficiency of tapping for subscriber service and to reduce system cost.

T   F   8. CATV "hub" system designs can improve the transmission quality and the reliability of service of large urban systems.

T   F   9. Larger size coaxial cables with lower transmission loss are seldom employed in coaxial CATV super-trunks.

T   F   10. It is possible to utilize a single optical fiber as an analog link to transport a single television channel.

T   F   11. It is possible to frequency division multiplex several television channels on a single optical fiber operating as an analog transmission system.

T   F   12. It is possible to time division multiplex several television channels on a single optical fiber operating as a digital transmission system.

T   F   13. An analog lightwave transmission link used for trunking to a CATV hub will be more susceptible to noise if it uses FM modulation than if it uses AM (vestigial sideband) modulation.

T   F   14. Analog lightwave transmission systems are not susceptible to any intermodulation distortion at all.

T   F   15. In any digital transmission system, the most critical adjustment is proper carrier modulation.

T   F   16. In most operating CATV systems today, all system transmission bandwidth is available at all subscriber taps within the system.

T   F   17. It is technically possible to utilize CATV technology (RF transmission on coaxial cables) to provide television, data, and telephone services to all subscriber service locations.

T   F   18. It is technically possible to utilize a single optical fiber (or a pair of fibers) to provide television services to all subscriber service locations in a completely lightwave CATV system.

T   F   19. Most of the lightwave links providing service in CATV systems today are in the subscriber loop plant portion of those systems.

T   F   20. There are no significant technical or economic advantages to be gained by using one common transmission technology to provide both CATV and telephone subscriber services.

CHAPTER 15

Network Considerations in Trunking

In the 40 or 50 lightwave projects that the author has been associated with, the cost of the interconnecting optical fibers alone has been approximately 50 percent of the total project cost. Accordingly, the efficient utilization of all fibers placed should be a primary objective of all lightwave system design efforts.

In many of the earlier projects that the author reviewed, the designer did not optimize the usage of all the fibers that were placed. In several instances, extra fibers were placed without compelling technical necessity and often without any rational justification at all. When operating experience with lightwave systems was limited and the level of confidence in the reliability of such plant was relatively low, this was quite understandable and may even have been commendable as prudent. All of the early installations worked well and did provide the predicted transmission capacity. With maturation of the technology, however, and with the extensive operational history of this type of plant now available, there is no justification for uneconomical designs today, and we are obliged to reexamine the basic philosophy that we bring to the design effort.

The earlier applications the author had opportunity to review were not really complex networks at all. Some had relatively high traffic requirements, perhaps, but they were all rather straightforward trunking applications between two discrete service points. These could all be addressed as individual transmission links with little or no consideration for consolidation into a network at all. At the same time, the author became involved in a number of lower traffic-density applications. But again, these were simple transmission links between two points, and had been addressed from this point of view by the designers.

With this orientation toward system designs, the question of digital multiplexing to increase traffic capacity was quite obvious and almost self-resolving. System designers made practically no distinction between point-to-point applications using lower fiber-count cables and high density point-to-point systems using higher fiber-count cables. All new applications were addressed with a sort of standard solution, generated without any special effort or thought.

CONVENTIONAL PAIRED CABLE APPLICATION

Let us review the basic evolution of a link or system from the perspective of how and why it fits into and relates with a network, regardless of the scope of that network. We shall examine a hypothetical application in which four switching centers, designated *A, B, C,* and *D,* are geographically located in a line along an existing cable route. We assume that the traffic requirement between any switching center is only one twisted pair. Quite obviously expanding to 50 pairs or hundreds of pairs would be possible, but using one pair for each link will present the design philosophy simply and clearly.

Figure 15–1 shows the four switching locations and their geographical relationship. For the moment we shall ignore the cable route length between locations, since it is not relevant to examining the design philosophy. Note that the traffic links required (one pair to service each link) are *A* to *B, A* to *C, A* to *D, B* to *C, B* to *D,* and *C* to *D.*

This hypothetical system is structured to resemble an actual application that the author reviewed, a rather complex network of ten switching locations (offices), all within a rather small geographic area. This was a densely populated suburban area, and the trunk requirements between all switching centers were quite heavy even though the cable route lengths between offices were only seven or eight miles or so. The geographical relationship of the various offices was not a star-type configuration conveniently arranged around a central toll office. There were already existing cable routes with switching locations positioned along the routes like beads on a string, sometimes four or five such offices in a row, just as we have proposed for our example here.

In place and requiring relief was a lot of high-pair-count (300 and 400 pairs) cable with a lot of T–1 carrier systems, including a multiplicity of intermediate cable repeaters, already working and awkward to expand. In addition, several of the exchanges were electromechanical (step-by-step) switches with digital replacements scheduled. It was proposed to initially place fiber cables along some (not all) of the interconnecting cable routes, and to handle not only new circuit additions but some paired-cable relief as well with digital transmission through the new lightwave plant.

Returning to our example, note that in the interconnecting cable routing the most significant factor is the physical availability of the interconnecting pairs at all switching locations (see Figure 15–1). The pair serving the *A* to *D* traffic link, for example, is spliced through or interconnected at switching locations *B* and *C.*

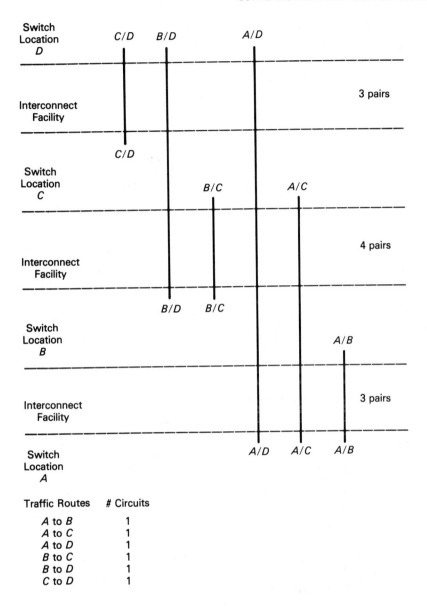

FIGURE 15–1
Conventional
Paired Cable

Traffic Routes	# Circuits
A to B	1
A to C	1
A to D	1
B to C	1
B to D	1
C to D	1

Should the traffic requirements change, we could easily reassign and utilize the pairs differently. For example, if traffic requirements between D and C increased, but requirements between B and D did not, we could open up the pair between B and D at location C, and provide an additional pair for C-to-D service.

This flexibility in usage is inherent in the appearance of each pair at each switching location. Although the initial thinking was in terms of discrete interconnections (*A* to *C, A* to *D,* etc.), flexibility can be very useful if traffic patterns develop differently than anticipated.

The philosophy applied in Figure 15-1 could very well be used in establishing pair counts for all interconnecting cables.

CONVENTIONAL T-1 CARRIER ON PAIRS

Figure 15-2 presupposes the traffic requirements for every interconnecting link (*A* to *B, A* to *C, A* to *D,* etc.) actually to be one T-1 carrier system. Thus, the single lines interconnecting switching locations represent two pairs, together accommodating a bidirectional T-1 system. No T-1 intermediate repeaters are shown since we are reviewing the basic design philosophy only. In Figure 15-1 each such line represented only a single pair.

In Figure 15-2 the plant is somewhat less flexible than the pairs depicted in Figure 15-1. Thus, if, as before, we were to experience a requirement for an additional T-1 system between locations *D* and *C,* we could accommodate this in Figure 15-2 by opening the two pairs at location *C,* as we did earlier, but now we have to relocate some T-1 terminal equipment also. And without some additional equipment or sophistication in intermediate terminal locations, we cannot manipulate circuits in increments smaller than 24 equivalent voice circuits, which is the capacity of the basic T-1 carrier system.

All of this still reflects the basic philosophy of providing a discrete interconnecting link for each identifiable traffic routing or requirement.

CONVENTIONAL LIGHTWAVE DESIGN

Now let us extend the discussion to optical fibers. We shall assume that the basic increment of service provided is to be four T-1 carrier systems, although it is obviously possible to multiplex up to a much higher digital level than this. Four T-1 systems presents a 6.3 Mb/s digital signal providing 96 equivalent voice circuits.

Figure 15-3 shows an interconnecting optical fiber facility for each identifiable traffic link, as before, but now each single line represents four fibers, two for the service facility and two for a protection-switched, paralleling redundant facility. If the traffic requirements develop in an orderly fashion as predicted, the provision of discrete connecting facilities for each requirement (again *A* to *B, A* to *C, A* to *D,* etc.) may be perfectly satisfactory. Of course here, the basic increment of service provided is four T-1 systems (96 voice channels) as a minimum.

Now when we consider some unpredicted traffic developments as we did before—for example, a requirement for additional capacity between *C* and *D*—the facility (the optical fiber "quad" of two fiber pairs) can still be opened at location *C,* but the rearrangement of optical terminal equipment required is much more ex-

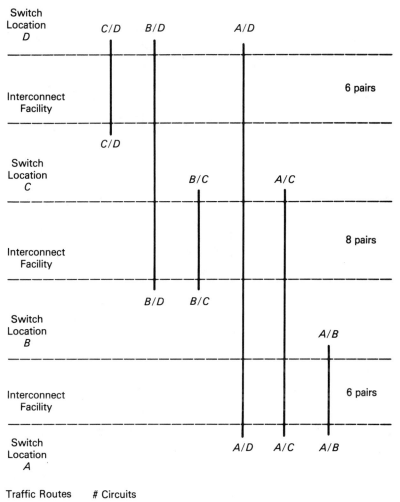

FIGURE 15–2
Conventional
T–1 Carrier on
Pairs

Traffic Routes	# Circuits
A to B	24
A to C	24
A to D	24
B to C	24
B to D	24
C to D	24

tensive. And the penalty of sacrificing a 96-channel facility between *B* and *D* to gain a 96-channel capability between *C* and *D* may be a bit more restrictive than we would wish. Due to the higher inherent capacity of the lightwave systems, if we configure as we did for pairs alone, or as we did for T-1 carrier on pairs, the plant placed becomes less flexible in use.

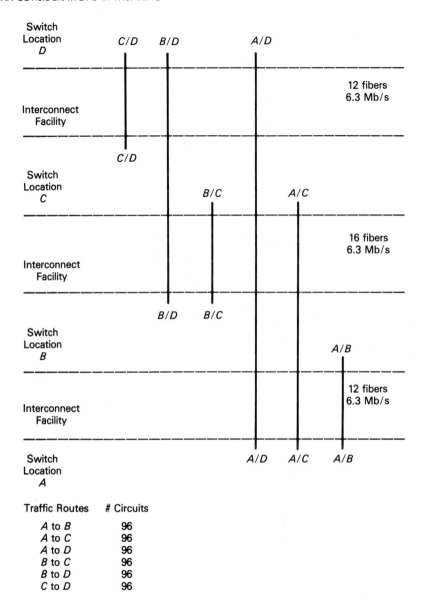

FIGURE 15–3
Conventional
Fiber Optic
Design

Traffic Routes	# Circuits
A to B	96
A to C	96
A to D	96
B to C	96
B to D	96
C to D	96

COMPARING DESIGNS

In our initial discussion, which dealt with pairs alone, the convenient accessibility of the pairs at intermediate locations provided a highly flexible usage of those pairs as traffic requirements changed. Escalating to 24-channel increments in the T-1 carrier application reduced that flexibility somewhat, and the restrictive availability of 96-channel increments as a minimum in the optical plant became even more awkward. Now consider the optical plant if each discrete interconnecting link is at the DS-3 transmission rate of 45 Mb/s (672 equivalent voice circuits). This is a very cumbersome increment indeed if we are forced to reallocate the traffic capacity within the local network.

RECONFIGURING THE LIGHTWAVE SYSTEM

Perhaps, however, we could construct the optical plant in a different manner right from the start. It might then be possible to provide some smaller increments of flexibility in the use and reassignment of the services provided. Indeed, we might even go down to the level of individual voice circuits at each switching location.

Figure 15-4 does just that. To provide 96 voice circuits for each traffic route, 288 circuits are needed between *C* and *D,* 384 circuits between *B* and *C,* and 288 circuits between *A* and *B.* The figure shows the same optical fiber capacity and transmission bit rates as shown in Figure 15-3, but each equivalent voice circuit is terminated and made available at voice frequency, at each fiber end.

Now we have the same flexibility for interconnection at voice frequency that we had in the example using pairs alone: there no longer exists a dedicated facility for each traffic route, even though sufficient voice circuits are provided throughout the network to interconnect and establish such traffic facilities.

Although this approach is the most flexible in usage of individual voice circuits, it surely will not be the most cost effective. For example, consider each voice circuit provided for traffic route *A* to *D*. In Figure 15-4 these will all require demultiplexing down to voice frequency and then back up to digital carrier at intermediate locations *C* and *D*. Altogether, there are twenty terminals in the figure, each of which requires T-1 channel banks and digital multiplexing and demultiplexing up to, and down from, the DS-2 (6.3 Mb/s) level. Surely, there is a more practical level of interconnection at these intermediate locations, perhaps one that does not go all the way down to voice frequency for each and every circuit.

INTERCONNECTION AT T-1 LEVEL

In Figure 15-5 we have not changed either the optical fiber count or the transmission bit rate, but we have terminated all fibers at all locations, and are providing a multiplexing/demultiplexing capability down to the DS-1 digital level (1.544 Mb/s) transmission rate. The system provides the flexibility of usage shown in that of Figure 15-2—not as flexible as the configurations of Figures 15-1 or 15-4, of course, but substantially less expensive than the latter.

DESIGN PHILOSOPHY

The preceding discussion was predicated upon providing discrete lightwave transmission capacity between locations rather than dedicating an individual lightwave system for each required traffic route. There may be disadvantages to such an ap-

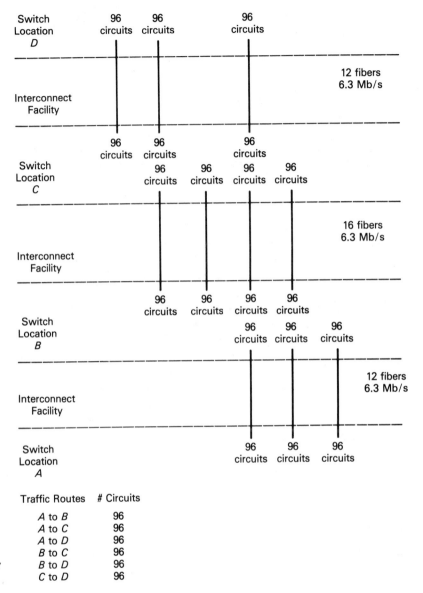

FIGURE 15–4
Voice Frequency
Interconnection

proach, however. For example, consider Figure 15–3. The transmission link provided for the traffic route A to D is an optical transmission system end to end. If the interconnecting facilities between all points on the link were not excessively long, we might even have end-to-end (A to D) transmission without any intermediate repeaters at all. In such a case, the fiber would be spliced through at all intermediate locations.

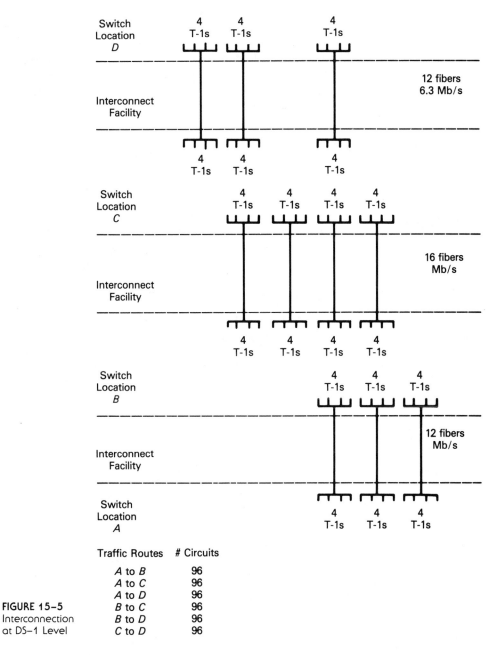

Switch Location D

4 T-1s 4 T-1s 4 T-1s

Interconnect Facility

12 fibers
6.3 Mb/s

4 T-1s 4 T-1s 4 T-1s

Switch Location C

4 T-1s 4 T-1s 4 T-1s 4 T-1s

Interconnect Facility

16 fibers
Mb/s

4 T-1s 4 T-1s 4 T-1s 4 T-1s

Switch Location B

4 T-1s 4 T-1s 4 T-1s

Interconnect Facility

12 fibers
Mb/s

Switch Location A

4 T-1s 4 T-1s 4 T-1s

Traffic Routes	# Circuits
A to B	96
A to C	96
A to D	96
B to C	96
B to D	96
C to D	96

FIGURE 15–5
Interconnection at DS–1 Level

Now contrast this approach with the optical facilities shown in Figures 15–4 and 15–5. In these figures, each interconnecting facility is shown as an individual lightwave link that involves an optical transmitter and receiver for each fiber. Recall that each single line in these drawings was for four fibers to be used in a protection-switched configuration. The interconnections at all intermediate locations in both of the figures are electrical, not optical.

It can be argued that these configurations would provide lower reliability of service between locations *A* and *D* than would the continuous lightwave transmission between those same two points shown in Figure 15–3. There is some merit to this argument, although in Figures 15–4 and 15–5, all lightwave links are protected with redundant fiber facilities and automatic transfer switching. A counterargument is that by installing individual lightwave links between all locations the individual fiber lengths are significantly reduced, which might permit less expensive fibers to be used; or less sophisticated, less costly LED light sources might be practical instead of lasers.

MAXIMIZING FIBER USAGE

Up to this point, we have not applied any higher level digital multiplexing at all. All optical transmission, whichever design was under consideration, has been limited to DS–2 and 6.3 Mb/s. This was intentional, to demonstrate a logical development, but it is somewhat academic.

In practice, when we work with single twisted pairs, we must add interconnect facilities to increase transmission capacity. When we work with T–1 carrier on pairs, we can add increments of 24 circuits when we add interconnecting facilities, but we cannot avoid all additions in physical facilities if we want more capacity. This is not so with optical fiber plant, of course. Consider again the North American digital hierarchy:

	Transmission Rate	Equivalent Voice Circuits
DS-1	1.544 Mb/s	24
DS-1C	3.152 Mb/s	48
DS-2	6.312 Mb/s	96
DS-3	44.736 Mb/s	672
DS-3C	90.148 Mb/s	1,344

Standard multiplex equipment is available to effect the transition from one level of digital signal to another. This may be easier to understand if it is referenced to the basic T–1 carrier system we are all familiar with, which provides 24 equivalent voice circuits from a DS–1 level (1.544 Mb/s) digital signal.

Multiplex	Transition From	Transition To
M1C	2 DS-1	1 DS-1C
M12	4 DS-1	1 DS-2
M13	28 DS-1	1 DS-3
M23	7 DS-2	1 DS-3

A DS–2 (6.312 Mb/s) signal does not tax a lightwave system at all. Operations at 45 Mb/s on optical fibers are readily available, inexpensive, and reliable. Accordingly, we can easily satisfy our circuit requirements in the application we have been addressing if we simply upgrade all transmission bit rates shown in Figure 15–5 to those shown in Figure 15–6.

This provides the potential for 672 equivalent voice circuits between all switching locations, which is well in excess of our original requirements. One might even

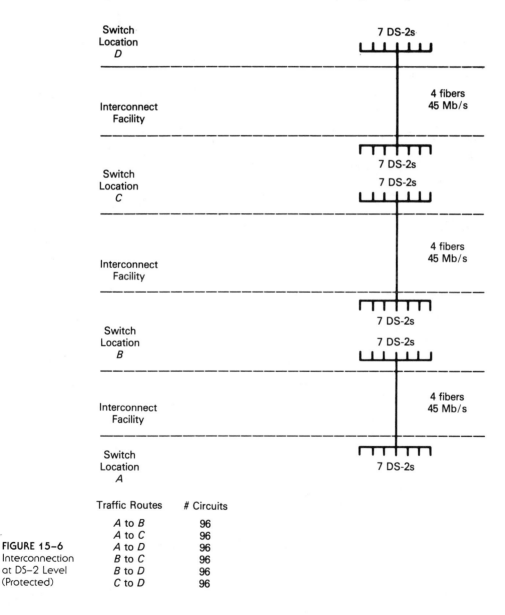

FIGURE 15–6
Interconnection at DS–2 Level (Protected)

Traffic Routes	# Circuits
A to B	96
A to C	96
A to D	96
B to C	96
B to D	96
C to D	96

assume this level of capacity could adequately provide for foreseeable growth requirements.

Note that in the figure we have reduced the number of interconnecting optical fibers. Each straight line represents four optical fibers in a protection-switched configuration. The number of circuits provided for any traffic route is still 96, but the capacity of the plant is actually substantially higher than this.

Note also that we have escalated the interconnection level at intermediate locations from the T-1 level shown in Figure 15-5 to DS-2 (four T-1 systems). This then becomes the basic interconnection increment, but it would still be possible to interconnect at either the DS-3 or the DS-1 level by using other terminal equipment.

In sum, we have drastically reduced the number of fibers in the system, while at the same time preserving some reasonable degree of flexibility for utilization of the system. We do have an uncomfortable level of commonality for all transported information, however: there are M23 multiplexers at the end of each discrete fiber link, and these units are common to all circuits carried. The arguments to protect each M23 unit with automatically transferred redundant fibers and lightwave terminal units (transmitters and receivers) are persuasive, and in fact, such would be the conventional protection-switched configuration.

We might still consider an alternative, however.

INTERCONNECTION AT DS-2 LEVEL

In Figure 15-7, we deviate from our previous practice of using a single line for the interconnecting facilities to denote four fibers in a protected-switching configuration. Rather, here each single line denotes only two fibers, one for each direction of transmission. We might even employ WDM to utilize a single fiber as a bidirectional transmission link, but that is only a further refinement on the philosophy underlying the figure.

Note that we still show interconnection at the DS-2 level (it could be DS-1 or DS-3, of course), but if we examine Facilities *AA* and *BB,* we see that they are completely independent of each other, including using separate fiber pairs. There is no digital multiplexer or lightwave terminal equipment common to both facilities at all.

A gross disruption of either subsystem, be it fiber break, lightwave terminal failure, or multiplex equipment outage, will thus be limited to affecting only 50 percent of the services provided. If we judiciously divide the traffic routes across these independent subsystems, we can achieve a very respectable level of service protection without any protection sensing and switching at all. For example, half of the *A*-to-*D* traffic would be assigned to facilities *AA, CC,* and *EE,* with the other half carried on *BB, DD,* and *FF.*

Contrast this with the more conventional approach of multiplexing all traffic up to a higher bit rate—perhaps to 90 Mb/s in our example—and then providing redundant fiber pairs and terminal switching, and it is not difficult to see that there may be attractive cost advantages to the approach shown in Figure 15-7.

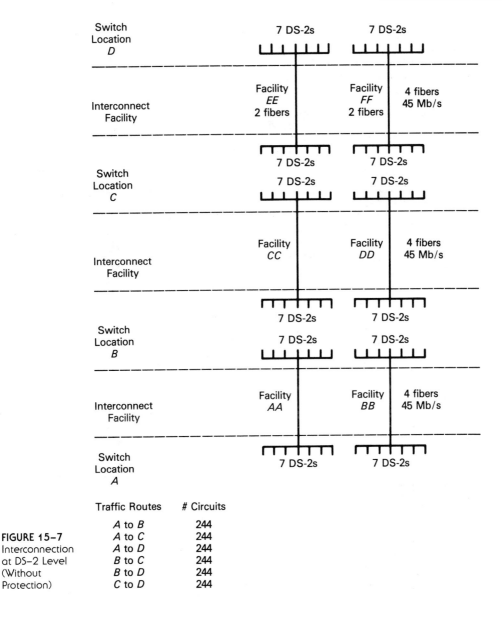

FIGURE 15–7
Interconnection
at DS–2 Level
(Without
Protection)

Traffic Routes	# Circuits
A to B	244
A to C	244
A to D	244
B to C	244
B to D	244
C to D	244

EXPANDING SYSTEM CAPACITY

What about subsequent expansion of the system's capacity? We addressed this whole subject initially because we anticipated that some awkward or unpredicted growth might develop in such a sophisticated local network. One approach might be simply to escalate by digital multiplexing to a higher level transmission bit rate, as in the last example, where we suggested possibly going up to 90 Mb/s. In this case, the block diagram in Figure 15–6 would apply, except that the bit rate for the interconnecting facilities would change to 90 Mb/s.

But perhaps there is a better way. Perhaps we can expand the capacity without increasing the commonality of terminal equipment and consequently without increasing the vulnerability to outage.

Figure 15–8 shows how, using WDM, a second 45 Mb/s subsystem can be superimposed on the initially installed fibers. Although the fibers are common to

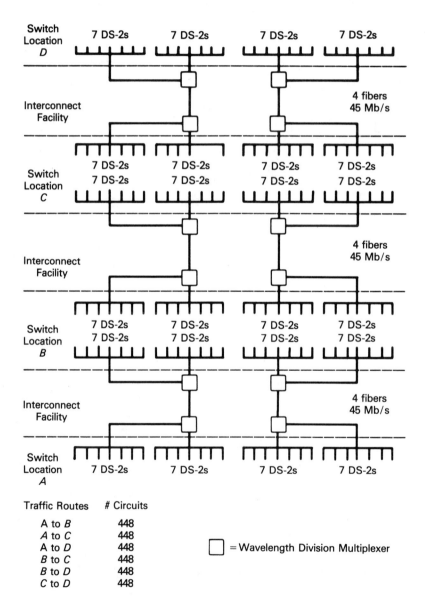

Traffic Routes	# Circuits
A to B	448
A to C	448
A to D	448
B to C	448
B to D	448
C to D	448

☐ = Wavelength Division Multiplexer

FIGURE 15–8
Expanding
System Capacit
by WDM

both subsystems, no terminal equipment or digital multiplexing units are common. And duplicating the entire arrangement on a second fiber pair, as shown in the figure, allows traffic on any particular route to be distributed across the two fiber pairs also.

We anticipate failures of terminal equipment or fibers to occur only a very small percentage of the time, and with this configuration, even when such an unlikely event occurs, the outage is limited to only 50 percent of the traffic. Perhaps this might be considered a reasonable level of protection even though no protection switching at all is provided in the system.

Capital invested in transmission capacity produces revenue, but capital invested in simply protecting the transmission capacity does not. Diversifying the traffic routing in the manner shown may well be an acceptable level of operational reliability. In any event, it is surely more efficient in terms of transmission capacity versus facility cost.

The WDM couplers shown in Figure 15–8 introduce transmission loss, but since the individual lightwave links are short, any such losses might be quite acceptable in many actual applications. The couplers could be installed initially, even if the terminal equipment to expand system capacity were deferred until the traffic growth required such expansion. Note that in Figure 15–8, the maximum transmission bit rate has been limited to 45 Mb/s. The use of stable, inexpensive LED light sources may be quite satisfactory under these conditions, even though two different lightwave frequencies are involved.

SUMMARY

For clarity, the drawings used in the examples we present do not show digital multiplex equipment or lightwave terminal units. They do show a logical development from twisted pairs through optical fibers, however, and we can see alternative methods of system design that could improve the efficiency of the system while adequately ensuring reliability. We have also shown full utilization of the various techniques for digital multiplexing or WDM.

In actual applications, one may find it neither necessary nor justifiable to escalate transmission capacity across the board in the manner shown. It is entirely possible to escalate to a lesser degree or incrementally, as circumstances dictate.

The twisted pair is a flexible but inefficient transmission facility. Many pairs in a system will be idle some part of the time, and thus will produce revenue only part of the time. The pair is the smallest transmission increment, however, and can be effectively used when only a few circuits must be provided or when several service points requiring few circuits each are involved.

Cable carrier improves the efficiency of the pairs, but introduces a larger increment in terminal size and thus is less flexible for utilization when few circuits are required.

Optical fibers are very efficient in their use of the interconnecting facility itself; thus, the physical plant costs are more acceptable. The tradeoff comes in requirements for higher sophistication in terminal equipment. Since terminal equipment costs can be postponed to some extent, that is, delayed until projected traffic requirements actually develop, optical transmission systems can permit a graceful growth that parallels or closely tracks traffic revenues.

How effectively this technology is applied will depend upon how logically we develop system designs. If we simply apply lightwave as a drop-in replacement for copper pairs or cable carrier systems and use the same design philosophy that paired cable responds to, we may negate some of the advantages that lightwave technology presents.

REVIEW QUESTIONS _____

True or False?

T   F   1. The cost of interconnecting optical fibers typically represents 25 percent of a lightwave transmission system's cost.

T   F   2. Multiple-pair cable technology provides substantial flexibility in the utilization of the facilities placed, but it does this at the cost of possible inefficiency in the use of individual pairs.

T   F   3. Multiple-pair cables directly relate system transmission capacity to pair count, that is, the number of individual transmission facilities that the cable presents for use.

T   F   4. When additional traffic requirements develop for an in-place multiple-pair cable, the only recourse is to place an additional cable to make more pairs available.

T   F   5. Cable carrier systems improve the efficiency of multiple pair cables by increasing their transmission capacity.

T   F   6. Trunking cable carrier systems are particularly efficient when very few circuits must be provided at several intermediate points along a cable route.

T   F   7. The T–1 carrier systems basically provide 32 voice circuits for use.

T   F   8. Cable carrier systems are somewhat limited in the expansion they can provide in a multiple-pair cable's transmission capacity.

T   F   9. Coaxial cable provides more transmission capacity than a twisted pair due to the greater transmission bandwidth it presents for use.

T   F   10. Optical fibers provide more transmission capacity than coaxial cables due to the smaller transmission bandwidth they provide for use.

T   F   11. Lightwave transmission systems usually require more intermediate repeaters than multiple-pair cable carrier systems do.

T   F   12. When traffic requirements are substantial, the dedication of a single transmission facility to serve each and every point within a network is usually inefficient.

T   F   13. If broadband transmission capacity is intelligently provided throughout a system, the system can usually be reconfigured to accommodate changes in traffic requirements.

T   F   14. Designing systems for broadband utilization of the transmission facilities can significantly reduce the optical fiber requirements in a system.

T   F   15. Optical fibers are most effectively employed when they are simply used as direct one-for-one replacements for metallic conductor pairs.

T   F   16. It is possible to design a lightwave system to operate at a high transmission data rate, but to operate it at a lower data rate initially.

T   F   17. If a link was designed to operate at a higher transmission data rate, an upgrade to that data rate would only require replacement or upgrade of terminal equipment.

T   F   18. Multiple-pair transmission systems are inherently inefficient in that some pairs may be idle some of the time.

T   F   19. It should not be necessary to reinforce optical cables by placing additional cables and fibers as often as it was necessary in the past to reinforce multiple-pair cables.

T   F   20. In lightwave transmission networks, we will probably utilize fewer physical interconnecting facilities (optical fibers) and less sophisticated electronic terminals than we did in multiple pair cable networks.

CHAPTER 16

Network Considerations in Distribution

Just as it was necessary to consider the implications of networking when we addressed trunking in lightwave transmission systems, it is necessary to consider the local network that is involved in distribution or exchange type of plant if we are to apply lightwave technology effectively in such plant. At first glance it might appear that in lower level plant applications, like exchange and distribution, the problems will be less complex or more readily resolved than in higher level applications, but this is not necessarily the case. Consider the chief differences between the two applications.

A trunking application is clearly defined and is most generally restricted to only two terminal locations, although occasionally more service points may be involved. The options for system configuration are somewhat limited, and the optimum solutions tend to apply in many cases.

In exchange plant, on the other hand, the nature of the applications to be addressed will cover a much broader range, and the economic evaluation will be more complex. Also, we are obliged to consider the impact of lightwave plant on several other aspects of the exchange network, whereas in trunking we had only to evaluate the technology against the established and conventional alternatives, e.g., pairs, coaxial cables, and microwave.

Moreover, since most lightwave installations to date have been on major traffic route systems, there will be less previous experience in exchange design to fall back on. We may find that these applications offer both the greatest opportunity and the greatest challenge to utilize the technology effectively.

Defining the term "distribution" is a difficult task. When telephone plant was less complex and fewer alternatives were available in the subscriber loop, the line of demarcation between trunking and distribution was much more clearly defined. As carrier came into wider use, primarily because cable pairs or open wire (individual conductors spaced on insulators and crossarms) was inefficient and labor intensive, a variety of relatively sophisticated devices began to intrude into the lower level plant. Line concentrators, grouped carrier terminals, and ultimately remote switching terminals, were all developed to utilize the fixed transmission facilities—the metallic conductors themselves—more efficiently.

But where does the trunk application end?

THE LOCAL NETWORK

It is not possible to define trunks simply as carrier-derived circuits, since this technology is used more and more widely in the exchange level of plant. It is not uncommon to find large numbers of circuits provided via carrier to the premises of a single subscriber, such as in PBX or PABX installations.

Even the local network switching function is less clearly definable today. When all subscriber services basically had their origin at a relatively few circuit switching centers, be they small unattended mechanical switches or a major telephone exchange, the telephone loop and the instrument itself were very unsophisticated. The telephone instrument was much more an electromechanical unit than an electronic device. The transistor, and the integrated circuit, however, have changed all that. Sophistication in the network is now much more widely distributed throughout the service area, the present trend to use remote digital switches extensively being a case in point.

Some of the changes in network structure were the direct result of less expensive, more efficient transmission techniques such as digital carrier systems. By consolidating more circuits on a transmission facility, it became economically practical to do more and more "exchange" trunking, that is, to establish smaller switch service areas by locating more switching functions closer to subscriber loops. Fed from such localized switches, the subscriber loops time share a limited number of "exchange" trunks that connect to the higher level "host" switch.

There are compromises involved in such a network design. As the circuit switching function is performed at many more locations within the service area, the sophistication of the switching is more broadly distributed across the area, and the sophistication of the required maintenance is also more widely distributed throughout the network. These are reflected in rather elaborate systems to test conventional subscriber loops through such a switch from a centralized remote location.

The nature of digitalized switching is such that the cost per circuit switched tends to be lower if more circuits are switched at the same location. This suggests that given a less expensive multiple-circuit transmission facility, the philosophy of widely distributing the switching function, at any level at all, may be less appropriate now than was previously the case. Lightwave transmission, using relatively broadband optical fibers and providing random access to substantial transmission bandwidth at any point along an installed facility, would permit bringing many subscriber lines from a number of serving area interfaces back to a single centralized switch location on a line-per-terminal basis.

It is highly speculative, perhaps, but we may find that the trend toward distributing switching widely throughout a service area will be reversed, and a new trend which consolidates even more subscriber lines within a single switch location will develop. Certainly, as profound a change in transmission technology as lightwave transmission presents, and with due regard to the economics of both transmission and switching, the equations which encouraged distributed switching will have to be reexamined.

BROADBAND VERSUS DEDICATED FACILITIES

The problem is compounded by a more basic debate. There is some industry support for an ultimate optical facility to each and every point of service. This would involve running an optical fiber or fiber pair from the network switching location to the physical premises where the service is utilized. In the public switched telephone network, the latter is the subscriber's home or place of business.

Without question, such a network structure could provide the maximum transmission capacity to every point of service. In effect, the philosophy simply substitutes the glass medium for the present copper pair.

The other side of this argument proposes a wideband utilization of the inherently broad-bandwidth lightwave plant to service a multiplicity of subscriber "clusters." The extension of subscriber service from these cluster locations might be effected by copper pairs or optical fibers; indeed, in some unique instances coaxial cables might be employed for the final plant extension.

Although the technique might limit the transmission bandwidth made available at each and every network terminal, it could still provide substantially more bandwidth than the presently installed and operating copper paired plant does. And given the recent developments in bandwidth utilization, the technique cannot be easily discredited on this basis alone.

The multiplicity of fibers involved in a design that essentially dedicates a physical facility (a fiber or a fiber pair) for each and every point of service is not particularly attractive. Indeed, it is essentially the same technique that has been employed in multiple-paired cable plant for a long time. The inefficiency of individual exchange pairs being idle for the majority of the time, and long lengths of such pairs actually providing no service at all, are not unfamiliar to the telecommunications engineer. And however refined the prediction process has become, we are all familiar with pairs placed on speculation and subsequently not used for long periods, or not enough pairs being placed initially and cable relief becoming necessary within a very short time.

Although it might be presumptuous to endorse any particular theory of evolution for future networks, we are obliged to try to address the requirements as we see them, or as we think they may evolve. The question facing us is, Can we usefully proceed to discuss lightwave transmission in exchange or distribution plant when the definition of just what exchange or distribution plant is is somewhat uncertain?

UNDERSTANDING LIGHTWAVE TECHNOLOGY

A logical response to the dilemma is to try to understand the basic technology of lightwave transmission well enough that, as the industry incorporates it more broadly into the national network, we are able to follow the developments intelligently and make those individual judgments that will be required as the necessity arises. Of course, objectivity may be difficult to come by, since previous experience may predispose us in one direction or another and there will be subtle pressures exerted in larger operations which have greater financial resources. These may give heavier weight to some aspects of system design, such as protective system configurations, since they are better able to bear the costs of such configurations.

There is a tendency among "experts" today to endorse almost all technically sophisticated solutions simply because those approaches are possible, or because they are "state of the art." Engineering, as defined in the American Heritage Dictionary, is "the application of scientific principles to practical ends, such as the design, fabrication, or operation of efficient and economical equipment, structures, or systems." Note the action words "practical," " economical," and "efficient." If we design systems without due regard to these, we are simply not performing engineering functions at all, according to this definition.

Accordingly, let us attempt to understand as fully as possible the technology itself, so that we can apply it effectively to any situation in which it presents sufficient advantages to justify its use.

PROFILING THE DISTRIBUTION APPLICATION

Without committing ourselves to any particular design philosophy, we can perhaps still characterize distribution applications with some measure of confidence. As used here, the term "distribution" implies the transmission facility that connects a discrete point of service to the national switched telecommunications network at a given level of circuit switching. Thus, a distribution facility includes the transmission facility from the last circuit switching function in the network to the equipment on the subscriber's premises, regardless of the sophistication of that equipment. We shall qualify the term to include transmission facilities between a so-called host switch location and a so-called remote switch. This may not be consistent with other definitions, but the fluidity of the plant demarcation, as pointed out earlier, makes the approach justifiable.

Distribution plant might be characterized as being distinctly limited in length; that is, the majority of installations will not involve distances greater than 25 km or so. Of equal significance in the definition of such plant is that it requires relatively low transmission data rates. We shall suppose the limitation is 45 Mb/s or less in the majority of cases.

Note that whether such plant connects a remote switch terminal with a host switch is irrelevant in our discussion. Even if the facility were intended for such service, the circuit capacity inherently available with a 45 Mb/s transmission rate should be more than adequate for most, if not all, such applications. After all, a 45 Mb/s rate can support 672 equivalent voice circuits.

Conversely, at any discrete point, such as a line concentrator or a grouped carrier terminal where a cluster of subscriber loops might be served, a 45 Mb/s signal would generally provide ample capacity, and duplicate installations or WDM implementation could accommodate instances where it did not. Our characterization of this plant as 45 Mb/s links may be challenged, of course.

In any event, some arbitrary data rate must be selected for purposes of examination and analysis, and if a logical methodology can be developed, then upgrading to a higher data rate may not completely disqualify the process established. Within this definition of distribution plant, then, how might lightwave technology be applied?

FIBERS AS DEDICATED FACILITIES

If lightwave transmission is applied in the form of a single dedicated transmission facility for every point of service in the distribution plant, then the problem is rather easily resolved. We have already, by definition, established the transmission data rate to be 45 Mb/s or less. Earlier, we learned that there is little, if any, significant cost advantage in specifying performance for data rates lower than 45 Mb/s. If the philosophy of dedicating a fiber facility as an individual service loop actually prevails, then a substantial amount of lower grade fiber cable might be involved, and this statement on fiber cost might not be valid. In any event, we can explore the provision of a 45 Mb/s subscriber loop with some confidence that any cost adjustments would most probably be reductions rather than increases.

The question that is most difficult to resolve is whether it is more probable that a single fiber, operating in a bidirectional mode, will be extensively used in the subscriber loop rather than a fiber pair (two discrete fibers). In the opinion of the author, a single bidirectional fiber would be both technically adequate and significantly less expensive, so that such an approach is likely to prevail. This should not be considered an authoritative endorsement of this type of distribution design, however, since there may well be new developments which might change things.

FIBERS AS BROADBAND FACILITIES

As regards broadband applications, the transmission design problems are easily resolved, but the possible configurations of the plant are more complex. Given a transmission data rate qualified to be only 45 Mb/s or less and an extreme link length of 25 km or so, a quick review of Chapter 10 resolves the question of whether such transmission is practical in the affirmative. How such a facility might be designed to retain flexibility in utilization is less clear, however.

Figure 16–1 shows, in block diagram form, a representative distribution system that is 25 km in length between terminal points *A* and *B*. These points might be a "host" switch at one end and a "remote" switch at the other, but the application is irrelevant to the discussion. Let us assume that the 45 Mb/s transmission capacity must be provided and reserved for traffic requirements between these two terminals. What other services, and at what other service points, can the facility support?

Figure 16–2 shows two new service points that presumably developed as requirements at some later time, that is, after the initial system was placed into service between points *A* and *B*. One of the new locations, point *C*, is between the original two terminals. The second location, point *D*, is actually past the original point *B* terminal. Could the original link design have been done in such a manner that the subsequent addition of these two new service points could be accommodated without compromising any of the service links, original or subsequent?

In Figure 16–3 the two additional service points have been established, and the drawing identifies the transmission losses that have been introduced into the

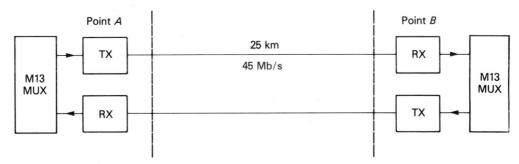

FIGURE 16–1 Initial Requirements

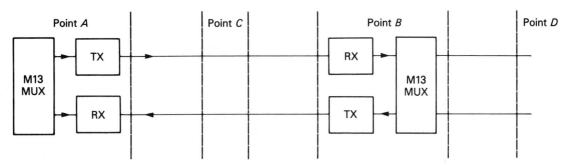

FIGURE 16–2 Additional Service Points

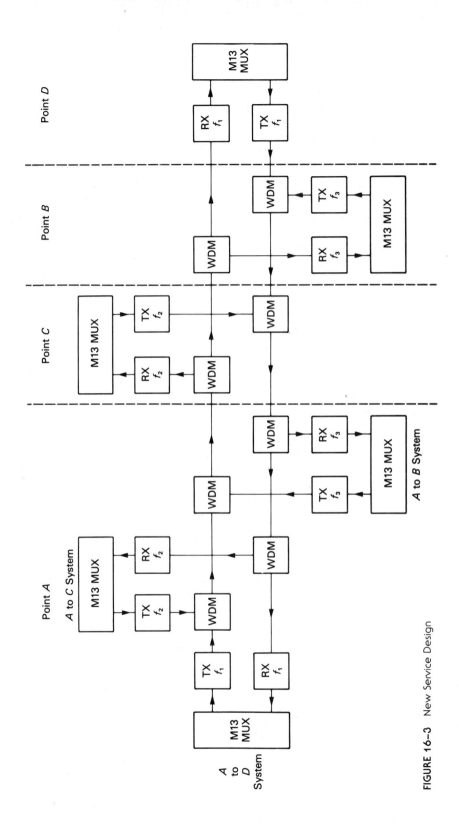

FIGURE 16–3 New Service Design

initial link, as well as the new transmission losses presented by the additional fiber that was required to extend the system out to point D. From the two WDM filters that have been inserted into the link at point A, we will accept a prudently conservative 4 dB of transmission loss. Moreover, at the new service location, point C, an additional coupler insertion loss is incurred which we will assume to be 2 dB. Similarly, at original terminal point B, we expect another 2 dB of loss from the coupler located there.

If the original design provided a reserve margin of only 8 dB—a quite conservative number—then the original system performance will remain unchanged, and service quality between the original terminals, points A and B, will be just as satisfactory as before the additions were made.

An 8 dB margin might seem to be quite an optimistic assumption, but recall that this link was originally only 25 km long. From Chapter 10, the transmission losses involved in the original design were by no means inhibitively large, so that we may very well be able to accept these additional losses. In any event, during the initial design work, we were free to select the transmitter output, fiber loss, and receiver sensitivity such that the link could well have been designed to accommodate these additional losses. In fact, we might even be able to change the transmitter light sources from LEDs to higher output lasers now, or change to more sensitive receivers, to make the new system additions possible, if this were actually necessary.

Can we accommodate the longer distance required to serve the new terminal at point D? This will depend upon how much fiber is added, or the distance involved between points B and D, and the loss of the fiber chosen for this extension. If necessary, we could employ higher output light sources for this new service alone, or a more sensitive receiver.

TRANSMISSION CAPACITY

If the initial system design was predicated upon supporting a 45 Mb/s bit rate, then any of the new service links that we create can function at this same data rate also. Not only will the original transmission capability between points A and B remain unchanged, but all new services could be implemented at this same 45 Mb/s rate or lower.

How many new service locations can be established on a lightwave facility will depend not only on the insertion losses incurred in optical couplers, but also on the frequency separation and isolation that those couplers can provide.

There has been only limited use of WDM in telecommunications installations to date, primarily because of the nature of the applications themselves, which have largely been high-traffic-density trunking routes. High fiber-count cables were extensively placed, and consequently there has been little need to use WDM, although many of these systems included some provision for future WDM operation on the facilities. At this time, the maximum practical use of WDM is debatable, but it seems reasonably certain that four separate lightwave carriers may be multiplexed in this manner.

IMPRACTICALITY OF DIGITAL MULTIPLEXING

Although the practicality or suitability of WDM could be compared directly to the efficiency or cost of higher level digital multiplexing in trunking applications, no such comparison is meaningful now, because the physical separation of the various service points precludes digital multiplexing. Multiplexing requires all derived service capacity to be produced and employed at one single location, unless back-to-back electronic installations are implemented at intermediate service points. However, it would appear that this level of sophistication is unsuitable for lower level plant installations.

On the other hand, it would be perfectly feasible to handle the service extension to point D in a configuration employing higher level digital multiplexing. Figure 16–4 shows the original system terminals, equipped with M13 multiplexers, but with the new service location at point C, which is irrelevant, omitted for clarity. The service extension to point D is handled as a discrete new lightwave link, and its data input is derived from one of the outputs of the M13 multiplexer that is part of the point A-to-B transmission system.

A DS 2 signal, or several such signals, could be extended as electrical input signals to the new lightwave link that connects points B and D as a separate transmission facility. The reliability of the service from A to D would then depend upon the reliability of the two lightwave links in tandem, the one connecting A to B, and the other connecting B to D.

The figure intentionally does not identify the service requirements at the new locations C and D, so that the reader is not predisposed to a particular application requirement. Strictly from a transmission point of view, it really does not matter what service these points require. They might be factories or industrial parks that were built subsequently, or they might be new shopping malls or new residential developments. Nor does it matter whether the service derived between points A and D was network trunking or not. Within the transmission link being addressed, it is just another signal to be transported.

The last point deserves further discussion. Studies of previous transmission media tended to differentiate between services such as exchange or trunk services. Such distinction within a lightwave transmission system is unnecessary and unimportant. Any circuit, or any group of circuits within a system, will enjoy the same quality of transmission and require no special handling or design effort. Any optical fiber can simultaneously transport trunk, special service circuits, or distribution or exchange level circuits interchangeably and equally well. Today, a trunk facility using fibers may be designed to link two switching centers with trunks, but it is inherently capable, today or tomorrow, of also serving other grades of service anywhere along the length of the physical facility itself.

The system designer must accordingly expand his thinking beyond the response to a specific requirement for facilities, and think instead in terms of the bandwidth—the transmission capacity—that the new facility inherently introduces throughout the geographical area it traverses.

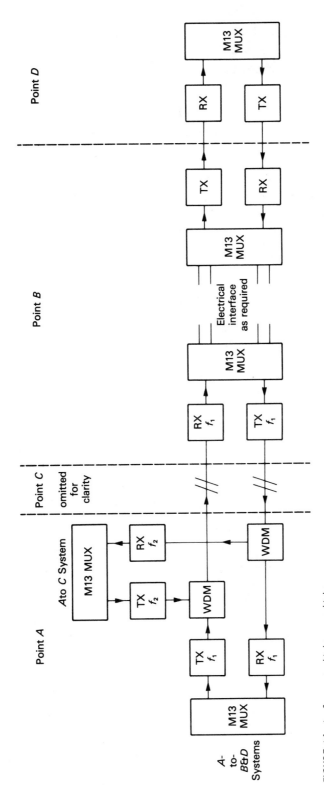

FIGURE 16-4 Separate Lightwave Links

FLEXIBLE FIBER UTILIZATION

The preceding example should not be taken to exclude side-lead extensions of transmission capacity. Figure 16–5 shows three locations, with one (point C) "off route" on a cable run between points A and B. The distances shown are arbitrary.

The design provided four separate fibers out to the side-lead branch point, shown in the drawing as point X, simply leaving those four fibers idle for a future extension through a new fiber cable down to point C. The remaining four fibers in an eight-fiber cable were to be extended through point X to serve the A-to-B link.

By using WDM couplers at point X and incorporating their insertion loss into the initial transmission design for the A-to-B link, we can reduce the fiber count in 10 km of cable from eight to four. Such a design change does not reduce the transmission capacity that was originally required to be 45 Mb/s for both services (A to

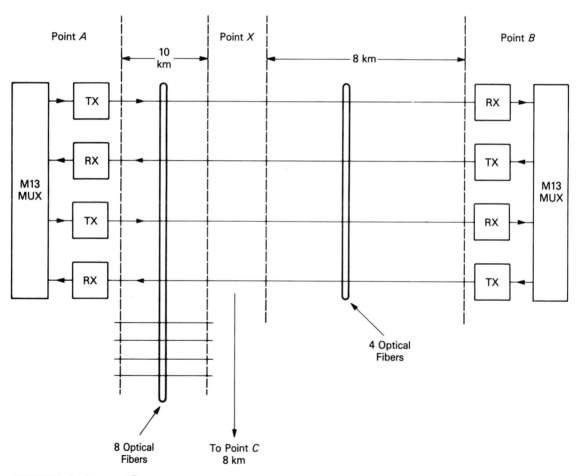

FIGURE 16–5 Proposed Design

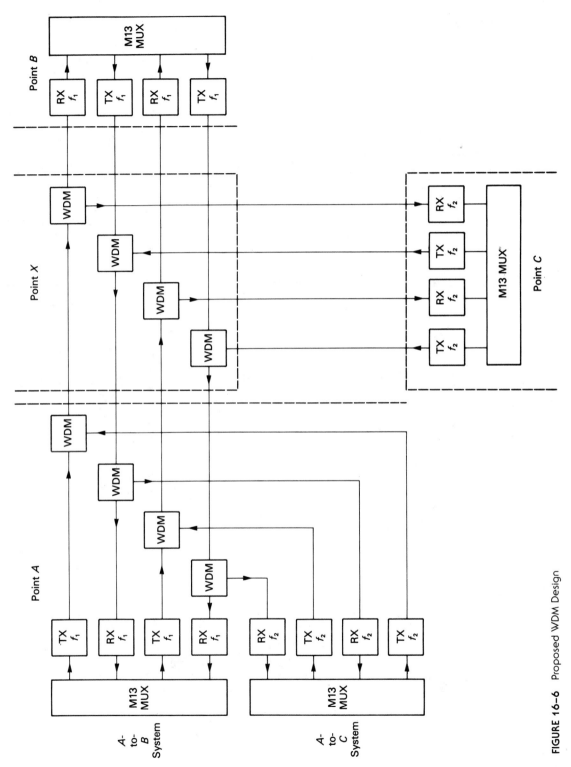

FIGURE 16–6 Proposed WDM Design

271

B and *A* to *C*), since the WDM-derived subsystem can also operate at that same transmission data rate. The revised system design is shown in Figure 16–6.

SUMMARY

The system design can substantially affect the cost of a particular installation. The depth of understanding that the designer brings to each problem will determine how cost effective installations in distribution plant will be, regardless of which basic network structure or design philosophy prevails.

The technology is sufficiently mature that any competent engineer can easily design and construct systems that perform well. The measure of good engineering in distribution applications will therefore depend not entirely on whether the system functions adequately, but on whether the installation is cost effective as well.

New developments in switching technology and new service concepts encourage new network architectures. Circuit switching is now distributed much more widely throughout the local network. Cost-effective transmission into and out of a multiplicity of such switches is both a natural and essential element of this trend.

The line of demarcation between trunk and exchange services is no longer so clearly defined, and we can anticipate more multipurpose utilization, or mingling of services, within a single transmission facility or system.

It is still unclear whether optical fibers will evolve in distribution and exchange plant as individual dedicated transmission facilities per point of service or not. There is a persuasive argument to utilize them as broadband, multiple-service arterial feeds, with some lower level of loop plant supplying the final subscriber connection link.

It may seem incongruous, but the design of cost-effective lightwave facilities for exchange plant service presents a more difficult design challenge than the design of a trunking application. In either case, previous experience with paired cable plant design could be a handicap for the designer if it encourages only a straight substitution of fibers for pairs.

REVIEW QUESTIONS

True or False?

T   F   1. Designing lightwave systems for telephone trunking is generally more complicated than designing lightwave systems for exchange or distribution service.

T   F   2. The cost of lightwave transmission facilities is usually less significant in exchange plant applications than in trunking applications.

T   F   3. Telephone carrier systems are not frequently employed in exchange plant using metallic conductor cables.

T   F   4. More "exchange circuit switching" is now being provided in smaller switching increments and at more switching locations than was generally the case in earlier years.

T　　F　　5. Lightwave transmission links are unsuitable for trunking to small, remote circuit switches within the exchange area.

T　　F　　6. There is no consensus as yet on just how optical fibers might best be applied in telephone subscriber loop plant.

T　　F　　7. The technical and economic characteristics of lightwave technology may influence the architecture of future exchange networks within the public telephone system.

T　　F　　8. It is not possible to utilize a single lightwave link for both telephone trunking and telephone exchange services.

T　　F　　9. It is possible to serve intermediate points along a lightwave system cable route without adding optical fibers.

T　　F　　10. Several completely independent transmission links can operate simultaneously over the same optical fibers at either the same or at different transmission data rates.

T　　F　　11. Although higher level digital multiplexing can increase the transmission capacity of a lightwave link, it is usually of little help in establishing multiple service points along a lightwave cable route.

T　　F　　12. By simply dedicating additional optical fibers for every point of service along a lightwave cable route, the designer will automatically assure an efficient application of lightwave technology.

T　　F　　13. Optical fibers are less reliable than metallic conductors.

T　　F　　14. Optical fibers present more transmission bandwidth for use than metallic conductors do.

T　　F　　15. In exchange plant applications, the use of WDM can significantly improve the cost effectiveness of lightwave transmission facilities.

T　　F　　16. The most cost effective applications of lightwave technology in exchange plant will be assured if we simply directly transfer the techniques of applying the technology in high-traffic-density trunking systems.

T　　F　　17. Lightwave transmission facilities lend themselves equally well to trunking or exchange applications within the local telecommunications network.

T　　F　　18. Bi-directional transmission on optical fibers presents a broader range of options for applying lightwave transmission in exchange plant.

T　　F　　19. Wideband transmission facilities, such as optical fibers, will have only limited impact on the way we presently develop and configure telecommunications networks.

T　　F　　20. Lightwave transmission facilities are more susceptible to lightning and electromagnetic interference in exchange plant applications than twisted-pair, metallic conductor cables are.

CHAPTER 17

Lightwave System
Testing and Maintenance

In any transmission system there will be a variety of tests that can be conducted. Some of these may be quite sophisticated and require expensive and specialized equipment. Engineering a transmission system is undertaken to produce the transmission performance and capacity that are required for use. Upon completion of the project, some level of testing is obligatory to determine whether the system actually presents the quality and capacity it was designed for. The tests involved may be quite extensive and, as proof-of-performance or acceptance tests, may even be a part of contractual obligations.

A different level of testing might be developed for long-term system operation and maintenance. The tests involved here may be periodically conducted as part of a routine maintenance program or to effect restoration of service after a system failure or outage. This level of testing may be quite defensibly less sophisticated.

Testing methods and test equipment can logically be divided into two categories: those associated with active terminal or repeater equipment and those more directly related to the interconnecting physical facilities. In the case of lightwave systems, the latter would be the optical fiber cable itself.

Although the higher level, more sophisticated proof-of-performance type of tests may be desirable initially, or even periodically in some cases, there is little need or justification for this sophistication in all cases. Some transmission characteristics are much less apt to change in the normal operation of the facility.

In a lightwave system, for example, we may need to positively establish the usable bandwidth or rise time of a new facility when it is initially put in service. There is little probability, however, that these characteristics will change significantly over an extended period of system operation. We might anticipate some changes in transmission level, from equipment aging perhaps, or temperature variations in the fiber's environment, but these would not usually affect bandwidth or rise time. It is difficult to envision any change in the system that would affect bandwidth or rise time and not be evidenced as a change in transmission level also.

Consequently, the condition of any particular link or system might be adequately determined by measurements of transmission level alone if there were advantages or economies to be gained by doing so. This is especially so if the link or system had been tested initially to verify that its bandwidth or rise time was satisfactory. Fortunately, measurements of transmission levels (i.e., optical signal amplitude) are not particularly difficult or expensive to perform. There are persuasive arguments for limiting system tests to measurements of transmission levels insofar as this is possible.

In many cases it is possible to verify system bandwidth or rise time performance without the use of specialized or sophisticated test equipment. Since, in digital lightwave systems, anomalous bandwidth or rise time characteristics would be evidenced by pulse distortion (broadening) and higher BER in transported signals, we might consider simply applying digital signals to the system. We could even apply the traffic that the system was originally designed to carry, and then measure the BER in the recovered digital signals themselves. This technique would offer the advantage of using test equipment and techniques that we are already quite familiar with and have at our disposal.

As we develop more experience with lightwave technology and apply it in lower density applications closer to the subscriber loop plant, the cost of more exotic test techniques, both in equipment required and operator sophistication, becomes less and less affordable. At the same time, the level of confidence we develop in lightwave technology grows stronger with each installation. Accordingly, we should not be intimidated by lightwave test requirements, and there is little merit in sophistication merely for its own sake. We are obliged to establish beyond doubt that the facility constructed performs as it was designed to do, but there is no necessity for testing beyond this level of obligation.

COMPARING MICROWAVE AND LIGHTWAVE SYSTEMS _____

Lightwave transmission introduces a new, much higher frequency spectrum than some of us may be familiar with. A useful analogy can be drawn with microwave technology, which also thrust us into a range of frequencies that were alien to previous experience. If we examine the evolutionary process that microwave followed, we may see similarities in the present development of lightwave. We may even be able to usefully apply some of our earlier experience with microwave.

Initially, the emphasis on microwave testing was heavily oriented toward sophistication. For example, in many early installations, it was general practice to make return loss and sweep frequency measurements of all waveguide and antenna systems. Many installations even utilized tunable waveguide connectors that could be individually adjusted to optimize the system.

This technique and level of sophistication are perhaps still employed in very high-traffic-density systems, but in lighter traffic installations, particularly more robust digital systems, the practice is rarely this complicated. The simpler methods now in general use were not adopted arbitrarily; they were the natural result of operating experience that clearly showed the more sophisticated techniques to be technically unnecessary and unjustifiably expensive.

It is also uncommon in working microwave installations to make many measurements of signal level directly at RF frequencies. Rather, most transmitters provide some RF output level measurement function either on a front panel meter or as a pin jack measurement of a correlated DC voltage. In microwave receivers where a wide range of automatic gain control (AGC) is required due to anticipated variations in atmospheric attenuation, RF input levels are generally referenced to a simple DC voltage measurement of the AGC bus.

A rather common test technique for microwave installations is to insert a variable attenuator into the waveguide system and establish, by introducing an artificial RF fade, the achieved fade margin of the system either to a level of receiver squelch in analog systems or to a measurable level of BER in digital applications. If provision is made for RF level measurements in a microwave installation, an arrangement for permanently coupling some RF energy out of the waveguide assembly is incorporated into the terminal equipment. It would be most unusual to mechanically disconnect the waveguide in an operating system simply to measure RF signal levels.

This level of logical practicality is not nearly as evident in lightwave operations as yet. The requirements and physical considerations of a lightwave system are not identical, of course, but some features of our microwave experience could contribute to developing less expensive lightwave techniques that might be perfectly acceptable technically.

OPTICAL CABLE AND SPLICE TESTING

Although more sophisticated tests are possible, the practical limitations of field testing generally restrict optical cable tests to measuring attenuation. A known level of lightwave energy is coupled into the fiber being tested and the level of energy coupled out of the fiber at the distant end is measured. The difference between these two levels is the attenuation introduced by the fiber.

The attenuation of a fiber may be significantly different for carriers of different wavelength, of course, so fiber attenuation tests should be made at or near the wavelength at which the system will actually be operating. The transmitting terminal associated with the system is a very convenient source of signals for fiber attenuation testing, but this convenience is not always available. For example, it is often necessary to measure fiber loss through individual lengths of cable before the cable has been placed in the field or before several cable lengths have been spliced together. In such instances, a portable source of lightwave signal is essential.

Several manufacturers offer highly portable, battery-powered testing sets that combine a lightwave signal source with a lightwave energy-measuring capability. These devices are not prohibitively expensive and are very easy to use. Relative calibration is simply a matter of directly coupling the transmitter output into the detector/receiver input. Then, if both ends of a cable are readily available, the attenuation of any fiber within that cable is easily measured. This technique is particularly useful in verifying the loss of cables on the reel in the warehouse before the cables have been placed in the field.

After placement, both ends of the cable are no longer available at a single location. Accordingly, measurement of fiber attenuation now requires two discrete units, a signal source at one end of the cable and a detector/meter at the other. Two of the combined-function units discussed earlier can be employed in such tests.

Regardless of the testing units employed, an interconnection with the optical fiber being tested is required. Unfortunately, the mechanical uncertainties of temporary optical fiber connections, the requirements for cleanliness, and similar considerations all introduce some variable element into the testing procedure. These shortcomings are unavoidable, but if reasonable care is taken in making the connections, the margin of error in the test results should be acceptably small. If gross variations in measurement are experienced, it is prudent to examine the temporary optical connections carefully and perhaps even remake them entirely.

TESTING PRECISION

From the preceding considerations, we might conclude that absolute precision in field measurements of fiber attenuation and signal levels may not be possible, and to some degree this is true. It is equally true, however, that absolute precision is not essential.

In engineering any transmission system, one must base all calculations on the specified performance characteristics of the elements composing the system. For example, a paired cable, a coaxial cable, or an optical fiber will be specified by the manufacturer to introduce a stated amount of transmission loss for some unit of length. The manufacturer presumably will not ship any of the product that introduces greater transmission loss than this specification. The manufacturer may, however, ship a product that introduces less loss than that.

The stability, calibration, or precision of test equipment under field measurement conditions might be questionable, of course, and the error inherent in cable route distances as taken off maps or staking sheets is unavoidable. Thus, one cannot expect, nor is it essential, that constructed systems evidence transmission levels or attenuation values that agree precisely with the design predictions. In this respect, lightwave systems do not introduce any unique problems or require any special considerations that we would not encounter in testing other transmission systems.

TESTING OPTICAL CABLE

A generally accepted practice in constructing a lightwave system is to measure and record the attenuation of each individual fiber, in each discrete length of cable, on the reel as it is received from the manufacturer. Not only does this allow for verification of the measured attenuation data provided by the manufacturer, but it correlates the test equipment and technique used locally with the equipment and technique that the manufacturer employed in the original measurements made on the same cable before shipment. In a sense, then, one can calibrate one's own equipment and method against the manufacturer's data. Such calibration also provides some practical experience in the use of the local testing equipment itself.

After placement of the cable, but before splicing, many people will remeasure the attenuation, again comparing the results with all previous attenuation data. This is a prudent practice, particularly if the project is the individual's first experience with optical cable. There are costs involved in such a procedeure, however, and it might be better to combine this step with other testing that is associated with fiber splicing operations. Cable damage which may have been introduced during placement will become just as apparent during these other tests.

It is important that testing equipment used in splicing operations be calibrated frequently. Since we are really interested in verifying attenuation rather than precisely measuring lightwave energy levels, we need not concern ourselves with calibration in its generally understood sense of being a reference to some standard device or level. Rather, we can consider calibration adequate if the same meter indicates the same level from the same signal source all or nearly all of the time, or if two or more meters all indicate the same level all or nearly all of the time. We can satisfy this requirement by simply checking all meter indications against the same signal source periodically.

OPTICAL TERMINAL TESTS

Optical transmitters can be tested for light energy output level by simply connecting the output into the power meter. Many transmitters on the market today also include a capability for self-monitoring the output level. This often includes an alarm feature that indicates reduced light energy output and relates it to laser aging. In such cases, there may be an alarm sensitivity adjustment whereby the alarm will be activated when the light energy output is reduced to some discrete, preestablished level.

The alarm feature may be verified or the alarm level adjusted by measuring the output level with the power meter and gradually reducing the light source output by internal transmitter adjustments. When the output level is reduced to the preestablished alarm level, the alarm should activate or should be readjusted to activate. In units provided with this feature, the manufacturer's recommended adjustment procedure should be reviewed and followed carefully.

Optical receiver tests are more difficult to perform. Since we must verify not only the optical signal input level but the quality (BER) of the electrical output signal at that optical signal input level, we need an (optical) input signal that is actually modulated with information. This requires a more sophisticated light energy test signal source than that previously discussed, or else we could use the system transmitting terminal itself.

By employing a variable optical attenuator inserted into the interconnecting optical fibers, we can set the receiver optical input level to whatever amplitude we wish. In this manner, if the transmitter and its associated receiver at any single terminal are operating on the same lightwave frequency, we can loop the local transmitter output back through the variable optical attenuator to the local receiver input. Then, by modulating the transmitter with digital information, we can manipulate the receiver optical input level and measure the BER in the receiver electrical output signal.

By means of the test setup shown in Figure 17–1, we can verify the specified operating threshold of the receiver by reducing the receiver optical input until the receiver electrical output signal deteriorates to the BER specified by the manufac-

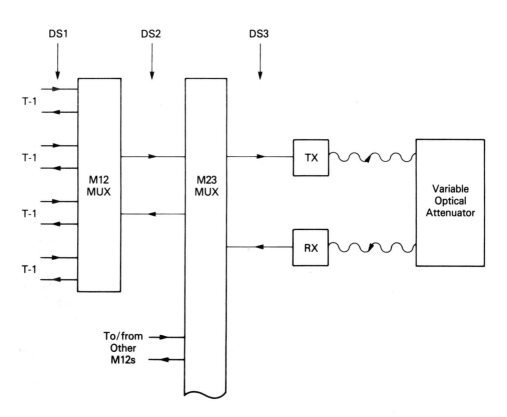

FIGURE 17–1
Loop-Back Test

turer as the threshold. Also, perhaps we can remove sufficient attenuation for the receiver BER to become unacceptable. In this manner, we can exercise and verify the dynamic range of the receiver.

The digital signal tests (on receiver electrical output) can be made at any convenient digital level available, that is, at DS-1, DS-2, DS-3, etc. If lower data rates are used to establish the BER performance, the test period will have to be longer, of course.

The use of an optical attenuator in the manner described is convenient and effective. However, the test setup could be technically improved somewhat by inserting an optical splitter or coupler on the output of the variable optical attenuator, as shown in Figure 17–2.

This would permit coupling a power meter onto the attenuator output and would give a constant measurement of the optical input level to the receiver, rather than relying entirely on the attenuator calibration. Of course, the insertion loss of this splitter or coupler must be known and included in any signal-level measurements. The configuration may be useful whenever an optical attenuator is employed in any test setup.

Essentially the same technique shown in Figure 17–2 can be employed in end-

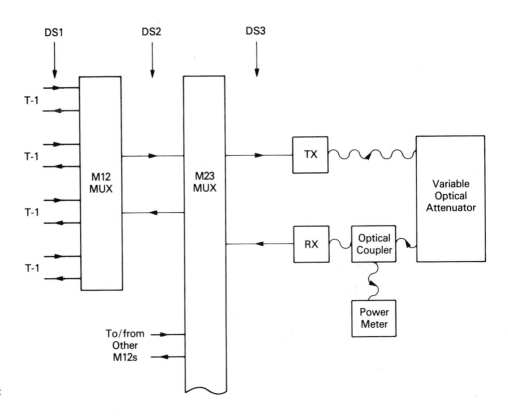

FIGURE 17–2
Loop-Back Test

to-end system testing, as shown in Figure 17–3. The use of an optical splitter or coupler to permit monitoring of lightwave signal levels is again optional.

LIGHTWAVE SYSTEM TESTS

System performance tests with a setup such as that of Figure 17–3 would consist of applying digital signal input to the system and making BER measurements at a convenient digital signal level at the distant terminal. The optical signal input to the receiver would be progressively attenuated (using the variable optical attenuator) until the BER degraded to the system minimum performance specification. Typically this might be a BER of 10^{-9}. When this level of degradation in transmission quality is reached, the receiver optical input signal level should be recorded.

Removing any attenuation (and the splitter/coupler if used) and measuring the optical input directly available on the interconnecting fiber demonstrates the normal operating input signal of the system. This can be directly related to the previously measured receiver threshold, and a judgment can be made as to whether or not the system margin satisfies the specified margins or whether the operating input level is optimally positioned within the dynamic range of receiver input levels.

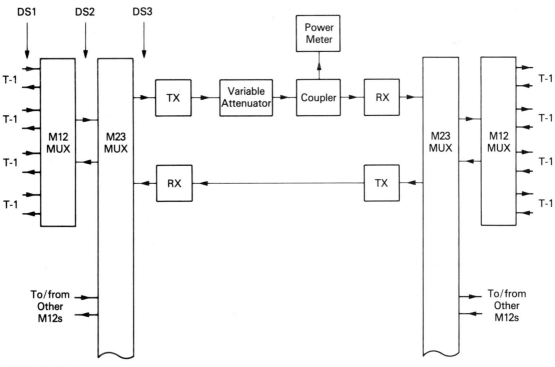

FIGURE 17–3 System Tests

If, during the performance of these measurements, the BER in the digital signals transported is found to be unacceptable at receiver optical input levels where it was predicted to be acceptable, we should suspect that the system bandwidth is not adequate, and more sophisticated investigating techniques would be necessary. In a well designed system using quality fibers and terminal equipment, this condition should be rarely encountered.

If, during the performance of the measurements, the BER in the digital signals transported is found to be acceptable at normal receiver optical input signal levels where it was predicted to be acceptable, we may safely assume that the system bandwidth is adequate. In such instances, the end-to-end system performance should be considered fully in conformity with system performance specifications.

It is beyond the scope of this text to develop in depth any unusual test procedures that might be required if the system were found not to be in conformity with performance specifications. Such tests might include remeasuring end-to-end fiber loss or bandwidth, or optically looping back signals at either or both system terminals. Suffice it to say that an unusual trouble condition would have been identified, and more extensive tests may be necessary to determine and correct the root cause.

Any number of other tests may be necessary, such as protection-switching operation or alarm-reporting functions. These are unique to the individual installation, however, and will be dictated largely by the manufacturer's manuals and specifications. It is beyond the scope of this text to cover all possible variations.

Although more sophisticated test techniques are possible, there is little real need for such tests as standard maintenance practices. It has been the author's experience that most operating systems actually rely on system fault reporting and alarm systems rather than on any scheduled periodic test procedures.

SYSTEM FAULT LOCATION

Optical fibers are dielectric (nonmetallic) in nature, and consequently some optical cables may contain no metallic elements at all. Under these conditions, test techniques required to locate a fault or irregularity cannot depend upon metallic continuity or resistance measurements.

A commonly employed technique uses *time domain reflectometry* (TDR). A pulse of lightwave energy is launched into the optical fiber being tested, and the level of lightwave energy reflected back out of the fiber is measured. The display may be a digital readout (often calibrated in distance) or a trace line on an oscilloscope. In the latter case irregularities such as splice losses or reflections will show up as changes in reflected energy level, and the distance to such irregularities can be determined by the position of the indication along the time baseline of the instrument.

This technique has found general acceptance in coaxial cable fault location, but can be adapted to other transmission media such as waveguides or twisted pair-cable, as well. In lightwave systems, the test unit is referred to as an *optical time domain reflectometer* (OTDR). Since the reflected energy is at a relatively low level and is attenuated by the fiber as it propagates back toward the test unit, the OTDR does have some limitations in locating faults in very long cable routes. These can be offset, however, by isolating shorter sections of fiber for test purposes.

Although OTDRs are sometimes employed to measure fiber splice losses because they are convenient to use, the results can be confusing. It is possible, for example, for an OTDR display to indicate splice gain in an actual case of splice loss. Hence, the accuracy of splice losses measured with an OTDR should be considered relative rather than precise. It can tell you which of several splices has the lowest loss, and it will indicate a splice with an excessive amount of loss, but the display does require some interpretation. The devices were designed to locate discontinuities in a fiber, and they do this with adequate precision in most cases, but they should not be considered an ideal unit for measuring splice losses.

Although optical fibers themselves are dielectric, many optical cables include a protective metallic shield or armor. In some cases this metallic shield can be useful in locating cable faults in the same manner that faults are located in conventional paired-cable plant.

The importance of documenting test procedures and recording and preserving all measured data cannot be overemphasized. When system irregularities are experienced, either initially or subsequently due to mechanical plant damage, the use of such information as fiber loss through sections of the system and precise splice locations can be extremely helpful. All data should be carefully filed for future reference.

Similarly, when new optical cables are placed, all data measured, such as individual fiber losses between points along the cable route, should be recorded and carefully preserved. Such information may be very useful in detecting or locating subsequent cable damage.

ROUTINE MAINTENANCE PRACTICES

To date, industry experience with lightwave systems has been very satisfactory. This history strongly suggests that very little routine maintenance will be required, and inquiries directed at the operators of current installations indicate that even after several years of operation, few of these people have found it necessary or useful to institute any extensive maintenance procedures.

Most of the equipment available on the market today includes features that provide alarms when performance degrades for any reason. For example, most lightwave transmitters sound an alarm for a reduction in, or loss of, light energy output level. Alarms of this type certainly can be considered maintenance aids, since corrective action may often be taken before an actual interruption of service is experienced.

In telecommunication systems, other terminal equipment, such as digital multiplexers, will certainly be equipped with status-monitoring and alarm-reporting functions.

A multiplicity of alarm levels and functions can be provided by the equipment manufacturer, and many of these are unique to a specific manufacturer's product. The reader should become thoroughly familiar with all features of the particular equipment he or she services or owns. The levels of alarm reporting available have a great deal to do with the development of a particular maintenance program.

More conventional test procedures, such as BER tests on digital signals or routine noise measurements on derived voice frequency circuits, can logically be considered as maintenance practices for lightwave transmission links just as they are for paired-cable or microwave installations. Such tests would obviously provide some indication of system degradation.

Any maintenance program should avoid, insofar as possible, the unnecessary disruption of optical fiber connections, due to the uncertainties of satisfactorily restoring such connections. It might be more practical to simply depend upon the system alarm-reporting functions. However, periodic testing of the alarm reporting system itself would then be desirable and necessary.

A PRACTICAL MAINTENANCE PROCEDURE

Since most, if not all, lightwave receivers include AGC functions, an AGC bus is available within the receiver circuitry. This bus develops a DC control voltage that varies in direct proportion to the receiver light energy input signal level. If the AGC bus voltage is calibrated against light energy input level, either at the factory or in the field during installation, then a simple DC measurement of the bus voltage can be substituted for any actual light energy measurements on the input optical fiber itself. Such DC measurements could be made and logged periodically with no disruption of optical fiber connections and without taking any derived circuits out of service. Indeed, measurements of this type, requiring only a digital DC voltmeter, could even be effectively employed to measure the transmission loss of optical fiber splices made under emergency conditions to repair optical cables which suffer mechanical damage at some time subsequent to placement.

In equipment that provides front panel access to the AGC voltage, a log sheet of the AGC bus voltage could be maintained easily and inexpensively. Such readings taken periodically would present a useful maintenance record of system performance.

The author was involved in a program that field tested the practicality of this method. Although the particular equipment involved required that a connection be made on the internal circuit board itself, to gain access to the AGC bus, the measured DC voltage tracked linearly with reductions in optical input signal until the receiver threshold was reached and the system "crashed."

It is suggested that equipment manufacturers include in their receivers a front-panel, readily accessible test jack for DC measurement of the AGC control voltage. This feature would be extremely useful and should add no significant cost to the product itself.

SUMMARY

Most of the lightwave systems that the author has been exposed to either eliminate completely any planned or periodic maintenance or go to the other extreme and frequently perform a number of sophisticated and largely unnecessary measurements. Of the two approaches, the former would be least expensive and might introduce fewer problems. Disrupting optical connections that are functioning well simply to adhere to an arbitrary test schedule makes little sense and is asking for trouble.

The availability of simple DC measurements as a monitoring technique for system performance has a great deal of merit. Given this facility, a system operator could initiate periodic measurement and logging procedures that would produce a real understanding of the condition of a system across a long period of time. Gradual deterioration would be readily apparent, and the technique might even be used to verify the quality of a restoration splice after a serious cable or fiber disruption.

The in-service traffic carried on a system can constitute a continual transmission-monitoring function. Trunks or circuits that go out of service sound an alarm and are reported in the normal course of events, and such alarms are indicative of system irregularities. On the other hand, there are system features that certainly merit periodic checks, and in the author's experience, such checks are all too frequently omitted entirely.

Consider automatic service protection switching, for example. To ensure service continuity, the designer calls for redundant equipment and facilities, and they are provided at a significant cost penalty. Such facilities are required to sense irregular system operation and very quickly transfer service to a stand-by facility. This transfer should occur without dropping circuit connections or initiating alarms in other, unrelated sections of the network.

Recently, the author directed inquiries at a half-dozen system operators whose lightwave plant had been in service for at least two years or more. The following two questions were asked:

1. Since placing your system in service, how many times has protection switching been invoked? The answer from all six operators was *never,* which is most encouraging.

2. Do you have a program in place to exercise the protection-switching feature periodically, in order to determine if it is functional? The answer from all six operators was *NO!*

One might conclude from this admittedly limited survey that either a lot of money was being spent on protection switching unnecessarily, or the maintenance programs which were in place were poorly designed. The reader may draw his or her own conclusions on this point.

REVIEW QUESTIONS

True or False?

T   F   1. It is practical to simply adopt the initial proof-of-performance tests as standard, periodically performed system monitoring testing procedures.

T   F   2. Transmission characteristics of a lightwave transmission link will change drastically over an extended period of system operation.

T   F   3. Transmission level measurements alone provide a useful and practical indication of the condition of a system or link.

T   F   4. The measurement of lightwave system transmission levels is a very sophisticated test operation.

T   F   5. There are sound and persuasive arguments for limiting system tests to transmission levels alone after initial proof of performance tests have been conducted and recorded.

T   F   6. The traffic carried on a digital lightwave system cannot be employed as a test signal for measuring the quality of transmission through the system.

T   F   7. The most sophisticated test procedures possible are not automatically essential to the proper long term operation of lightwave facilities.

T   F   8. When microwave transmission was first being introduced into the telecommunications networks, some test procedures tended to be unnecessarily sophisticated.

T   F   9. The automatic gain control (AGC) voltage in a microwave receiver is used to indicate receiver RF input signal level.

T   F   10. It is possible to measure the relative level of an optical input signal to a receiver simply by measuring the DC voltage on the receiver AGC bus.

T   F   11. Disconnecting optical connectors to make routine lightwave signal measurements as part of a maintenance program is a sound practice.

T   F   12. Absolute precision in measuring lightwave energy levels is essential in testing optical fiber splicing losses.

T   F   13. When using OTDRs to measure fiber splice losses, some possibility for error is inherent in the use of the device itself.

T   F   14. Temporary connections between optical fibers and testing equipment can introduce some uncertainty in the precision of the test measurements.

T   F   15. Lightwave terminals can be tested by looping the transmitter output back into the receiver input, using some optical attenuation.

T   F   16. The detection threshold of a lightwave receiver can be demonstrated by using an optical attenuator and an unmodulated lightwave test signal.

T   F   17. Fiber bandwidth or rise time tests are difficult to perform in the field and require test equipment that is not widely available today.

T   F   18. When a system has been equipped for service protection, there is little point in periodically testing the protection switching operation.

T   F   19. If the transmission levels are found to be correct but the quality of transmission through a link is still unsatisfactory, a reasonable test would be to loop back each system terminal and measure the transmission performance through the terminal.

T   F   20. Most companies operating lightwave systems have instituted extensive routine testing and maintenance programs.

CHAPTER 18

System Design Case Histories

In this chapter we shall examine a number of actual system designs that the author had an opportunity to review over the past year or so. We shall examine the proposed design and present alternative designs, and compare them both technically and economically. The intent is to expose the reader to practical applications of the material presented throughout the text.

We shall use the cost data presented earlier in Chapters 6, 7, and 12, but we must qualify the data as representative only and subject to change. In the economic analysis presented, we shall not include any cable placement costs or costs for system equipment installation or testing. These costs would be difficult to estimate with any acceptable precision and, in any event, would be roughly the same for any of the designs discussed.

Even with these qualifications, the analysis should be a useful instructional exercise and will reinforce the reader's understanding of lightwave technology.

CASE HISTORY A

Application Requirements

The requirement was to connect a "host" telephone switch to three "remote" telephone switches. Although the circuit capacity required between all locations was not given, the service is obviously to provide trunking to each remote switch. We shall assume the circuit requirements to be four T-1 carrier systems for each remote switch location, which provides a total capacity of 96 equivalent voice circuits for each service location.

Proposed Design

Figure 18–1 shows the proposed system design as it was presented for review. Note that the total fiber length required to serve the most distant location, remote switch 3, is only 33 km. It was proposed to use single mode fiber and an operating wavelength of 1,300 nm, with a 45 Mb/s transmission data rate applied throughout, except in the link connecting remote switches 2 and 3, where a 6 Mb/s data rate was proposed.

As shown in the figure, the proposed design employed back-to-back terminals at the intermediate service locations (remotes 1 and 2) with electrical interconnection for all "through" services at both of these locations. All links provided four optical fibers in a fully protection-switched configuration to ensure continuity of service.

Transmission Considerations

At 1,300 nm, single mode fibers that introduce only 0.5 dB/km of transmission loss are readily available. Such a fiber would introduce a total transmission loss of only 16.5 dB in the longest fiber length, 33 km (from the host switch to remote 3). It is possible, using WDM, to consider direct optical transmission from the host switch to all three service locations.

Alternative Design 1

Figure 18–2 shows an alternative system configuration. Each location served is provided with a completely independent transmission link, except for the common utilization of the optical fibers using WDM.

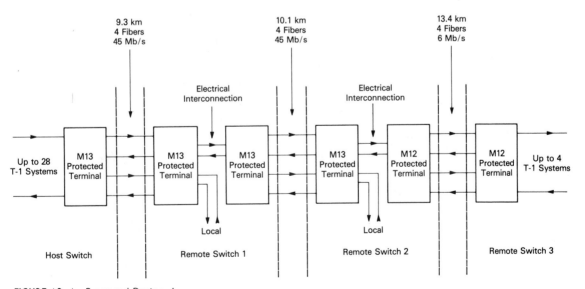

FIGURE 18–1 Proposed Design *A*

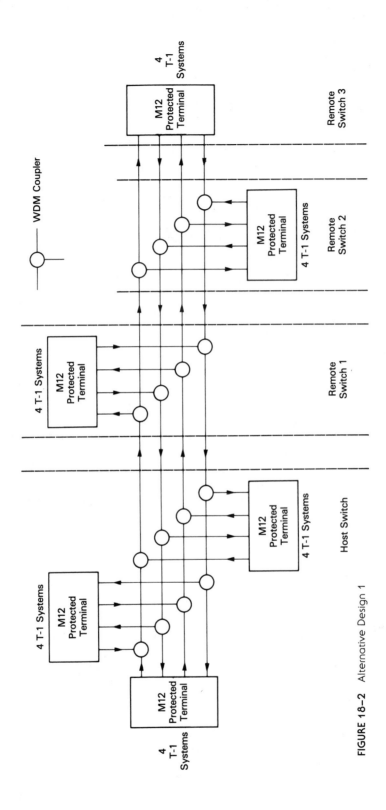

FIGURE 18–2 Alternative Design 1

In the design of Figure 18-1, all traffic to and from remote switches 2 and 3 was vulnerable to any failure of the terminal equipment at both the host switch and the remote 1 location. Remote switch 3 was further exposed to any equipment failure at the remote 2 location.

In Figure 18-2, the same cable route and the same number of fibers are employed, but with separate wavelengths to serve all three locations, each is provided with an independent transmission system. WDM couplers are present, of course, but they are completely passive in nature and should present a very high level of reliability.

In conformity with the original requirements, Figure 18-2 provides a fully protected link for all three service locations. Thus, it would be possible to use a 6 Mb/s data rate for all services in Alternative 1, employing less sophisticated (and less expensive) M12 digital terminal equipments throughout as shown in the figure.

Alternative 1 is technically sound and is completely responsive to the service requirements, providing four fully protected T-1 systems for each service location. The reader is encouraged to calculate the transmission losses introduced by the WDM couplers and ascertain whether this alternative design is technically acceptable. The necessary data were all presented earlier in the text.

Alternative Design 2

Another approach may merit consideration as well. All of the previously discussed terminal configurations (both M12 and M13 terminals) were fully protected. Each such unit consisted of two optical transmitters, two optical receivers, and two digital multiplexers. Protection was a function of sensing a failure of any particular system element, be it an optical fiber, transmitter, receiver, or multiplexer, and then switching to a backup system.

Figure 18-3, on the other hand, shows a configuration using WDM and unprotected terminal equipment. Each link represents a completely independent transmission system into and out of each location serviced. If the traffic (derived voice circuits) is equally divided between these two independent systems, a very respectable level of service protection can be provided without the sophistication and complexity of sensing and switching at all.

At first glance, it may appear that the configuration of Figure 18-3 requires more terminal equipment, but this is not actually so. As pointed out earlier, a protected terminal (M12 or M13) includes the same quantity and type of equipment that two unprotected terminals require, with additional equipment and cost for sensing and switching. Note also that the parallel unprotected configuration actually presents a transmission capacity of eight T-1 systems, although there is no necessity for equipping all eight initially. This is significantly more capacity than provided in Alternative 1, and the additional capacity can translate to a higher grade of service (all trunks are busy less frequently) if more circuits are actually equipped. The additional capacity might also eliminate the necessity, real or perceived, of utilizing a higher transmission data rate to ensure system expandability at some later date.

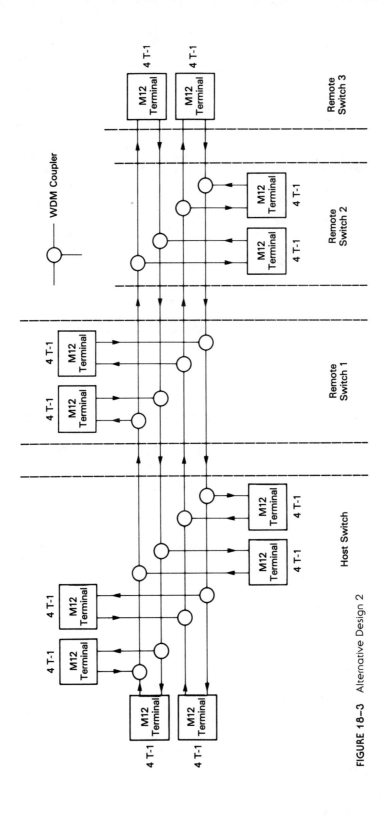

FIGURE 18–3 Alternative Design 2

Alternative 2 is, then, technically sound and exceeds the service requirements, providing a transmission capacity of eight T–1 systems for each service location.

Economic Considerations

The transmission data rates presented by all of the designs just discussed are quite nominal and will present no cost penalties or advantages in fiber and cable selection. Since all the designs employed the same number of fibers, there is no need to develop any cable costs for comparison purposes; economic comparisons rest entirely with terminal equipment costs alone.

The proposed design (Figure 18–1) required a total of four M13 terminals operating at a 45 Mb/s data rate and two M12 terminals operating at a 6 Mb/s rate. This equipment costs out as follows:

Item	Unit Cost	Quantity Required	Extended Cost
M13 Terminal	$18,000	4	$72,000
M12 Terminal	9,000	2	18,000
			Total $90,000

Alternative 1 (Figure 18–2) required six M12 terminals and 16 WDM couplers. This equipment costs out as follows:

Item	Unit Cost	Quantity Required	Extended Cost
M12 Terminal	$9,000	6	$54,000
WDM Coupler	300	16	4,800
			Total $58,800

Alternative 2 (Figure 18–3) required twelve unprotected M12 terminals and sixteen WDM couplers, and costs out as follows:

Item	Unit Cost	Quantity Required	Extended Cost
M12 Terminal (Unprot)	$4,000	12	$48,000
WDM Coupler	300	16	4,800
			Total $52,800

The lower cost of the M12 terminals in the last set of figures reflects the fact that they are unprotected.

General Discussion

Although cost may not be the primary consideration in system design, the designer is certainly obliged to produce as cost effective a system as is possible consistent with an acceptable level of reliability. The inclusion of sensing and switching sophistication in a facility that must be maintained for an extended period of service life may represent a subtle, but real, disadvantage. The point might also be made for any transmission facilities that unjustifiably introduce higher transmission data rates.

We leave the final analysis of the preceding cost data to the reader, but two points perhaps deserve special mention:

1. In Alternative 2, the initially provided transmission capacity is twice that provided by Alternative 1.

2. In both Alternatives 1 and 2, the reliability of service for the three locations might justifiably be considered to be higher than in the proposed design.

CASE HISTORY B

Application Requirements

The network geography of the next application is shown in Figure 18-4. Note the transmission data rates and fiber counts.

The initial traffic capacity required was a total of 50 T-1 systems, with all traffic terminating at the toll office from all service locations. The allocation of these 50 carrier systems is shown in Figure 18-5, along with the substantial traffic growth projection.

The proposed design used single mode fiber throughout but included two spare optical fibers in every cable link.

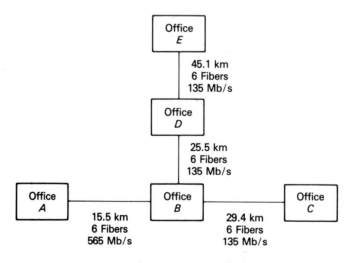

FIGURE 18-4
Proposed Design
B

Total: 100 km, 6 Fibers, 135 Mb/s
Total: 15.5 km, 6 Fibers, 565 Mb/s

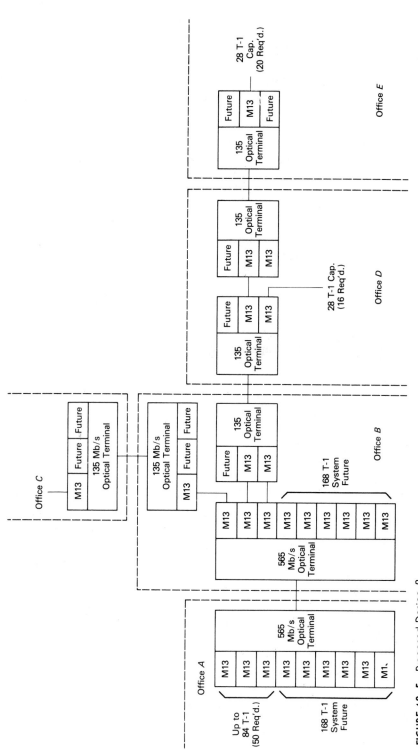

FIGURE 18–5 Proposed Design B

Special Considerations

The provision of two extra optical fibers in all optical cables was interesting. The system designer stated that this arrangement was intended to accommodate future traffic requirements in the network. This seemed rather implausible, however. The proposed design used a 565 Mb/s data rate for part of the system (capable of carrying 336 T-1 carrier systems, or 8,064 equivalent voice circuits) and 135 Mb/s in the balance of the network (capable of carrying 84 T-1 systems, or 2,016 equivalent voice circuits), making the probability of traffic growth above these already high capacities to be extremely low—in fact, hardly likely at all.

In any event, if extra fibers were desired, the provision of only two of them was inappropriate, for it was clear from a review of the proposed design that the designer followed a philosophy that four optical fibers are essential to any transmission facility. This is the conventional, fully protected facility that is so familiar, and the author takes no general exception to this basic approach. The proposed design would compromise the approach, however, in that only two extra fibers were made available, and these were presented as offering significant traffic growth potential. If the designer is committed to four fibers as an operating facility requirement, then two fibers cannot seriously be considered as constituting a second transmission facility. The designer might much more logically have provided four extra fibers.

If, on the other hand, the designer suggests that the two extra fibers are intended to improve the overall system reliability, we fail to see that two fibers, unequipped with terminal equipment or automatic switching, and installed under a common cable sheath with the other interconnecting fibers, represent any significant improvement in reliability at all.

At any rate, what cost penalty do the two extra fibers represent? An additional pair of fibers will cost approximately $.50 per meter of cable length. Since the project required a 115 km optical cable, the total cost of the extra fibers will be on the order of $57,000, not an insignificant sum.

These spare fibers do not serve any useful transmission purpose, and it is uncertain that they ever would. Furthermore, they fail to produce any offsetting additional revenues, and it seems quite doubtful that they ever will.

In including these spare fibers, the designer may have been influenced by previous experience with multiple-pair cables. With these cables, traffic carrying capacity can be related directly to pair count, even when telephone carrier systems are applied, since each carrier system requires a finite number of conductors. Perhaps the designer in this case simply felt more comfortable with extra "pairs," which he supplied in the form of the extra fibers. However, in lightwave transmission systems transmission bandwidth is the basic denominator of primary interest, and when adequate bandwidth has been provided, extraordinary justification should be required if additional fibers over and above this bandwidth capability are proposed.

Note in Figure 18-5 that the proposed design made no provision for service at the intermediate office *B* location. Inquiry as to why not revealed that this location was actually a local telephone switching exchange, and availability of service was provided in the alternative designs that were subsequently developed and that will be presented here.

Proposed Design

The following bill of materials and cost data are developed from the proposed design which is shown in Figures 18–4 and 18–5:

Item	Unit Cost	Quantity Required	Extended Cost
135 Mb/s Terminal	$23,000	6	$138,000
565 Mb/s Terminal	34,000	2	68,000
M13 Digital Mux	10,000	14	140,000
6-Fiber Cable	2.80/m	115 km	322,000
		Total	$668,000

As presented here, the system is implemented only to the levels of service shown in Figure 18–5. Note that no service is provided between offices *A* and *B*. Service between these locations could be implemented by simply adding an M13 digital multiplexer at both service locations. Since each such unit costs $10,000, the total to initiate this service would be an additional $20,000.

If we eliminate the unnecessary two extra optical fibers throughout the network, the total given above would be reduced from $668,000 to $611,000.

Alternative Design 1

Figures 18–6 and 18–7 show one alternative to the proposed design. In this alternative, we have eliminated the spare fibers and are operating the lightwave link be-

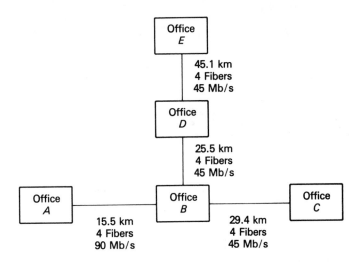

FIGURE 18–6
Alternative
Design 1

Total: 100 km, 4 Fibers, 45 Mb/s
Total: 15.5 km, 4 Fibers, 90 Mb/s

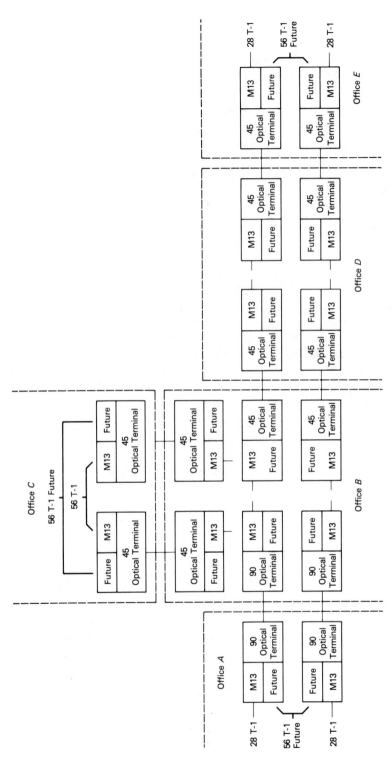

FIGURE 18-7 Alternative Design 1

tween offices *A* and *B* at a 90 Mb/s data rate. All other links are operating at 45 Mb/s, and all links are parallel unprotected systems rather than high-data-rate fully protection-switched installations.

The alternative divides all traffic on each link evenly across the two independent facilities provided, thus achieving a respectable degree of service protection without either the cost or the sophistication of switching systems.

From Figure 18–5, the original requirement was for only 50 T–1 systems. Alternative design 1 provides a system capacity of 56 T–1 systems, and they may be allocated for service in a variety of ways.

Although the initial transmission data rates are reduced, since there is probably very little, if any, cost advantage in fiber and cable due to these reduced data rates, system transmission engineering should be based upon a 90 or even a 135 Mb/s data rate throughout. This should allow an upgrade to higher rate transmissions at any future time.

The following bill of materials and cost data are developed from alternate design 1 which is shown in Figures 18–6 and 18–7:

Item	Unit Cost	Quantity Required	Extended Cost
45 Mb/s Terminal — (Unprot)	$11,000	12	$132,000
90 Mb/s Terminal (Unprot)	9,000	4	36,000
M13 Digital Mux	10,000	4	40,000
4-Fiber Cable	2.30/m	115 km	264,500
		Total	$472,500

Alternative 1 can be gracefully expanded in capacity by simply adding terminal equipment, and the system can even be converted to a high-data-rate protection-switched facility if the traffic growth ever demands it.

One possible weakness in the design is the fact that all traffic through the network must be demodulated and remodulated, with processing or electrical interconnection provided at the office *B* location. It can be argued that this drawback exposes all traffic to possible interruption, and this is equally true of the proposed design also. Another alternative might be directed specifically toward this point.

Alternative Design 2

Figures 18–8 and 18–9 show an approach that is similar to alternative 1, but instead of using a higher transmission data rate on the link between offices *A* and *B*, we have simply added four more optical fibers to the interconnecting cable in this link.

Now all traffic from office *A* to both offices *D* and *E* need not be processed at all at the office *B* location. This should improve the reliability of service for both office *D* and office *E*, and certainly eliminates entirely any initial requirement for a 90 Mb/s data rate anywhere within the network.

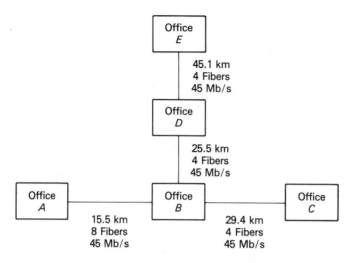

FIGURE 18-8
Alternative
Design 2

Total: 100 km, 4 Fibers, 45 Mb/s
Total: 15.5 km, 8 Fibers, 45 Mb/s

As in alternative 1, all terminals are unprotected, relying on traffic distribution across two independent, parallel links to provide service reliability. This system can also be expanded gracefully and may escalate to a fully protected configuration at some later date, but expansion may be delayed until traffic growth is actually experienced. Note that not only can we employ less sophisticated 45 Mb/s terminals throughout, but also, we do not require any digital multiplex equipment at any station beyond that which is inherently provided by the 45 Mb/s terminals themselves.

The following bill of materials and cost data are developed from alternate design 2, which is shown in Figures 18–8 and 18–9:

Item	Unit Cost	Quantity Required	Extended Cost
45 Mb/s Terminal (Unprot)	$11,000	16	$176,000
4-Fiber Cable	2.30/m	100 km	230,000
8-Fiber Cable	3.30/m	15 km	49,500
		Total	$455,500

Alternative Design 3

The increase in cable cost introduced by the eight fiber cable is not insignificant. It is possible, however, to eliminate this increase by employing WDM in the link connecting offices A and B. Figures 18–10 and 18–11 show an approach that is identical with alternative 2 except that only four fibers are provided in the link connecting offices A and B. The second independent transmission system is provided on this link by using two different optical wavelengths and WDM couplers.

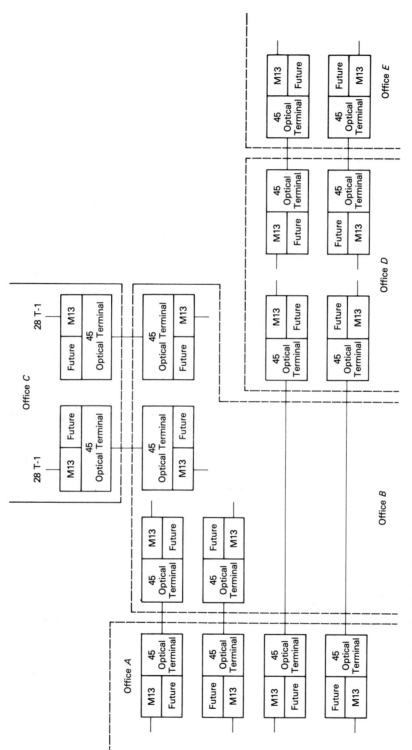

FIGURE 18-9 Alternative Design 2

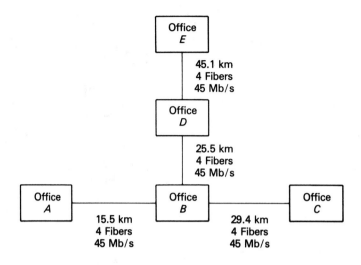

FIGURE 18–10
Alternative
Design 3

Total: 115 km, 4 Fibers, 45 Mb/s

In all other respects alternative 3 presents the same operating features and capacity of alternative 2, including the graceful growth capability. Note that traffic to and from offices D and E does not require any signal processing at all at the intermediate office B location, just as in alternative 2.

The following bill of materials and cost data are developed from alternative design 3 as shown in Figures 18–10 and 18–11:

Item	Unit Cost	Quantity Required		Extended Cost
45 Mb/s Terminal (Unprot)	$11,000	16		$176,000
WDM Couplers	300	8		2,400
4-Fiber Cable	2.30/m	115 km		264,500
			Total	$442,900

General Discussion

A comparison of the proposed design with all the alternatives may be facilitated by tabulating the cost data as follows:

System Design	Cost from Bill of Materials
Proposed Design	$668,000
Alternative 1	472,500
Alternative 2	455,500
Alternative 3	442,900

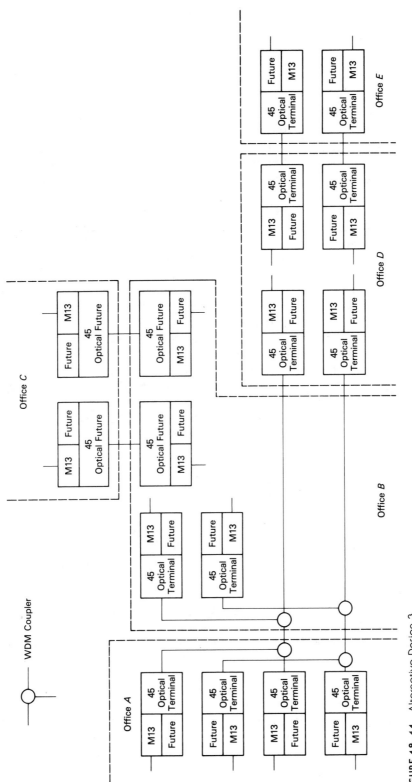

FIGURE 18–11 Alternative Design 3

303

In a project of this magnitude, a cost savings of $10,000 or even $20,000 may not be decisively significant. A network of this scope will be in service for many years, and the final judgment must include such factors as potential growth, long-term maintainability, and the like. On the other hand, a cost differential on the order of $200,000, such as exists between the proposed design and any of the alternatives, certainly cannot be ignored.

If the original transmission engineering provides a potential system bandwidth greater than that necessary for the initial traffic load, then subsequent system expansion can be orderly and efficient. The growth option can be preserved simply by selecting fibers with bandwidth greater than actually initially required, and there would be an insignificant, if any, cost penalty involved.

Note that in any of the alternative designs the entire system capacity could be doubled at some future date by merely applying WDM techniques, and that all of the alternatives do supply the capacity that was required initially.

The alternative designs can all accommodate traffic growth incrementally, that is, increased transmission capacity can be added only to those sections of the network that subsequently require additional capacity.

The proposed design committed the capital investment initially, with no assurance that the additional capacity would eventually be required, and certainly no assurance that it would be necessary throughout the entire network.

It is left to the reader to analyze the preceding figures and to evaluate all the alternatives, keeping in mind that some of them may present an arguably higher level of service reliability than the proposed design, which required all traffic to be processed at the intermediate location at office *B*.

CASE HISTORY C

Application Requirements

This project involved five electromechanical telephone switching offices that all "homed" their trunk traffic on the same larger electromechanical "host" switch. The in-place network geography is shown in Figure 18–12. Trunking for these facilities utilized digital telephone carriers (T–1 systems) applied to multiple-paired conventional telephone cables. The transmission distances involved required a substantial number of carrier repeaters in each cable run.

It was proposed that all five remote switches, and the host switch as well, be replaced with new digital switching equipment. For reasons of economy, it was decided to purchase "dumb" remote switches for all five remote installations. Such units have no stand-alone switching capability, and if umbilical trunk service to the host switch were totally interrupted, a complete loss of service for all subscriber telephone lines fed from each of the remote switch units would occur.

To reduce the probability of isolation of each of the remote switches, it was proposed that an entirely new trunking network using lightwave transmission facili-

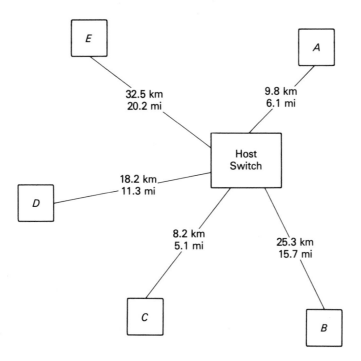

FIGURE 18–12
Existing Facilities
(Case History C)

ties be constructed. The system operator proposed to place the new optical cables as shown in Figure 18–13.

This configures the network as a "ring" rather than as a "star" configuration as was shown in Figure 18–12. The reasoning was that a ring configuration would provide alternative cable routing for all traffic to and from each of the remote units, thereby improving the reliability of service throughout the network. Traffic for each remote would be transmitted in both directions around the ring, and these independent transmission paths would improve reliability.

Although the traffic capacity required for each of the remote switches is only nominal, the proposed ring configuration consolidates all traffic into a single transmission signal at each location. The operator proposed a 135 Mb/s transmission data rate throughout the entire network to handle the consolidated traffic and to provide some capability for future traffic growth, which was projected to be limited in scope.

In Figure 18–14 the ring network is redrawn to show the terminal equipment requirements. Note that in this figure, and in Figure 18–13, a total of six transmission links are involved. In the star network in Figure 18–12, only five transmission links were required. The ring network requires 12 terminal installations, the star network only 10.

Note that for either the ring or the star design, the total cable requirements were essentially the same. Thus, there is not any decisive advantage for either design as far as cable and cable placement costs are concerned.

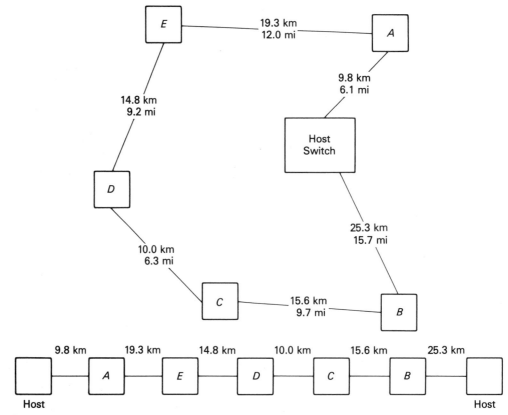

FIGURE 18-13 Proposed Design C

Special Considerations

In the narrative accompanying the proposed design, the designer noted that the existing digital carrier trunking on multiple-paired cables had imposed high maintenance costs, due to lightning damage during frequent thunderstorm activity in the area. This experience was certainly a major contributing factor in the choice of the ring network design. However, even given that the existing facilities may be unacceptably unreliable, and that the designer clearly identified the root cause to be the susceptibility of the existing plant to lightning, surely some of this susceptibility is due to the multiplicity of digital carrier repeaters that were required. In any event, no mention was made of a history of disruption of service due to mechanical cable damage or disruption.

A basic advantage of lightwave transmission systems is their inherent invulnerability to electromagnetic interference and lightning. Regardless which of the system designs prevails, we can realistically expect the service reliability of a lightwave net-

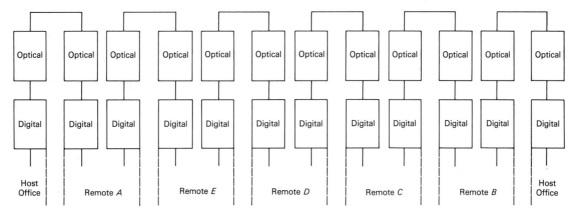

FIGURE 18-14 Proposed Design *C*

work to be much better than that of the existing paired-cable carrier systems in service now.

One might question the necessity and justification for any alternative cable routing at all. From Figure 18–14, it is clear that the ring design requires extensive processing of all transmission signals at each and every location. Not only have we introduced a much higher transmission data rate, and consequently a higher level of complexity in all terminal installations, but we are obliged to demodulate and remodulate all traffic at all locations, even if in a given case the traffic does not require access at the location.

For example, consider traffic between remote switch *A* and the host switch. Transmitted in both directions through the ring, in one direction the information must pass through ten intermediate electro-optical units (transmitters or receivers) and through ten intermediate digital multiplexers. It is difficult to believe that such a transmission path is substantially more reliable than the star network would have provided; indeed, it can be persuasively argued that this arrangement is less reliable. Thus, the ring configuration may be interesting, but only academically superior.

Proposed Design

The following bill of materials and cost data are developed from the proposed design as shown in Figures 18–13 and 18–14:

Item	Unit Cost	Quantity Required	Extended Cost
135 Mb/s Terminal	$23,000	12	$276,000
M13 Digital Mux	10,000	12	120,000
4-Fiber Cable	2.30/m	95 km	218,500
		Total	$614,500

In this configuration, the network is equipped to handle only one DS-3 signal. At each intermediate location, the local traffic must be extracted from and inserted into this DS-3 signal. If traffic growth develops, some quantity of additional digital multiplexers will be required.

As in the previous case histories, several alternative designs are possible.

Alternative Design 1

One alternative design would simply provide individual optical cables for each service location following the same star network configuration shown in Figure 18-12. Since the traffic capacity requirements for each individual link would be substantially reduced, it would be possible to satisfy the initial service requirements for each link using a transmission data rate of 45 or 90 Mb/s and still provide abundant capacity for growth.

It might even be practical to consider a lower data rate, such as 6 Mb/s (a capacity of 8 T-1 systems providing 192 equivalent voice circuits). This could reduce the initial system costs even further, and if transmission engineering satisfied the 45 or 90 Mb/s requirements initially, the system could easily be updated to the higher data rate and increased capacity at any later time. For brevity, we have not developed the cost for the 6 Mb/s alternative, but the reader is encouraged to do so to reinforce his understanding and apply the knowledge previously gained.

The following bill of materials and cost data are developed from alternative design 1 as shown in Figure 18-12:

Item	Unit Cost	Quantity Required	Extended Cost
45 Mb/s Terminal (Unprot)	$11,000	10	$110,000
4-Fiber Cable	2.30/m	95 km	218,500
		Total	$328,500

Expansion of any individual link can be discriminatory, requiring capital investment only where growth is actually experienced, and could be effected either through digital multiplexing or WDM.

Alternative Design 2

It might be instructive to reconsider the ring design and see whether it can be improved upon. Upon reexamining Figure 18-14, it is evident that one disadvantage of the design is that all signals, in passing through any intermediate location, have to be demodulated, digitally processed in some way, and remodulated again even though some of these signals are not utilized for service at that location at all. The effect on the reliability of transmission of the additional equipment and signal processing that have been introduced cannot be ignored.

A second disadvantage of the approach is the inflexibility of the network. Traffic growth at any particular location can be accommodated, but it may require additional equipment or rearrangements of equipment at a number of intermediate stations. In effect, the design forces us to adopt a high transmission data rate initially, and perhaps prematurely as well, simply to be able to handle potential growth that may never actually develop.

The design cannot expand gracefully, that is, by adding capacity incrementally and discriminately, thus deferring capital investments until such growth is required and some supporting new revenues are apparent. The basic design itself imposes these disadvantages.

How might the design be improved? Perhaps it would help if there were fewer intermediate stations in the system ring.

In Figure 18–15 we have broken out switch locations *B* and *C* and reconfigured them as a separate ring together with the host office. To do this we have added 8.2 km of additional optical cable between remote *C* and the host location, but we no longer need to place the 10.0 km of cable between remote *C* and remote *D*. The

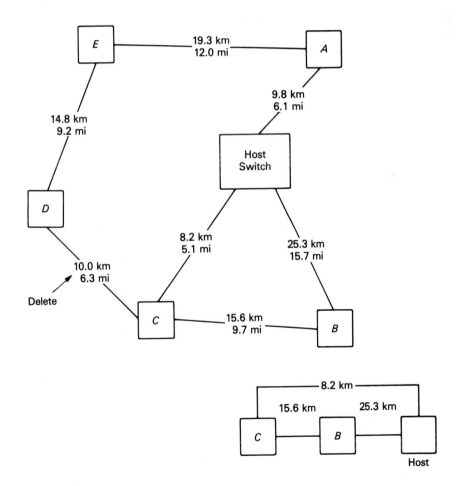

FIGURE 18–15
Alternative
Design 2

cable route lengths within this smaller ring are such that we can transmit between any two points in the ring without any optical repeaters at all. Rather, we shall simply add four more fibers in each cable link and provide alternative traffic routing between all stations on a continuous-fiber transmission link.

Each of these independent transmission links could be a fully protected configuration requiring four fibers and 8-fiber cable throughout, as shown in Figure 18–16. But since alternative cable routing is available, it is reasonable to use unprotected facilities for each independent transmission path, relying on the alternative path as the protection-switched facility.

Figure 18–17 shows the completed system configuration. This is the alternative which we have cost developed. We suggest the reader develop a bill of materials and cost data on the configuration shown in Figure 18–16 as an educational exercise.

Note that 6 Mb/s would provide sufficient transmission capacity for either of these two alternatives, and if transmission engineering was done for 45 Mb/s, substantial growth capability would be preserved. We have cost developed the system on the basis of using 45 Mb/s terminal equipment throughout, however.

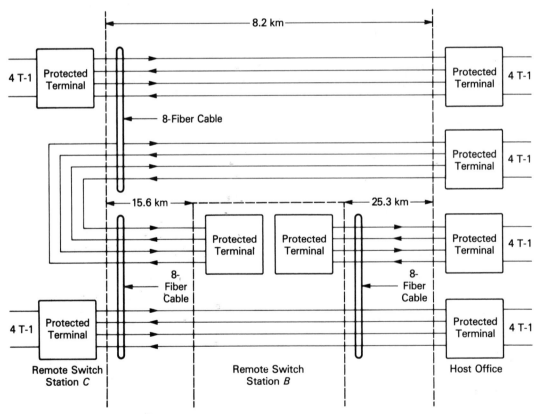

FIGURE 18–16 Alternative Design 2

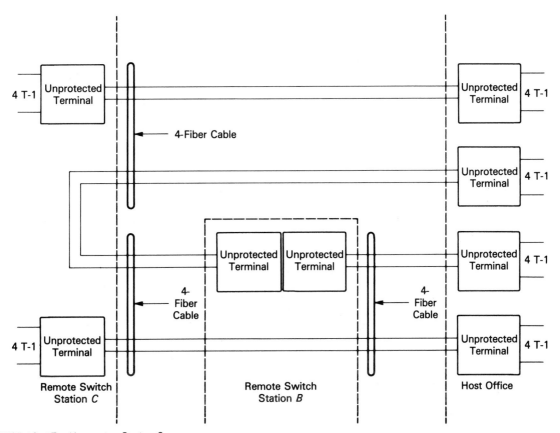

FIGURE 18-17   Alternative Design 2

The following bill of materials and cost data are developed from alternate design 2 as shown in Figures 18-15 and 18-17:

Item	Unit Cost	Quantity Required	Extended Cost
45 Mb/s Terminal (Unprot)	$11,000	8	$ 88,000
4-Fiber Cable	2.30/m	49 km	112,700
		Total	$200,700

But this development only addresses remote switches *B* and *C*, and we must make provision for the remaining offices as well.

In Figure 18-18 we show switch locations *A*, *D*, and *E* configured as a separate ring together with the host office. To do this we have added 18.2 km of additional optical cable between remote *D* and the host, but we no longer have to place the 10.0 km of cable between remote *C* and remote *D*. The cable route lengths within this smaller ring are such that we cannot transmit between any two points in the

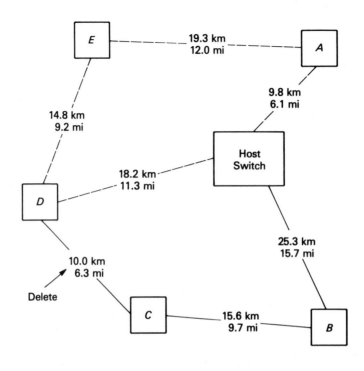

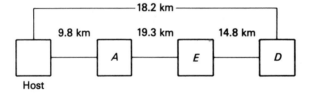

FIGURE 18–18
Alternative
Design 2

ring without any optical repeaters at all. We will need a single optical repeater in the alternate transmission path between remote A and the host office, and we will locate this repeater at the remote switch D location.

Figure 18–19 shows six fibers in each cable link and the 45 Mb/s unprotected terminals that are employed throughout. Service protection is inherent in the alternative traffic route provided by the alternate cable facilities.

The following bill of materials and cost data were developed from alternate design 2 as shown in Figures 18–18 and 18–19:

Item	Unit Cost	Quantity Required	Extended Cost
45 Mb/s Terminal (Unprot)	$11,000	12	$132,000
Optical Repeater	13,000	1	13,000
6-Fiber Cable	2.80/m	62 km	173,600
		Total	$318,600

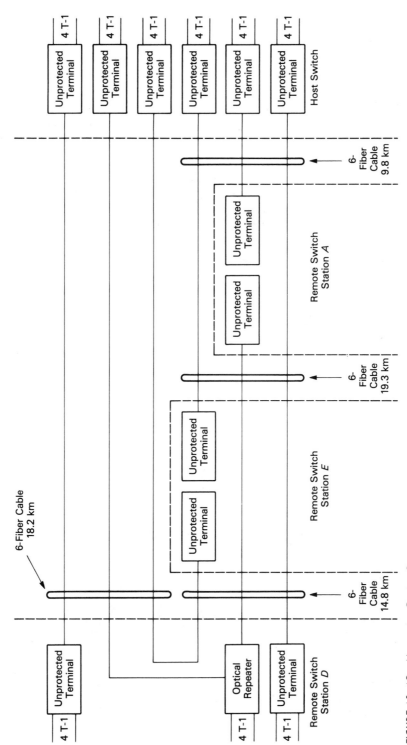

FIGURE 18–19 Alternative Design 2

Alternative design 2 must include both of the smaller ring systems, so the total cost of this alternative would be a total of $519,300. This does not include the additional cable placement costs that the approach would incur over and above the construction costs of both alternate 1 and the proposed design.

Alternative Design 3

It is possible to reduce the cable costs in alternative design 2 by applying WDM. Figure 18–20 shows just such a configuration.

In this modification, the separate ring established earlier for stations *B, C,* and the host office remains unchanged, since it only required a four fiber cable in the first place. The second ring, which includes the host switch and stations *A, D,* and *E,* has been redesigned in Figure 18–20 using WDM. This reduces the required fiber count in this portion of the network from the previous six fibers to four.

The following bill of materials and cost data are developed from alternate design 3 as shown in Figures 18–17 and 18–20:

Item	Unit Cost	Quantity Required	Extended Cost
45 Mb/s Terminal (Unprot)	$11,000	20	$220,000
Optical Repeater	13,000	1	13,000
WDM Coupler	300	12	3,600
4-Fiber Cable	2.30/m	111 km	255,300
		Total	$491,900

General Discussion

A comparison of the proposed design with all the alternative designs may be facilitated if we tabulate the cost data as follows:

System Design	Cost from the Bill of Materials
Proposed Design	$614,500
Alternative Design 1	398,500
Alternative Design 2	519,300
Alternative Design 3	491,900

The cost differential between the proposed design and alternative 1 is an attention-getting $216,000. As presented here, both designs are fully responsive to the initial service requirements.

Although alternatives 2 and 3 present smaller cost reductions, the reductions are significant, and note that both designs do provide a completely independent second cable route, just as the proposed design did, if this design feature is judged to be essential.

In all the alternative designs the service reliability should be at least as high, and arguably higher, than that provided by the proposed design.

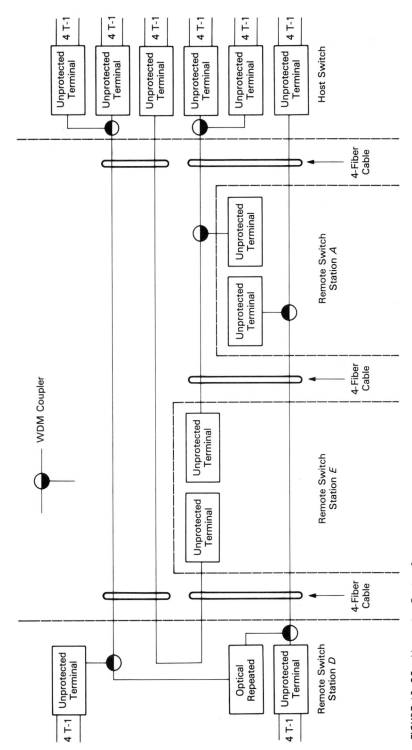

FIGURE 18–20 Alternative Design 3

By now, the reader should be knowledgeable enough to complete an evaluation of this project, so we shall not endorse any particular approach.

SUMMARY

The case histories presented in this chapter were real applications, and the designs proposed were the solutions actually developed by practicing telecommunications engineers.

The point of the exercises was not to criticize the proposed designs, but to take advantage of the opportunity they presented to challenge the reader's comprehension of lightwave technology, and to exercise the reader in practical applications of the technology.

Good engineering practice is not simply a question of technical adequacy alone; as we have seen, most problems lend themselves to multiple solutions.

Sound engineering is the judicious blending of technical adequacy with considerations of economy, efficiency, and practicality.

Understanding and Applying Decibels

If we can adopt units of measure that are equally compatible with both signal levels and transmission losses and gains, then the determination of the precise signal level at any point and time within a system would be a simple matter of addition or subtraction. The basic concept is to deal not with absolute power levels, but with ratios of power. How may this be accomplished?

RATIOS

A *ratio,* by definition, is the relation in degree or number between two similar things. The ratio between a 10-lb weight and a 1-lb weight, for example, is 10 to 1, or 10. Not 10-lbs, but a ratio of 10 to 1, since the 10-lb weight is 10 times heavier than the 1-lb weight. The ratio between a 100-lb weight and a 10-lb weight would also be 10 to 1, since the 100-lb weight is 10 times heavier than the 10-lb weight.

When we are dealing in ratios, we cannot express a specific weight as a ratio. In the example above, a ratio of 10 to 1 was equally accurate and applicable to both cases, even though the weights compared were significantly different in the two cases. A ratio is a very useful quantity, and we apply it extensively in lightwave transmission technology.

A basic unit to denote the power ratio of two levels of power, voltage, current, or sound intensity is the *Bel,* named after Alexander Graham Bell. Like any ratio, the Bel cannot be used to denote a specific level of power, voltage, current, or sound intensity. Since light energy is power, we can utilize both the ratio and the Bel in lightwave transmission systems.

By definition, 1 Bel will denote a ratio of 10, but only between two discrete levels of power. Let us be quite certain we understand this point.

If we had a signal input to a device of 10 watts (a watt is a unit of measure of power) and an output signal from that device of 100 watts, the ratio of the output to the input would be 10 to 1. Since the output level is greater than the input level, the device would be said to have gain, and the gain could be denoted as +1 Bel, since the ratio between the levels was 10.

The same would be true of another device whose input was 100 watts and whose output was 1,000 watts. Since the ratio of the output to the input is again 10 to 1, or 10, this device would also be said to have a gain of +1 Bel.

A Bel may have a positive or negative value, denoting gain or loss, respectively. For example, if the power input to a section of cable was 100 watts, and the signal recovered at the distant end of that cable was 10 watts, the ratio of the input to the output would be 10. Since this time the input is greater than the output, the cable would be said to have loss, and the loss could be denoted as –1 Bel.

Again, if the input level to a device was 10 watts and the output from that device was 1 watt, the ratio of input to output would be 10 to 1, and the transmission loss could be expressed as a negative ratio, –1 Bel, and similarly for a 1,000 watt input and a 100 watt output.

Gain and loss can be specified without any reference to the actual signal level that might be applied to, or passed through, a device. Thus, an amplifier with a gain of +1 Bel would have the same impact on an input signal regardless of whether the actual input signal were 1 watt or 2 watts—that is, there would be a tenfold amplification to 10 watts and 20 watts, respectively.

This is very convenient in working with networks or systems composed of many different items. We can, for example, specify cable loss without actually having the cable in place or in service, since we do not need to measure the actual system signal loss through the cable.

The Bel is a perfectly respectable, technically acceptable unit of ratio measurement, but was found through time to be awkwardly large. To avoid the necessity for computing with decimal values of the Bel, the decibel was adopted.

THE DECIBEL

The *decibel,* denoted dB (note lower case d and capital B), is one-tenth of a Bel. Thus, 10 dB = 1 Bel, 20 dB = 2 Bels, and 30 dB = 3 Bels. In all respects the decibel functions exactly as does the Bel. It may be negative or positive to denote loss or gain, but it cannot, of itself, be used to denote any specific or discrete power level, only the ratio between two such levels. The relationship between ratios, Bels, and decibels is shown in Table A-1.

TABLE A–1
Ratios, Bels, and
Decibels

Power Ratio	Value in Bels	Value in decibels
1 to 1	0	0
2 to 1	.3	3
10 to 1	1	10
100 to 1	2	20
1,000 to 1	3	30
10,000 to 1	4	40

LOGARITHMS

Both Bels and decibels are logarithmic. The logarithm, or simply, log, of a number is defined as the exponent to the base 10 that produces that same number. For example, the log of N is X if $10^X = N$. Note that in this expression, X is the exponent of 10, and not the other way round.

Example $N = 1,000$, or $N = 10^3$, or $N = 10 \times 10 \times 10$, or $N = 10 \times 100$, or $N = 1,000$. If $1,000 = 10^3$, then the exponent 3 is the logarithm of 1,000. ∎

Example What is the logarithm of 10,000?
$N = 10^4$, or $N = 10 \times 10 \times 10 \times 10$, or $N = 100 \times 10 \times 10$, or $N = 1,000 \times 10$, or $N = 10,000$. Then $10^4 = 10,000$ and the exponent 4 is the logarithm of 10,000. ∎

These examples all used multiples of ten for simplicity, but every number has a logarithm. Some random examples are given in the following table:

Number	Logarithm
2	.301
3	.477
7	.845
10	1.000
14	1.146
100	2.000
283	2.452

Note that for numbers between 0 and 10, the logarithm is a decimal number between 0 and 1; for numbers between 10 and 100, the logarithm is a decimal number between 1 and 2; for numbers between 100 and 999, the logarithm is a decimal number between 2 and 3; and so on. The logarithm for any number may be found in a log table in many technical books or can be taken directly off a slide rule or calculator.

POWER RATIOS

The formula to determine the ratio between two power levels in decibels is

$$dB = 10 \log \frac{P_2}{P_1}$$

This formula applies equally to all units of power including watts, milliwatts, microwatts, and even horsepower, which is 746 watts by definition.

Example

What is the power ratio in dB between a 400-horsepower engine and a 200-horsepower engine?

$$dB = 10 \log \frac{P_2}{P_1}$$

$$= 10 \log \frac{400}{200}$$

$$= 10 \log 2$$

It may be disappointing to the big-car owner, but the vehicle only has 3 dB more power than the smaller car. Perhaps this fact will make the small-car owner feel a little better.

We did not apply the preceding formula earlier in developing the decibel values given in Table A–1 simply because it was so easy to derive the ratios directly from the exponent used with the factor 10. We can rewrite Table A–1 as follows with 10 explicitly the base:

TABLE A–2
Power Ratios and Decibels

Power Ratio	Base 10	Value in Bels	Value in decibels
1 to 1	10^0	0	0
2 to 1	$10^{0.3}$	0.3	3
10 to 1	10^1	1	10
100 to 1	10^2	2	20
1,000 to 1	10^3	3	30
10,000 to 1	10^4	4	40

Note that the exponent of the base 10 is the ratio in Bels in each case. If we apply the formula, we easily see that the result is the same. For example, given an input to a device of 100 watts, and an output from the device of 1,000 watts, what is the power ratio in decibels? We have

$$dB = 10 \log \frac{P_1}{P_2}$$

$$= 10 \log \frac{1,000}{100}$$

$$= 10 \log \frac{10}{1}$$

$$= 10 \log 10$$

Since the log of 10 is 1, dB = 10×1, i.e., the power ratio is $+10$ dB. This agrees with the value given in both Tables A-1 and A-2 for a power ratio of 10 to 1.

We stated earlier that decibels can be either positive or negative in value. Utilizing the power ratio formula, let us do a calculation to familiarize ourselves with the process.

Example   If a cable input is 20 watts and its output is 2 watts, what is the cable loss in dB?

$$dB = 10 \log \frac{P_2}{P_1}$$

$$= 10 \log \frac{2}{20}$$

$$= 10 \log \frac{1}{10}$$

$$= -10 \log 10$$

Since the log of 10 is 1, dB = -10×1, i.e., the power ratio is -10 dB.

It is plain that the power ratio between any two power levels can be easily determined. It will be convenient if the more common power ratios are simply memorized, and this is easily done by reviewing Table A-2. Since a power ratio of 2 to 1 is 3 dB, it follows that when one power level is twice as great as another (a ratio of 2 to 1), the power ratio would be $+3$ dB. When one power level is only half the amplitude of another (a ratio of 1 to 2), the power ratio would be -3 dB.

Similarly, a 10 to 1 ratio, or ten times as much power, would be $+10$ dB, and one tenth as much power would be -10 dB. A 100 to 1 ratio would be $+20$ dB if it were gain, and -20 dB if it were loss.

How can we relate power ratio to discrete signal power levels?

THE DECIBEL-MILLIWATT _____

Since lightwave transmission systems operate with lightwave signal levels that rarely are as large as a tenth of a watt (0.1 watt), and that often are much lower than that, the use of the watt as a unit of power is awkward. A more convenient unit is the *milliwatt,* which is one thousandth (0.001 or 10^{-3}) of a watt.

A standard reference power level has been established to be one milliwatt, and is called the decibel-milliwatt, denoted as dBm. (Note the capital B and lower case d and m.) Zero decibel-milliwatts (0 dBm) is assigned the specific value of one milliwatt (0.001 watt). In lightwave transmission systems, optical signal power levels are denoted as power in dBm.

Like decibels, decibel-milliwatts may have a negative or positive value. A negative value or $-$dBm denotes a power level that is less than one milliwatt; a positive value or $+$dBm denotes a power level that is above one milliwatt. Zero dBm is, of course, a power level of exactly one milliwatt.

The numerical value of dBm (either negative or positive) denotes the ratio of the power level to the reference power level, that is, to one milliwatt. For example, in –3 dBm, the figure 3 means that the ratio is 3 dB, and since 3 dB is a power ratio of 2 to 1, the power level must be either twice the level of one milliwatt, or one-half the level of one milliwatt. The minus sign rules out the former possibility, so the power level that –3 dBm denotes is one-half a milliwatt, or 0.0005 watt, or 5×10^{-4} watt.

A value of $+10$ dBm would be 10 dB greater than 0 dBm. A ratio of 10 dB means ten times as great or one tenth as much as the reference value, which is one milliwatt. The positive sign in the term $+10$ dBm indicates that the stated power level is ten times the reference power level, or 10 times 1 milliwatt, or 0.01 watt, or 10^{-2} watt.

Note that 1 milliwatt can be written as 0 dBm or 10^{-3} watt or 0.001 watt.

Expressing optical (lightwave) signal power levels in dBm and transmission gain or loss in dB provides a very simple method of determining signal power at any point in any system. Manufacturers generally specify the power output at lightwave frequency of a light source or transmitter in either dBm or directly as milliwatts, and sometimes in both. Optical fiber and cable manufacturers generally specify fiber transmission loss in dB, and usually this is qualified in terms of fiber length, as dB/km, for example. All optical components or devices that might be inserted into the optical transmission path of the system, such as splices, fiber connectors, optical splitters, and couplers, will have their insertion loss specified in dB. The optical detector/receiver will have its detection threshold or sensitivity specified in terms of input optical power level in dBm.

USING dB AND dBm

Figure A–1 shows a simple lightwave system composed of a light source or transmitter, a 2 km length of fiber, an optical signal splitter which divides input equally across two outputs, a 5 km length of fiber, and an optical detector/receiver. The specifications for these components are as follows:

Transmitter output	-3 dBm
Fiber transmission loss	0.5 dB/km
Optical splitter insertion loss	3.5 dB

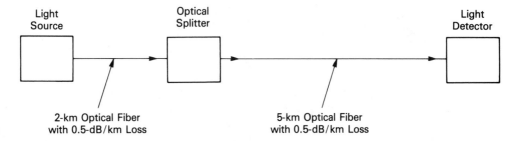

| Light Source | | Optical Splitter | | | Light Detector |

FIGURE A-1
System
Calculations

2-km Optical Fiber
with 0.5-dB/km Loss

5-km Optical Fiber
with 0.5-dB/km Loss

To determine the optical signal power input to the receiver, we need only calculate and add together all the transmission losses in the link.

Since the fiber is stated to have 0.5 dB/km, the first fiber section, 2 km long, will have 1.0 dB total loss. The 5 km fiber section will have 2.5 dB of transmission loss. Since the optical splitter was specified to have 3.5 dB of loss, the total link loss is $(-1.0) + (-2.5) + (-3.5) = 7.0$ dB.

Since the transmitter is specified to provide an optical output power of only -3 dBm, the receiver input power level will be $(-3.0) + (-7.0) = -10$ dBm.

We can reinforce our understanding of dBm with another example. Suppose a manufacturer specifies the optical power output of a light source (transmitter) to be 2 milliwatts. What is the output of that unit in dBm? By our earlier formulas,

$$dB = 10 \log \frac{P_2}{P_1}$$

$$= 10 \log \frac{0.002 \text{ (given power level)}}{0.001 \text{ (reference power level)}}$$

$$= 10 \log 2$$

Since the log of 2 is 0.301, dB $= 10 \times 0.301$, i.e., the power ratio is $+3$ dB. A positive value of 3 dB would be twice the reference power level of 1 milliwatt, or $+3$ dBm $= 2$ mw.

This same result can also be obtained by calculation. Since the transmitter optical output was given in mw, we can apply the formula

$$dBm = 10 \log Pwr \text{ (mw)}$$

$$= 10 \log 2$$

Since the log of 2 is 0.301, dBm $= 10 \times 0.301$, i.e., the power level is $+3$ dBm.

Note that we can state loss or gain in dB, but not in dBm. Note that we can state absolute power levels in dBm but not in dB. As we have seen, by simple addition or subtraction, we can combine dB and dBm in calculations.

OTHER APPLICATIONS OF dB

In other disciplines, dB is also used as a ratio, but a reference value other than dBm may be established as a standard. For example, the term dBw, where 0 dBw = 1 watt, is often encountered in radio transmission systems.

In cable television systems the term dBmV is commonly employed, where 0 dBmV = 1 millivolt measured across 75 ohms.

Decibels are also used to denote ratios between two voltages or currents, but in such cases the formula is altered to become

$$dB = 20 \log \frac{E_2}{E_1}$$

and

$$dB = 20 \log \frac{I_2}{I_1}$$

respectively.

Returning to power calculation, we can establish quite easily how many dBm a power level of 0 dBw is. We have

$$dBm = 10 \log \frac{P_2}{P_1}$$

$$= 10 \log \frac{0 \text{ dBw (given power level)}}{0 \text{ dBm (reference power)}}$$

$$= 10 \log \frac{1.000 \text{ (watt)}}{0.001 \text{ (watt)}}$$

$$= 10 \log 1,000$$

Since the log of 1,000 is 3, dBm = 10 × 3, i.e., the power level is +30 dBm. Thus, 0 dBw = +30 dBm.

Since we are talking about power in either case, we can even establish how many dBm 1 horsepower (746 watts) is:

$$dBm = 10 \log \frac{P_2}{P_1}$$

$$= 10 \log \frac{1 \text{ hp}}{0 \text{ dBm}}$$

$$= 10 \log \frac{746 \text{ (watts)}}{0.001 \text{ (watt)}}$$

$$= 10 \log 746{,}000$$

Since the log of 746,000 is 5.872, dBm = 10 × 5.872, i.e., the power level is +58.72 dBm. Thus, 1 hp = +58.72 dBm.

Glossary

Absorption In an optical fiber, loss of optical power resulting from conversion of that power into heat.

Absorption losses Losses caused by impurities, principally transition metals and neighboring elements (Cr, Mn, Fe, Co, Ni), by water, or by intrinsic material absorption.

Acceptance angle Half the vertex angle of that cone within which optical power may be coupled into bound modes of an optical waveguide.

Acceptance cone A cone whose included angle is equal to twice the acceptance angle.

Active Containing, or connected to and using, a source of energy.

Aerial cable Cable suspended in the air on poles or other overhead structures. Usually implies the use of a "messenger strand" to which the cable is lashed for support.

Alternating current (AC) An electric current which continually varies in amount and reverses its direction periodically. The plot of current vs. time is usually a sine wave.

Ampere Unit of electric current, or rate of flow of electricity. One coulomb per second. One volt impressed across a resistance of one ohm causes a current of one ampere to flow.

Amplification The act of increasing the amplitude or strength of a signal.

Amplitude modulation Process by which a continuous carrier wave is caused to vary in amplitude in accordance with the amplitude of the modulating signal.

Analog signal A signal which is continually variable and not expressed by discrete states of amplitude, frequency, or phase.

Angle of incidence The angle between an incident ray and the normal to a reflecting or refracting surface.

Angstrom (A) 10^{-10} meter. Use of the angstrom as a unit of optical wavelength has largely been supplanted in recent years by the nanometer (10^{-9} meter).

Antireflection coating A single or multiple layer of thin dielectric coating that reduces the reflectivity of an optical surface.

APD See **avalanche photodiode**.

Armored cable A cable having one or two layers of steel tapes or steel wires spirally applied to the sheath to provide mechanical protection.

Asynchronous Not synchronous.

Atmosphere The gaseous envelope surrounding the Earth, composed of 78 percent nitrogen, 21 percent oxygen, 0.9 percent argon, plus some carbon dioxide and water vapor. The atmosphere is divided into several layers, as follows:

Troposphere	0–10	miles
Stratosphere	10–50	miles
Ionosphere	50–370	miles
Exosphere	370+	miles

Attenuation The decrease in signal strength along a fiberoptic waveguide caused by absorption and scattering. Attenuation is usually denoted in decibels per kilometer (dB/km).

Attenuation-limited operation The condition prevailing when the received signal amplitude, rather than any distortions, limits performance.

Automatic gain control (AGC) A feature of some amplifiers and radio receivers which provides a substantially constant output even though the signal input varies over wide limits.

Avalanche effect The cumulative multiplication of carriers in a semiconductor caused by an electric field across the barrier region strong enough so that electrons collide with valence electrons, releasing new electrons which have more collisions, which release more electrons, etc.

Avalanche photodiode (APD) A photodiode designed to take advantage of avalanche multiplication of photocurrent. As the reverse-bias voltage approaches the breakdown voltage, hole-electron pairs created by absorbed photons acquire sufficient energy to create additional hole-electron pairs when they collide with substrate atoms; thus, a multiplicative effect is achieved. Amplification is almost noiseless, and this makes APDs 10 to 15 dB more sensitive than PIN photodiodes. Some problems with APDs are their temperature sensitivity, high reverse-bias voltages (200 to 400 V to achieve a hundredfold multiplication of current), and price, which is higher than that of PIN photodiodes.

Axial mode See **longitudinal mode**.

Axial ray A ray passing through the axis of an optical waveguide without any internal reflection.

Backscattering The scattering of light into a direction generally opposite to the original direction.

Bandwidth 1. The capacity of an optical fiber to transmit information expressed in bits of information transmitted in a specific time period for a specific length of optical waveguide (usually expressed as some number of megabits/sec/km). Bandwidth is limited by pulse spreading or broadening due to dispersion, so that adjacent pulses overlap and cannot be distinguished. 2. The range of frequencies within which a fiberoptic waveguide or terminal device performs at a given specification.

Bandwidth-limited operation The condition prevailing when the system bandwidth, rather than the amplitude (or power) of the signal, limits performance. The condition is reached when the system distorts the shape of the waveform beyond specified limits. For linear systems, bandwidth-limited operation is equivalent to distortion-limited operation.

Beam divergence The increase in beam diameter with distance from a light source's exit aperture. Measured in milliradians at specified points, usually where power density or energy density is $1/2$ or $1/e^2$ *or* e^{-2} the maximum value, and expressed as "half-angle" or "full angle" divergence.

Beam splitter A device that divides an incident beam into two or more separate beams. Prisms, thin films, sheets of glass, and partially silvered mirrors can be used as beam splitters.

Bel The fundamental division of a logarithmic scale for expressing the ratio of two powers which are in the ratio of one to ten. The Bel is an awkwardly large unit, so the decibel (one-tenth of a Bel) is used instead.

Bend loss A form of increased attenuation caused by allowing high-order modes to radiate from the side of a fiber. The most common types of bend losses are (1) those occurring when the fiber is curved around a restrictive radius of curvature, and (2) microbends caused by small distortions of the fiber imposed by externally induced perturbations, such as poor cabling techniques.

BER Acronym for bit error rate.

Bidirectional Having equal effectiveness in two directions which are separated by 180 degrees in azimuth.

Bidirectional transmission Signal transmission in both directions along an optical waveguide or other component.

Binary Having two possible states or values.

Binary digit One unit of information in binary (two-level) notation.

Binary state Either of the two conditions of a bistable device—the "one" state or the "zero" state.

Bipolar signal A type of direct current signal in which consecutive marks are of opposite polarity and a space is represented by zero voltage.

Bit An electrical or light pulse whose presence or absence indicates data. The capacity of the optical waveguide to transmit information through the waveguide without error is expressed in bits per second per unit length. Acronym for "binary digit."

Bit error rate In a digital communications system, the fraction of bits transmitted that are received incorrectly. If BER is specified at 10^{-9} (a typical value), then an average of one bit per one billion sent will be read incorrectly by the receiver.

Bit rate The speed at which digital information is transmitted, usually expressed in bits per second.

Brightness An attribute of visual perception in accordance with which a source appears to emit more or less light; since the eye is not equally sensitive to all colors, brightness cannot be a quantitative measure.

Buffer 1. A device used as an interface between two circuits or pieces of equipment to reconcile their incompatibilities or to prevent variations in one from affecting the other. 2. A circuit used for transferring data from one unit to another when temporary storage is required because of different operating speeds or times of occurrence of events.

Buried cable A cable installed directly in the earth without the use of underground conduits.

Cabling 1. The act of twisting together two or more wires, pairs, or pair groups by machine to form a cable. 2. The act of installing distribution cable, particularly in a new area.

Carrier frequency 1. The frequency of an unmodulated carrier wave. 2. Any of the frequencies which are suitable for use as carriers.

Carrier system A method of transmitting electrical intelligence by modulating it onto a higher frequency carrier wave, and then, at the receiving end, recovering the original intelligence by the reverse process of demodulation. Useful because many channels of intelligence can be modulated on one carrier wave and carried on a single transmission channel.

Carrier transmission A means of transmitting information electrically in which the transmitted wave is a wave resulting from the modulation of a single-frequency sinusoidal wave by a complex modulating wave.

Carrier wave The sinusoidal single-frequency wave which is modulated by a complex wave (called the modulating wave) to obtain a modulated wave capable of carrying much information over a single channel.

Cavity The volume (resonator) which provides feedback for laser oscillations. The most common configuration consists of an active medium between two plane or curved mirrors, called cavity mirrors or end mirrors.

Chemical vapor deposition (CVD) technique A process in which deposits are produced by heterogeneous gas-solid and gas-liquid chemical reactions at the surface of a substrate. *Note:* The CVD method is often used in fabricating optical waveguide preforms by causing gaseous materials to react and deposit glass oxides. Typical starting chemicals include volatile compounds of silicon, germanium, phosphorus, and boron, which form corresponding oxides with oxygen or other gases after heating. Depending upon its type, the preform may be processed further in preparation for pulling it into an optical fiber.

Chopping Modulating input to a detector at a known frequency to improve response.

Chromatic dispersion A measure of the pulse broadening due to the source spectral width (the combined effect of the material and waveguide dispersions).

Circuit reliability The percentage of time a circuit is available to the user during a specified period of time.

Cladding The low refractive-index material which surrounds the core of the fiber and protects against surface contaminant scattering. In all-glass fibers the cladding is glass. In plastic-clad silica fibers, the plastic cladding also may serve as the coating.

Cladding mode A mode of light propagating in an optical fiber that is confined to the cladding and core by virtue of a lower index medium surrounding the cladding. Cladding modes correspond to cladding rays in the terminology of geometrical optics.

Cladding mode stripper A device that encourages the conversion of cladding modes to radiation modes, thereby stripping the cladding modes from the fiber. Often, a material having a refractive index equal to or greater than that of the waveguide cladding.

Clock circuit A circuit that provides accurately timed pulses of uniform length which can be used to control and synchronize other circuits.

Coherence A fixed phase relationship among various points of an electromagnetic wave in space (spacial coherence) or in time (temporal coherence).

Collimation The process by which a divergent or convergent beam of radiation is converted into a beam with the minimum divergence possible for a given system (ideally, a parallel bundle of rays).

Common equipment   Any equipment which is used in some manner by a number of channels or pieces of equipment. Common equipment is usually provided in duplicate for greater reliability.

Complementary metal oxide semiconductor (CMOS)   A technology used in the manufacture of logic integrated circuits by combining N-channel and P-channel MOS transistors.

Concatenation (of optical waveguides)   The linking of optical waveguides end to end.

Conduit   A pipe or tube, of tile, asbestos-cement, plastic, or steel, which is placed underground to form ducts through which cables can be passed.

Connector   A reusable device for making temporary junctions between two fiberoptic cables.

Conservation of radiance   Formerly called the conservation of brightness. A basic principle which states that optical paraphernalia cannot increase the radiance of a source; in other words, the radiance of an image cannot exceed that of the object that produces the image.

Continuous-wave (cw)   Laser operation in which radiation is emitted continuously.

Core   The light conducting portion of an optical waveguide. The core is composed of a high-refractive-index material typically made of silicon tetrachloride ($SiCl^4$). The addition of germanium tetrachloride ($GeCl^4$) increases the refractive index of the core and creates an index gradient along a waveguide.

Core diameter   The diameter of the circle that circumscribes the core area.

Couplers   In fiberoptics, devices which link three or more fibers, providing two or more paths for the transmission signal. In an active coupler, a switching mechanism selects among several routes; in a passive coupler, routing is determined by the geometry of the device.

Coupling loss   The amount of power in the fiber optic link lost at discrete junctions such as source-to-fiber, fiber-to-fiber, or fiber-to-detector [expressed in decibels (dB)].

Critical angle   The angle made by the reflected ray in an optical waveguide when the ray enters the core at the maximum half-angle of the acceptance cone or maximum acceptance angle.

Crosstalk   The undesired coupling of energy due to signals traveling on one conductor interfering with signals on another conductor. Optical waveguides that are intact eliminate crosstalk.

Cutback technique   A technique for measuring fiber attenuation or distortion by performing two transmission measurements, one at the output end of the full length of the fiber, and the other within 1 to 3 m of the input end, access being had by "cutting back" the test fiber.

Cutoff wavelength   The wavelength greater than that at which a particular waveguide mode ceases to be a bound mode. In a single mode waveguide, concern is with the cutoff wavelength of the second-order mode.

CVD   Abbreviation for chemical vapor deposition.

Dark current   A detector's output current in the absence of incident illuminating radiation, caused by emission of thermionic electrons. Dark current limits the minimum radiation intensity that can be detected.

Data buffer   A data storage device used to compensate for a difference in data rate, or time of occurrence, of signals.

dBm   Decibels below or above one milliwatt (mW). A positive number indicates a value greater than 1 mW, a negative number a value below 1 mW. Zero dBm = 1 mW.

dbw or dBw   A unit of power. Decibels referenced to a unit of one watt (W). Zero dBw = 1 W.

Deci-   A prefix meaning one-tenth.

Decibel (dB)   The standard logarithmic unit used to express gain or loss of optical power.

Demultiplexer   A device that separates a multiplexed signal into its components.

Detect   To rectify a modulated carrier wave and thereby recover the original modulating wave.

Detection   The process by which a wave corresponding to the modulating wave is obtained from a modulated wave.

Detector noise-limited operation   Used to denote operation when the amplitude, rather than the width, of the pulse limits the distance between repeaters. In this regime of operation, the losses are sufficient to render the amplitude of the pulse too small to allow an intelligent decision on whether a pulse is present or absent.

Dichroic filter   An optical filter designed to transmit light selectively according to wavelength (most often, a high-pass or low-pass filter).

Dichroic mirror   A mirror designed to reflect light selectively according to wavelength.

Dielectric   A nonconducting (insulating) material, such as glass.

Differential quantum efficiency   In an optical source or detector, the slope of the curve relating output quanta to input quanta.

Diffraction   Deviation of light rays from the paths predicted by geometrical optics.

Diffraction, angle of   Deviation of part of a beam, determined by the wave nature of radiation and occurring when the radiation passes the edge of an opaque obstacle.

Diffraction grating   An array of fine, parallel, equally spaced reflecting or transmitting lines that mutually enhance the effects of diffraction to concentrate the diffracted light in a few directions determined by the spacing of the lines and the wavelength of the light.

Digital   Referring to the use of digits to formulate and solve problems, or to encode information.

Digital data   Any data expressed in (usually binary) digits.

Digital signal   A signal expressed by discrete states, e. g., the absence or presence of a voltage, the level of amplitude of a voltage, or the duration of the presence of a voltage. Information to be transported may be assigned value or meaning by combinations of the discrete states of the signal using a code of pulses or digits.

Digitizing   The process of converting an analog signal to a digital signal.

Diode laser   Synonym for injection laser diode (ILD).

Direct-burial   Said of telephone cable or wire which, because of its ruggedness, moisture protection, and resistance to rodent attack, is suitable for installation directly in the ground, without conduit or other protection.

Discontinuity   A point of abrupt change in the impedance of a circuit, where wave reflections can occur.

Dispersion   A term used to describe the chromatic or wavelength dependence of a parameter, as opposed to the temporal dependence which is referred to as distortion. The term is used, for example, to describe the process by which an electromagnetic signal is distorted because the various wavelength components of the signal have different propagation characteristics.

Dispersion-limited operation   Used to denote operation when the dispersion, rather than the amplitude of the pulse, limits the distance between repeaters. In this regime of operation, waveguide and material dispersion are sufficient to preclude an intelligent decision as to whether a pulse is present or absent.

Distort   To change the natural shape of, said particularly of a communication waveform during its transmission through a circuit.

Distortion   A change of signal waveform shape due to the temporal dependence of some parameter. *Note:* In a multimode fiber, the signal can suffer degradation from multimode distortion. In addition, dispersive mechanisms like waveguide dispersion, material dispersion, and profile dispersion can cause signal distortion in an optical waveguide.

Distortion-limited operation   The condition prevailing when the distortion rather than the amplitude of the received signal limits performance. The condition is reached when the system distorts the shape of the waveform beyond specified limits.

Distributed feedback   A condition which causes certain wavelengths of light in a laser cavity to resonate more strongly than others. Distributed feedback can be established by spatially periodic variations in some optical parameter of the active medium (e.g., index of refraction, or gain), by a periodic variation in the cavity shape, or by a grating used as an end mirror. The result is distributed, rather than uniform, feedback of resonating modes.

Double window   An optical fiber having desirable transmittance characteristics in both the first and second window regions.

Driving current   In diode lasers, the threshold current, e.g., the minimum electrical input (amperes) required to initiate lasing. The peak driving current is the maximum amperage the diode can accept before failing, usually from heat damage.

Duobinary coding   A signal design technique that codes and shapes binary data signals into a special waveform, characterized by three voltage levels. The process results in a two-to-one bandwidth compression, thus providing twice the data capacity for a given bandwidth. Duobinary coding also permits the detection of errors without the addition of error-checking bits to characters.

Duty factor, pulse (dimensionless)   The ratio of average pulse duration to average pulse spacing.

Dynamic range   Said of a transmission system, the difference in decibels between the noise level of the system and its overload level.

Electroluminescence   Nonthermal conversion of electrical energy into light. An example is the photon emission resulting from electron-hole recombination in a pn junction such as in a light-emitting diode.

Electromagnetic wave   A wave capable of propagating energy through space at the speed of light, consisting of electric and magnetic fields at right angles to each other and to the direction of propagation. Depending upon its frequency, the wave may be known as a radio wave, a light wave, or an x-ray.

Electronic   Describing devices which depend upon the flow of electrons in a vacuum or in semiconductors such as electron tubes and transistors.

Electron volt   The amount of energy gained by one electron in passing from a point to another point which is one volt higher in potential.

Equilibrium mode distribution   The condition in a multimode optical waveguide in which the relative power distribution among the propagating modes is independent of length.

Excess insertion loss   In an optical waveguide coupler, the optical loss associated with that portion of the light which does not emerge from the nominally operational ports of the device.

Extinction ratio   The ratio of light transmitted when a modulator is turned on to that transmitted when it is turned off.

Extrinsic joint loss   Loss caused by imperfect alignment of fibers in a connector or splice. Contributors include angular misalignment, lateral offset, end separation, and end finish. Generally synonymous with insertion loss.

Far infrared   A part of the spectrum often defined as containing the wavelengths greater than 2.5 μm.

FDM   Abbreviation for frequency division multiplexing; in optical communications, one also encounters wavelength division multiplex (WDM); WDM involves the use of several distinct optical sources (lasers), each having a distinct center frequency. FDM may be used with any or all of those distinct sources.

Ferrule   A mechanical fixture, generally a rigid tube, used to confine the stripped end of a fiber bundle or a fiber.

FET photodetector   A photodetector employing photogeneration of carriers in the channel region of a field effect transistor structure to provide photodetection with current gain.

Fiber bandwidth   The lowest frequency at which the magnitude of the fiber transfer function decreases to a specified fraction of the zero frequency value. Often, the specified value is one-half the optical power at zero frequency.

Fiber buffer   A material that may be used to protect an optical fiber waveguide from physical damage, providing mechanical isolation and/or protection.

Fiber-optic link   Any optical transmission channel designed to connect two end terminals or to be connected in series with other links.

Fiber optics (FO)   The branch of optical technology concerned with the transmission of radiant power through fibers made of transparent materials such as glass, fused silica, or plastic.

First window   Characteristic of an optical fiber having a region of relatively high transmittance surrounded by regions of low transmittance in the wavelength range of 800 to 900 nm.

Free-electron laser   A type of laser that differs from all others in that the optical energy comes from unbound (''free'') electrons rather than from electrons which are bound to an atom or molecule, or from molecular vibrations. Although still experimental, these lasers offer the potential to be efficient, tunable, and very powerful sources of coherent radiation.

Frequency modulation (FM)   A process whereby the frequency of a single-frequency carrier is varied in accordance with the instantaneous value of a modulating wave.

Frequency response   The transmission gain or loss of a system, measured over the useful bandwidth, compared to the gain or loss at some reference frequency.

Fresnel reflection   The reflection of a portion of the light incident on a planar interface between two homogeneous media having different indexes of refraction. Fresnel reflection occurs at the air-glass interfaces at entrance and exit ends of an optical fiber. The resultant transmission losses (on the order of 4 percent per interface) can be virtually eliminated by the use of antireflection coatings or index matching materials.

Fused quartz   Glass made by melting natural quartz crystals; not as pure as vitreous silica.

Fused silica   Synonym for vitreous silica. See also **fused quartz**.

Fusion splice A splice accomplished by the application of localized heat sufficient to fuse or melt the ends of two lengths of optical fiber, forming a single continuous fiber.

Fusion splicer An instrument which permanently joins two optical fibers by welding their cores together with a brief electric arc.

Gain The amplification of a signal's intensity as it propagates through an active medium.

Gallium aluminum arsenide (GaAlAs) The compound used to make most semiconductor lasers that operate at 800 to 900 nm in wavelength.

Giga A prefix used to represent one billion, or 10^9, or 1,000,000,000; abbreviated as G, as in GHz, one billion Hertz per second.

Gigahertz (GHz) One billion hertz. One billion cycles per second.

Gopher-protected cable A cable for buried use having extra steel tape protection to discourage gophers from gnawing on it.

Graded index fiber An optical fiber which has a refractive index that gets progressively lower away from the center. This characteristic causes the light rays to be continually refocused by refraction in the core. In general, a fiber type wherein the core refractive index decreases almost parabolically radially outward toward the cladding.

Grade of service The ability of a telephone system to connect one subscriber with another (1) without having the call blocked by busy trunks or (2) without causing the call to wait longer than an acceptable time, both expressed as the probability that the blocking or unacceptable delay will occur.

Guided wave A wave that is concentrated between materials having different properties, and is propagated within those boundaries.

Hertz (Hz) Frequency of periodic oscillations, expressed in cycles per second.

Heterodyne 1. Combining two carriers to generate a new carrier which may be either the sum or difference of the original frequencies. 2. To shift a carrier frequency to a new frequency by combining it with another carrier which is locally generated.

Heterodyne detection The detection of a signal at the beat frequency of two component signals. If one of the component signals is from a powerful "local oscillator," the second signal can be detected with great sensitivity.

Heterodyne frequency Either of the two frequencies, the sum and the difference, which result from an amplitude modulation process.

High loss Sometimes defined as optical waveguide with attenuation of more than 50 dB/km.

Horsepower A unit of mechanical power equivalent to 550 foot-pounds per second, or 745.7 W.

Hydroxyl ion absorption Absorption of optical power in optical fiber due to hydroxyl (OH) ions. This absorption has to be minimized for low fiber loss.

ILD Injection laser diode.

Index matching material A material, often a liquid, gel, or cement, whose refractive index is nearly equal to an optical element index. Material with an index nearly equal to that of an optical fiber's core is used in splicing and coupling to reduce reflections from the fiber end face.

Index of refraction The *relative* index of refraction is a fraction or ratio of the velocity of light in one medium compared to the velocity of light in another medium. The *absolute* index of refraction is a fraction or ratio of the velocity of light in a given medium compared to the velocity of light in a vacuum. Because the density of air is so low, it is convenient to consider the velocity of light in air the same as its velocity in a vacuum.

Index profile   A characteristic of an optical fiber which describes the way its index of refraction changes with its radius.

Infrared   Electromagnetic radiation with wavelength between 0.7 μm and about 1 mm. Wavelengths at the shorter end of this range are frequently called near infrared, and those longer than about 20 micrometers, far infrared.

Injection laser diode (ILD)   A laser employing a forward-biased semiconductor junction as the active medium. Synonyms are diode laser and semiconductor laser.

Insertion loss   The total optical power loss caused by the insertion of an optical component such as a connector, splice, or coupler. Losses intrinsic to a coupler or connector include angular misalignment loss, gap loss, and lateral offset loss. Losses extrinsic to a coupler or connector are caused by mismatches in fiber parameters or dimensions.

Integrated circuit   A functional circuit whose components and interconnecting leads are chemically formed on a single chip of semiconductor material.

Integrated optics   Devices in which several optical components are "integrated" onto a single substrate; analogous to integrated electronic circuits. Although still in the research phase, integrated optics has potential for use in optical signal processing and in fiberoptic communications.

Intensity   The square of the electric field amplitude of a light wave. Intensity is proportional to irradiance and may be used in place of the term when only relative values are important.

Intermodal distortion   Synonym for multimode distortion.

Ion exchange technique   A method of fabricating a graded index optical waveguide by an ion exchange process.

Ionization   The process of giving net charge to a neutral atom or molecule by adding or subtracting an electron. Can be accomplished by radiation or by creation of a strong electric field.

Jacket   A layer of material, generally plastic, that surrounds an optical fiber to protect it from physical damage. Unlike the cladding, the jacket is physically distinct from the fiber core.

Jitter   Time-related, abrupt, spurious variations in the duration of any specified related interval.

Joule   An international unit of work or energy. The work required to maintain a current of one ampere through one ohm for one second. A watt-second.

Kilo-   A prefix for one thousand (1,000, or 10^3).

Kilobit   One thousand bits.

Kilohertz (kHz)   1. One thousand hertz. 2. One thousand cycles per second.

Laser   Acronym for "light amplification by stimulated emission of radiation." A device which generates or amplifies electromagnetic oscillations at wavelengths between the far infrared (submillimeter) and ultraviolet. Like any electromagnetic oscillator, a laser oscillator consists of two basic elements: an amplifying (active) medium and a regeneration or feedback device (resonant cavity). A laser's amplifying medium can be a gas, semiconductor, dye solution, etc.; feedback is typically from two mirrors. Distinctive properties of the electromagnetic oscillations produced include monochromaticity, high intensity, small beam divergence, and phase coherence. As a description of a device, "laser" refers to the active medium plus all equipment necessary to produce the effect called lasing.

Laser head   The enclosure containing the active medium, resonant cavity and other components and accessories of a laser except for the power supply.

Lashed cable   An aerial cable fastened to its supporting messenger by a continuous spirally wrapped steel wire.

Lasing threshold   The lowest excitation level at which a laser's output is dominated by stimulated emission rather than spontaneous emission.

Launch numerical aperture (LNA)   The numerical aperture of an optical system used to couple (launch) power into an optical waveguide.

Lay   1. The fashion in which wires are twisted together to form a cable. 2. The axial distance along the cable that a conductor advances in one spiral turn around the cable core.

LED   Acronym for light-emitting diode.

Light   1. In a strict sense, the region of the electromagnetic spectrum that can be perceived by human vision, designated the visible spectrum and nominally covering the wavelength range of 0.4 μm to 0.7 μm. 2. In the laser and optical communication fields, custom and practice have extended usage of the term to include the much broader portion of the electromagnetic spectrum that can be handled by the basic optical techniques used for the visible spectrum. This region has not been clearly defined, but, as employed by most workers in the field, it may be considered to extend from the near-ultraviolet region of approximately 0.3 μm, through the visible region, and into the mid-infrared region to 30 μm.

Light emitting diode (LED)   A pn junction semiconductor device that emits incoherent optical radiation when biased in the forward direction.

Lightguide   Synonym for optical waveguide.

Light source   A generic term that includes lasers and LEDs.

Lightwave   Any electromagnetic radiation having a wavelength in the range from 800 to 1,600 nm in the near-infrared region.

Linewidth   The frequency or wavelength range over which most of the laserbeam's energy is distributed.

Long wavelength   As applied to fiberoptic systems, operation at wavelengths in the range of 1,100 to 1,700 nm.

Loss   See **attenuation**.

Low-loss   Sometimes defined as optical waveguide with attenuation of less than 10 dB/km.

Macrobending   In an optical waveguide, all macroscopic deviations of the axis from a straight line; distinguished from microbending.

Material dispersion   1. Light impulse broadening due to differential delay of various wavelengths of light in a waveguide material. This group delay is aggravated by broad-linewidth light sources.   2. That dispersion attributable to the wavelength dependence of the refractive index of the material used to form the waveguide.

Mechanical splice   A fiber splice accomplished by fixtures or materials rather than by thermal fusion. Index-matching material may be applied between the two fiber ends.

Mega-   A prefix for one million (1,000,000, or 10^6).

Megabit   One million bits.

Megahertz (MHz) One million hertz. One million cycles per second.

Micro- a prefix for one millionth (10^{-6}).

Microbending In an optical waveguide, sharp curvatures involving local axial displacements of a few micrometers and spatial wavelengths of a few millimeters. Such bends may result from waveguide coating, cabling, packaging, installation, etc.

Micron The unit used for specifying the wavelength of light, equal to one millionth of a meter.

Milli- A prefix for one thousandth (10^{-3}).

Milliwatt One thousandth of a watt.

Modal dispersion That component of pulse spreading caused by differential optical path lengths in a multimode fiber.

Mode In any cavity or transmission line, one of those electromagnetic field distributions that satisfies Maxwell's equations and the boundary conditions. The field pattern of a mode depends on the wavelength, refractive index, and cavity or waveguide geometry.

Mode coupling In an optical waveguide, the exchange of power among modes. The exchange of power may reach statistical equilibrium after propagation over a finite distance that is designated the equilibrium length.

Mode field diameter A functional representation of the energy-carrying region of the fiber. Also referred to as spot size.

Mode filter A device used to select, reject, or attenuate a certain mode or modes.

Modem A single unit of equipment which combines the functions of modulator and demodulator. The modem is an economical arrangement, since the two circuits can use common elements.

Modified chemical vapor deposition (MCVD) A process for making low-loss optical waveguide. The process involves passing various reactant gases down the center of a glass tube and heating the tube with a burner that moves back and forth outside the tube. This results in oxidation of some of the gas vapor. The oxidized glassy flecks adhere to the walls of the tube. In practice, the tube is rotated at a rate of at least 100 rpm to achieve uniformity of reaction and uniform deposition. After a sufficient buildup of glassy particulates has been achieved, the tube is heated to the melting point and collapsed to form a rod. The rod is fire polished and inserted in a furnace, and the fiber is drawn.

Modulation A controlled variation with time of any property of a wave for the purpose of transferring information.

Modulator A device that produces a controlled variation of a laserbeam's intensity or phase. Common types of modulators are electro-optical and acousti-optical.

Monochromatic Consisting of a single wavelength or color. In practice, radiation is never perfectly monochromatic, but, at best, displays a narrow band of wavelengths.

Monomode optical waveguide Synonym for single mode optical waveguide.

Multifiber cable An optical cable that contains two or more fibers, each of which provides a separate information channel.

Multimode A term that describes optical waveguide that permits the propagation of more than one mode.

Multimode distortion In a multimode optical fiber, that pulse distortion resulting from differential mode propagation rates.

Multimode fiber   A fiber that supports propagation of more than one mode of a given wavelength.

Multimode laser   Simultaneous emission at several wavelengths. Also called multiline in gas lasers.

Multiplexer   A device which combines two or more optical signals onto one communications channel. The signals can be of different wavelengths (wavelength division multiplexing) or can occupy different time slots (time division multiplexing). A multiplexer combines information signals from several channels into one single optical channel for transmission.

NA   Abbreviation for numerical aperture.

Nano-   A prefix meaning one thousandth of a millionth. One billionth (10^{-9}).

Nanosecond   One billionth of a second.

NEP   Acronym for noise equivalent power.

Neutral density filter   A filter which reduces the intensity of light without affecting its spectral character.

Noise   1. Any random disturbance in a communication system which tends to obscure the clarity and validity of a signal in relation to its intended end use. 2. Any signal having random fluctuations and frequency components. 3. An unwanted signal in a communication system; strictly speaking, if not random, it should be called interference.

Noise equivalent power (NEP)   At a given modulation frequency, wavelength, and for a given effective noise bandwidth, the radiant power that produces a signal-to-noise ratio of 1 at the output of a given detector.

Noise, intrinsic   Thermal noise which is normally present in a transmission path or device and which is neither caused by nor affected by input level or system loading.

Noise, thermal   Noise produced by the random motion of free electrons in all electrical conductors. The movement of an electrical charge (electron) through the resistance of the conductor produces a (noise) voltage. Thermal noise is white noise.

Noise, white   A noise whose power per unit of frequency is essentially independent of frequency over a specified frequency range. White noise is a broadband noise having constant energy per hertz, per 100 hertz, etc. Its amplitude-frequency curve slopes upward at 3 dB per octave.

Nomogram   A chart having a series of curved or straight-line scales across which a straight-edge can be laid to give a graphical solution of an equation involving three variables. Also called a nomograph.

Non-return to zero   See **NRZ signal**.

NRZ signal   A signal which is continuous through mark and space elements. Examples are multilevel signals and frequency-shift signals. NRZ means non-return (to) zero.

Numerical aperture (NA)   Measure of light acceptance of an optical waveguide.

Optical axis   The axis of symmetry of an optical system.

Optical cable   A fiber, multiple fibers, or a fiber bundle in a structure fabricated to meet optical, mechanical, and environmental specifications.

Optical cable assembly   An optical cable that is connector terminated. Generally, an optical cable that has been terminated by a manufacturer and is ready for installation.

Optical cavity   A region bounded by two or more reflecting surfaces, referred to as mirrors, end mirrors, or cavity mirrors, whose elements are aligned to provide multiple reflections. The resonator in a laser is an optical cavity.

Optical combiner A passive device in which power from several input fibers is distributed among a smaller number (one or more) of input fibers.

Optical detector A transducer that generates an output signal when irradiated with optical power.

Optical fiber Any filament or fiber, made of dielectric materials, that guides light, whether or not it is used to transmit signals.

Optical fiber preform An optical material structure that is an intermediate step in optical fiber manufacturing. It is basically a scaled-up version of the desired fiber. The optical fiber preform has the form of a rod or tube which has layers of similar materials as the desired fiber. The optical fiber is drawn from the preform.

Optical link Any optical transmission channel designed to connect two end terminals or to be connected in series with other links.

Optical power Colloquial synonym for radiant power.

Optical repeater In an optical waveguide communication system, an opto-electronic device or module that receives a signal, amplifies it (or, in the case of a digital signal, reshapes, retimes, or otherwise reconstructs it), and retransmits it.

Optical spectrum Generally, the electromagnetic spectrum within the wavelength region extending from the vacuum ultraviolet at 40 nm to the far-infrared at 1 mm.

Optical time-domain reflectometer An instrument which locates faults in an optical fiber by sending a short pulse of light through the fiber and then timing the arrival of backscattered signals which originate at discontinuities in the fiber.

Optical time domain reflectometry A method for characterizing a fiber wherein an optical pulse is transmitted through the fiber and the resulting light is scattered and reflected back to the input and is measured as a function of time. Useful in estimating attenuation coefficient as a function of distance and identifying defects and other localized losses.

Optical waveguide 1. Any structure capable of guiding optical power. 2. In optical communications, generally a fiber designed to transmit optical signals.

Opto-electronic Pertaining to a device that responds to optical power, emits or modifies optical radiation, or utilizes optical radiation for its internal operation. Any device that functions as an electrical-to-optical or optical-to-electrical transducer. *Note:* Photodiodes, LEDs, injection lasers, and integrated optical elements are examples of opto-electronic devices commonly used in optical waveguide communications.

Opto-electronic device A device which is responsive to electromagnetic radiation (light) in the visible, infrared, or ultraviolet spectral regions; emits or modifies noncoherent or coherent electromagnetic radiation in these same regions; or utilizes such electromagnetic radiation for its internal operation.

Output power Radiant power, expressed in watts.

Paired cable Cable in which the conductors are combined in pairs, i.e., two wires which are twisted about each other. Each wire of the pair has its distinctive color of insulation.

Parity Describing a self-checking code employing binary digits in which the total number of "ones" or "zeros" is always even or always odd. Also, the state of being equal or equivalent.

Passive Describing a device which does not contribute energy to the signal it passes. Also, a term sometimes used to describe devices which do not consume operating power to perform their designed functions.

Photoconductor A semiconductor detector in which the intensity of incident radiation is measured by observing light-induced changes in device resistance while applying an external voltage to the detector.

Photocurrent The current that flows through a photosensitive device (such as a photodiode) as the result of exposure to radiant power. Internal gain, such as that in an avalanche photodiode, may enhance or increase the current flow, but is a distinct mechanism.

Photodetector Any device which detects light, generally producing an electronic signal with intensity proportional to that of the incident light.

Photodiode A diode designed to produce photocurrent by absorbing light. Photodiodes are used for the detection of optical power and for the conversion of optical power to electrical power.

Photoelectric effect Historically, an effect referring to all changes in material electrical characteristics due to photon absorption. More recently, descriptive of the emission of electrons as the result of the absorption of photons in a material. This definition is quite broad, since the photons can be of any energy and the electrons can be released into a vacuum or into a second material. The material itself may be solid, liquid, or gas. With this broad definition, photoconductive, photoelectromagnetic, photoemissive, and photovoltaic effects are all photoelectric.

Photoemission The process by which light incident on certain materials in a vacuum causes emission of electrons.

Photometer An instrument that measures light intensity. Strictly speaking, a photometer measures light intensity in photometric units (lumens) corresponding to the intensities of light perceived by the human eye, but the term has been applied to systems that measure intensity in watts at wavelengths outside the visible region.

Photon A quantum of electromagnetic energy.

Photovoltaic effect A mode of detector operation in which light incident at a semiconductor junction generates an electric potential.

Pico- A prefix denoting one millionth of a millionth; one trillionth (10^{-12}). Pronounced "pie-ko."

Pigtail A short length of optical fiber, permanently fixed to a component, used to couple power between it and the transmission fiber.

PIN photodiode A diode with a large intrinsic region sandwiched between p-doped and n-doped semiconducting regions. Photons absorbed in this region create electron-hole pairs that are then separated by an electric field, thus generating an electric current in a load circuit.

Plant A general term applied to any of the physical property of a telephone or other transmission company which contributes to the furnishing of communication or power services.

Plastic-clad silica (PCS) fiber An optical fiber with a glass core and a plastic cladding.

Polar signal A signal whose information is contained in current reversals in the circuit, one direction of flow being considered a marking signal and the opposite direction a spacing signal.

Power Energy per unit time.

Power amplifier An amplifier designed to produce a gain in signal power, as distinguished from a voltage amplifier.

Power density Power per unit area (watts per square meter).

Power efficiency (dimensionless)   The ratio of emitted optical power of a light source to the electrical input power.

Power peak   In a pulsed laser, the maximum power emitted.

Power level   The amount of power at a point in a circuit compared with some reference power.

Preform   A tube or rod composed of glass compounds that is inserted into a furnace and from which optical waveguide is drawn.

Pulse broadening   An increase in pulse duration. *Note:* Pulse broadening may be specified by the impulse response, the root-mean-square pulse broadening, or the full-duration-half-maximum pulse broadening.

Pulse decay time   The time required for the instantaneous amplitude of an electrical wave to go from 90 percent to 10 percent of the peak amplitude.

Pulse dispersion (pulse spreading)   The separation or spreading of the input characteristics of the optical signal that appears along the length of the optical fiber and limits the useful transmission bandwidth of the fiber. Expressed in units of nanoseconds per kilometer. Three basic mechanisms for dispersion are the material effect, the waveguide effect, and the multimode effect.

Pulsed laser   A laser that emits light in pulses rather than continuously.

Pulse length   The time duration of the burst of energy emitted by a pulsed laser; also called pulse width. Usually measured at the "half-power" points (0.707 times the full height of a voltage or current pulse).

Pulse rise time   The time required for the instantaneous amplitude of an electrical wave to go from 10 percent to 90 percent of the peak amplitude.

Pulse spreading   The increase in pulse width in a given length of fiber due to the cumulative effect of material dispersion and modal dispersion.

Pulse train   A succession of pulses which follow each other closely, usually at equal intervals.

Quantum efficiency   In an optical source or detector, the ratio of output quanta to input quanta. Input and output quanta need not both be photons.

Quiescent   Inactive; without an input signal.

Radiant energy   Energy (joules) that is transferred via electromagnetic waves; there is no associated transfer of matter.

Radiant flux   The time rate of flow of radiant energy, measured in watts.

Radiant power   The time rate of flow of radiant energy, expressed in watts. The prefix is often dropped and the term "power" is used.

Radiation angle   Half the vertex angle of the cone of light emitted by a fiber.

Radiometer   An instrument for measuring incident radiation in radiometric units (watts). Radiometric measurements can be made at any wavelength, but the spectral range of a particular instrument may be limited to a narrow range.

Radiometric units   The system of units defined for measurement of the intensity of electromagnetic radiation; the basic unit is the watt.

Ray   A geometric representation of a light path through an optical device; a line normal to the wave front indicating the direction of radiant energy flow.

Rayleigh scattering   Scattering of a lightwave propagating in a material medium due to the atomic or molecular structure of the material and variations in the structure as a function of distance. The scattering losses vary as the reciprocal of the fourth power of the wavelength. The distances between scattering centers are small compared to the wavelength. Rayleigh scattering is the fundamental limit of fiber loss in the operating wavelength region (0.8–1.6 μm) of optical fiber systems.

Receiver   A unit including a detector and signal-processing electronics that converts optical input into electronic output; often used in communications.

Redundant   1. Exceeding what is necessary or normal. 2. Containing more information than is needed for intelligibility. About 75 percent of the information content of normal speech is redundant. 3. Said of the elements of equipment which exists in duplicate so that if one fails the second continues operation without interruption.

Reflection   The abrupt change in direction of a light beam at an interface between two dissimilar media so that the light beam returns into the medium from which it originated.

Refraction   The bending of a beam of light at an interface between two dissimilar media or in a medium whose refractive index is a continuous function of position (graded index medium).

Refractive index n (dimensionless)   The ratio of the velocity of light in a vacuum to the velocity of light in the specified medium.

Regeneration   The process of receiving distorted signal pulses and recreating new pulses from them at the correct repetition rate, correct pulse amplitude, and correct pulse width.

Regenerative repeater   A device which receives distorted digital signals and then reshapes and retimes the signals before retransmitting them.

Repeater   A signal amplification device, often used along fiberoptic cables to extend transmission distances.

Replication   Production of optical components by casting epoxy against an optically finished master to which coatings have been applied. The epoxy, which adheres to an optically unfinished substrate, picks up the coatings and copies the surface finish of the master.

Resonator   A cavity in which radiation of certain discrete frequencies can set up standing waves. In a laser resonator, the cavity is bounded at two ends by mirrors, one totally reflective and one partially reflective.

Responsivity   The ratio of an optical detector's electrical output to its optical input, the precise definition depending on the type of detector; generally expressed in amperes per watt or volts per watt of incident radiant power.

Ribbon cable   A cable whose conductors lie side by side in a single plane. Usually has a molded polyethylene insulation.

Rise time   For an emitter, the time it takes for light intensity to rise from 10 percent to 90 percent of peak output. Detector rise time—also called response time—is the time during which the detector output goes from 10 percent to 90 percent of peak.

RZ signal   A discontinuous signal in which mark signals are separated by spaces which contain no signal. RZ signifies return to zero.

Scattering   The change in direction of light rays or photons after striking a small particle or particles. Also, the diffusion of a light beam caused by the inhomogeneity of the transmitting medium.

Second window   Characteristic of an optical fiber having a region of relatively high transmittance surrounded by regions of low transmittance in the wavelength range of 1,200 to 1,350 nm.

Semiconductor   A material whose resistivity is between that of conductors and insulators, and whose resistivity can sometimes be changed by light, an electric field, or a magnetic field. Current flow is sometimes by movement of negative electrons and sometimes by transfer of positive holes. Used in transistors, diodes, photodiodes, photocells, and thermistors. Some examples are silicon, germanium, selenium, and lead sulfide.

Semiconductor laser   A laser in which lasing action occurs at the junction of n-type and p-type semiconductor materials. Most such lasers used in fiberoptic systems are made of GaAlAs. The wavelength of the emitted light may be made to vary from 730 nm to 925 nm by varying the aluminum content. The InGaAsP laser emits light at wavelengths from 1,020 nm to 1,700 nm.

Semiconductor, n-type   A semiconductor material, such as germanium or silicon, which has a small amount of impurity, such as antimony, arsenic, or phosphorus added to increase the supply of free electrons.

Semiconductor, p-type   A semiconductor material which has been doped so that it has a net deficiency of free electrons.

Sensitivity   Imprecise synonym for responsivity. In optical system receivers, the minimum power required to achieve a specified quality of performance in terms of output signal-to-noise ratio or some other measure.

Serial   Said of pulses which are sent separately, one after the other. The opposite of parallel.

Service   1. The aggregate of all the acts performed by a telephone company in providing communications to its customer. 2. The quality of that which is provided. 3. A measure of the traffic adequacy of a communications facility. See also **grade of service**.

Signal-to-noise ratio (S/N ratio)   The difference in amplitude of a signal (before modulation or after detection of a modulated carrier) and the noise present in the spectrum occupied by the signal when both are measured at the same point in the system.

Silica—silicon dioxide, SiO_2   Naturally occurring as rock crystal, or quartz. When SiO_2 is used as a chemical compound, it is melted and forms fused silica.

Silicon tetrachloride ($SiCl_4$)   The major constituent of optical waveguides. $SiCl_4$ is usually furnished as a liquid, and oxygen is bubbled into the container. This results in the formation of a gas that is directed into a fused quartz tube, which becomes the preform from which optical waveguide is drawn.

Single mode fiber   An optical waveguide through which only one mode will propagate. Single mode waveguide is produced by reducing the diameter of the core of the waveguide to 2 to 10 microns. The diameter of the core is dependent on the difference in the refractive indexes of the core and cladding. As the difference in the refractive indexes of the core and cladding decreases, the diameter of the core increases. Theoretically, the core could be infinitely large as the difference in the indexes become infinitely small. Single mode operation is desirable because all modes except the lowest and simplest mode are excluded. This reduces the time distortion of signals propagating in unwanted modes, retains phase relationships, and reduces dispersion to the lowest possible value.

SNR   Abbreviation for signal-to-noise ratio.

Solid-state   Denoting the use of semiconductors, such as diodes and transistors, instead of vacuum tubes.

Solid-state laser   A laser whose active medium is glass or crystal.

Source   A device that, when properly driven (with electrical energy), will produce information-carrying optical signals.

Source efficiency   The ratio of the emitted optical power of a source to the input electrical power.

Spectral width   A measure of the wavelength extent of a spectrum. One method of specifying the spectral linewidth is the full width at half maximum (FWHM), i.e., the difference between the wavelengths at which the magnitude drops to one-half of its maximum value.

Spectrum   A continuous range of frequencies, wide in extent, within which waves have some specified characteristic in common, e.g: audio spectrum, radio spectrum, etc.

Splice   A permanent connection of two optical fibers.

Splitter   A passive device which divides optical power among several output fibers from a common input.

Star coupler   A passive device in which power from one or several input waveguides is distributed among a larger number of output optical waveguides.

Step index fiber   A type of fiber which has an abrupt change in index of refraction at the core-cladding interface. Generally, such fibers have larger cores, higher losses, and lower bandwidths than graded index types.

Step index profile   A refractive index profile characterized by a uniform refractive index within the core and a sharp decrease in refractive index at the core-cladding interface.

Subscriber's loop   Circuit between a local office and a subscriber's telephone set.

Substrate   The material onto which an optical coating is deposited to form components such as filters, mirrors, and beamsplitters. Also, the base layer for a semiconductor component.

Suspension strand   Cable made of steel wires twisted together, supported by poles. Used to carry an aerial cable which is lashed to it.

Synchronous operation   A method of on-line transmission of telegraphic or encrypted signals in which the sending and receiving terminals are kept in step by a timing device, whether traffic is being passed through or not.

T-1 carrier system   A 24-channel, transistorized, time-division, pulse-code modulation voice carrier generally used on exchange cable to provide short-haul trunks. Uses two pairs, in one or two cables, for two directions of transmission. Requires regenerative pulse repeaters at approximately 6,000 foot intervals.

Tap   A device for extracting a portion of the optical signal from a fiber.

Tensile strength   The maximum stress that a material subjected to a stretching load can withstand without permanent damage. Tensile strength is a common way of specifying the strength of optical fibers and cables.

Terahertz (THz) 10^{12} Hz.

Thermal noise-limited operation   Operation wherein the minimum detectable signal is limited by the thermal noise of the detector and load resistance, and by amplifier noise.

Thermoelectric cooling   A method of cooling in which an electric current is passed through two dissimilar metals joined at two points; heat is liberated at one junction and absorbed at the other junction.

Thin film circuit A circuit whose elements are films formed on an insulating substrate. A thin film is one which is only several molecules thick.

Threshold current For a diode laser, the current above which optical output is coherent. When driven at currents below threshold, the laser behaves as an incoherent LED.

Timing signal Signal used to ensure synchronization of interconnected digital equipment. Usually a symmetrical square-wave signal.

Total internal reflection The total reflection that occurs when light strikes an interface at angles of incidence (with respect to the normal) greater than the critical angle.

Transducer A device which receives a wave from one transmission system or medium and transmits a wave containing equivalent information to a different system or medium.

Transistor A three-terminal electrical device made of semiconductor material and capable of performing amplification or switching of electrical signals.

Transmission loss Total loss encountered in transmission through a system.

Transmitter In a fiberoptic system, the device which converts a modulated electrical signal into an optical signal for transmission through a fiber. A transmitter typically consists of a light source (LED or diode laser) and driving electronics.

Trunk One telephone communication channel between (1) two ranks of switching equipment in the same central office, (2) central office units in the same switching center, or (3) two switching centers. A trunk is for the common use of all calls of a given category between its two terminals.

Tunable laser A laser or a parametric oscillator whose emission can be varied across a broad spectral range.

Two-way Describing a transmission system which can transport signals in both directions simultaneously.

Ultraviolet Electromagnetic radiation with wavelengths between about 40 and 400 nm. Radiation between 40 and 200 nm is termed vacuum ultraviolet because it is absorbed by air and travels only through a vacuum. The near-ultraviolet has wavelengths close to those of visible light; the far-ultraviolet has shorter wavelengths.

Underground cable Cable installed in subsurface conduits terminating at intervals in manholes, thus permitting the placement, replacement, or removal of cables at will.

Vapor-phase axial deposition (VAD) A process for making optical waveguide wherein the preform results from gases directed at the end of a rotating glass or rod mandrel. Flame hydrolysis causes soot to form both the core and cladding. The rod is pulled at a given rate, depending on the diameter of the preform. After the soot is deposited, the preform moves into an oven and is heated to remove any water. The preform is then treated in an atmosphere of thionyl chloride ($SoCl_2$), sintered, collapsed into a rod, and drawn into optical waveguide.

Velocity of light The velocity of light in a vacuum is 2,997,925 meters per second or 186,280 miles per second. For rough calculations, the figure of 3,000,000 meters per second is generally used.

Video 1. Pertaining to the signal which carries a television picture. 2. Describing the four-megahertz-wide band of frequencies which constitutes a television signal.

Visible light That part of the spectrum to which the human eye is sensitive, usually defined as wavelengths between 390 and 780 nm.

Vitreous silica Glass consisting of almost pure silicon dioxide (SiO_2).

Voice channel   A transmission path suitable for carrying analog voice signals, covering a frequency band of 250 to 3,400 Hz.

Watt   The unit of electric power equal to the rate of work when a current of one ampere flows under a pressure of one volt. For direct currents, a watt is equal to the product of the voltage and current, or the product of circuit resistance by the square of the current. For alternating currents, a watt is equal to the product of effective volts and effective current times the circuit power factor.

Wave   1. A periodic variation of an electric voltage or current. 2. A wave motion in any medium—mechanical as in water, acoustical as sound in air, electrical as current waves on wires, or electromagnetic as radio and light waves through space.

Waveguide   Any device which guides electromagnetic waves along a path defined by the physical construction of the device.

Wavelength   The distance between three consecutive nodes of a wave, equal to 360 electrical degrees. Wavelength is equal to the velocity of propagation divided by the frequency, when both are in the same units.

Wavelength division multiplexing (WDM)   The provision of two or more channels over a common optical waveguide, the channels being differentiated by optical wavelength.

WDM   Wavelength division multiplex.

Wideband   1. Passing a wide range of frequencies without distortion. 2. Having a bandwidth of 20 kHz or more. 3. Describing digital circuits or equipment capable of handling 50-kilo-bits-per-second signals.

Zero state   The condition of a binary memory cell when a "zero" is stored.

Reference Tables

Dimensional
Relationships

	A	nm	μm	μ	m
1 angstrom (A)	1	10^{-1}	10^{-4}	10^{-4}	10^{-10}
1 nanometer (nm)	10	1	10^{-3}	10^{-3}	10^{-9}
1 micrometer (μm)	10^4	10^3	1	1	10^{-6}
1 micron (μ)	10^4	10^3	1	1	10^{-6}
1 meter (m)	10^{10}	10^9	10^6	10^6	1

Numerical
Prefixes

Prefix	Symbol	Value	Notation	
tera	T	one million million	1,000,000,000,000	10^{12}
giga	G	one billion	1,000,000,000	10^9
mega	M	one million	1,000,000	10^6
kilo	k	one thousand	1,000	10^3
hecto	h	one hundred	100	10^2
deka	da	ten	10	10^1
deci	d	one tenth	.1	10^{-1}
centi	c	one hundredth	.01	10^{-2}
milli	m	one thousandth	.001	10^{-3}
micro	μ	one millionth	.000 001	10^{-6}
nano	n	one thousandth of one millionth	.000 000 001	10^{-9}
pico	p	one millionth of one millionth	.000 000 000 001	10^{-12}
femto	f	one quadrillionth	.000 000 000 000 001	10^{-15}

Wavelength vs Frequency

$$\text{Frequency} = \frac{\text{Speed of Light}}{\text{Wavelength (m)}}$$

$$\text{Frequency} = \frac{3 \times 10^8 \text{ m/s}}{\text{Wavelength (m)}}$$

Wavelength	Frequency
0.85 μm	3.53×10^{14} Hz
0.90 μm	3.33×10^{14} Hz
1.20 μm	2.50×10^{14} Hz
1.30 μm	2.31×10^{14} Hz
1.50 μm	2.00×10^{14} Hz

Numerical Aperture (NA) vs Acceptance Angle

Core NA	Full Acceptence Angle
0.10	11.48 degrees
0.15	17.25 degrees
0.20	23.07 degrees
0.25	28.95 degrees
0.30	34.91 degrees
0.35	40.97 degrees

Equivalent Power Values of dBm

dBm	Power	dBm	Power
0	1.00 mW	0	1.00 mW
−1	.794 mW	+1	1.26 mW
−2	.631 mW	+2	1.58 mW
−3	.501 mW	+3	1.99 mW
−4	.398 mW	+4	2.51 mW
−5	.316 mW	+5	3.16 mW
−6	.251 mW	+6	3.98 mW
−7	.199 mW	+7	5.01 mW
−8	.158 mW	+8	6.31 mW
−9	.126 mW	+9	7.94 mW
−10	.100 mW	+10	10.0 mW
−11	.079 mW	+11	12.6 mW
−12	.631 mW	+12	15.8 mW
−13	.051 mW	+13	19.9 mW
−14	.039 mW	+14	25.1 mW
−15	.031 mW	+15	31.6 mW
−16	.025 mW	+16	39.8 mW
−17	.019 mW	+17	50.1 mW
−18	.015 mW	+18	63.1 mW
−19	.012 mW	+19	79.4 mW
−20	10.0 μW	+20	100.0 mW

dBm	Power	dBm	Power
–21	7.94 μW	+21	125.0 mW
–22	6.30 μW	+22	158.0 mW
–23	5.01 μW	+23	200.0 mW
–24	3.98 μW	+24	251.0 mW
–25	3.16 μW	+25	316.0 mW
–26	2.51 μW	+26	398.0 mW
–27	1.99 μW	+27	.50 W
–28	1.58 μW	+28	.63 W
–29	1.25 μW	+29	.79 W
–30	1.00 μW	+30	1.00 W
–31	.794 μW	+31	1.25 W
–32	.630 μW	+32	1.58 W
–33	.501 μW	+33	1.99 W
–34	.398 μW	+34	2.51 W
–35	.316 μW	+35	3.16 W
–36	.251 μW	+36	3.98 W
–37	.199 μW	+37	5.01 W
–38	.158 μW	+38	6.30 W
–39	.125 μW	+39	7.94 W
–40	.100 μW	+40	10.00 W

Answers to Odd-Numbered Review Questions

Quest.#	Answer	Comment
Chapter 1		
1	True	The tactile theory was discredited by the 11th Century.
3	False	In 1864, James Clark Maxwell combined the studies of electricity, magnetism, and light into one theory.
5	False	The photon is the unit of measure of light energy. When light is absorbed the photon is converted to some other form of energy, such as heat, for example.
7	True	Light is electromagnetic energy and as such it moves in a wavefront just as sound and radio waves do.
9	False	Since the velocity of propagation of light is different in different media, the distance it will travel during a full cycle will be different also.
11	False	The velocity of propagation of light is different in different media. It is fastest in a vacuum and slower in air, for example.
13	True	See Figure 1–3 in the text.
15	False	See Figure 1–4 in the text.
17	True	This is called the photoelectric effect. Light may also cause chemical changes in some substances.
19	False	As shown in Figure 1–5 in the text, the angle of incidence always equals the angle of reflection from a smooth surface.
Chapter 2		
1	False	See Figure 1–4 in the text.
3	False	The Law of Inverse Squares applies, and the level of illumination will vary inversely with the square of the distance from the light source.
5	True	Those wavelengths of light that are absorbed will not be observable as colors to a viewer, since they will not be re-radiated to the viewer.
7	False	From the Greek words "chromo" and "monos," monochromatic refers to "having or exhibiting only one color."

Quest.#	Answer	Comment
9	False	Opacity is the quality or state of a body that makes it impervious to light rays.
11	False	Rayleigh's theory applies to particles that are no larger than about one-tenth of the wavelength of light.
13	False	The primary causes of light attenuation are absorption and scattering.
15	True	Radiant electromagnetic energy does travel in a wavefront.
17	True	This technique of measuring light energy is perhaps the one most commonly encountered.
19	True	The decibel, a tenth of a Bel, is a common term to denote the ratio between two levels of power, and lightwave energy is power.

Chapter 3

Quest.#	Answer	Comment
1	False	Since an optical fiber presents a relatively small cone of acceptance for light energy, a source that presents a narrower beam of light would couple more energy into the fiber.
3	True	Since a digital bit is a two state unit, being either one or zero value, any two state condition of the light, on or off for example, could represent modulation.
5	False	Chromatic dispersion may be the limiting factor in a single mode fiber system, but it would not be in a multimode system.
7	False	Modal dispersion is endemic in multimode fibers where more than one mode of propagation is supported.
9	True	The term "refractive index" denotes the velocity of propagation of a medium, and the indices of more dense media are higher, reflecting the slower transmission velocity that will be experienced.
11	True	See Figure 3–10 in the text.
13	False	Graded index fibers depend upon the refractive optical mechanisms of the fiber to minimize pulse dispersion.
15	False	Mode Field Diameter defines the active, light propagating cross section of a single mode fiber, and this cross section is always somewhat larger than the core diameter itself.
17	True	This is so since the Numerical Aperture defines the cone of acceptance for light rays presented as fiber input. If the cone of acceptance is narrower, it is more difficult to efficiently couple light into the fiber.
19	False	There is a direct correlation between the transmission data rate that can be supported and the bandwidth that a transmission facility presents for use.

Chapter 4

Quest.#	Answer	Comment
1	True	This is true of all types of fiber, single mode, step index multimode, and graded index multimode.
3	False	The attenuation peaks in optical fibers are due to contaminants in the glass itself. Techniques for removing impurities from glass are constantly being improved and such attenuation peaks are being drastically reduced if not eliminated entirely.
5	False	Fiber attenuation is denoted as a power loss ratio, but the term is dB/km. The term dBm denotes a discrete power level, not a power ratio.
7	False	Bipolar signals cannot be transmitted through lightwave systems. There is a modification of bipolar format, called pulse bipolar, that can, however.
9	False	Modal dispersion is caused by different propagation modes, such as reflective or refractive modes, and not by different signal wavelengths.
11	False	A design objective and a major advantage of single mode fibers is the reduction, if not total elimination, of modal dispersion.
13	False	In a single mode fiber, only one mode of propagation is supported, the axial mode, and no reflective or refractive propagation is present.

Quest.#	Answer	Comment
15	False	The velocity of propagation is different for different wavelengths of light traveling in the same medium.
17	False	A single mode fiber can pass light signals at several wavelengths within the transmission characteristics of the particular fiber.
19	True	Higher bit rate transmission involves shorter pulse periods, and such systems are particularly vulnerable to pulse dispersion. Thus such installations must strictly limit all dispersion effects from any cause.

Chapter 5

Quest.#	Answer	Comment
1	False	Zero dBm equals a discrete power level of one milliwatt. A power ratio of +3 dB would be twice as much power. Thus +3 dBm equals a power level of two milliwatts.
3	False	The term Signal to Noise Ratio is most generally employed in describing the quality of an analog signal.
5	True	To truly define the quality of the output signal, it is necessary to define the level of the input signal that will assure that quality.
7	True	The "bit" is the fundamental unit of information in a digital system and may have one of two values, such as a "zero" or a "one," for example.
9	True	Since each bit may contain logic information as a value, "zero" or "one" for example, more sophisticated intelligence can be represented by formatting a number of bits serially.
11	False	The bit period is the time alloted for each bit. Whether or not the pulse period is the same length of time as the bit period depends upon format. The statement is only true for an NRZ signal. See Figure 5-1 in the text.
13	False	The transmission bandwidth is equal to the reciprocal of the minimum pulse period.
15	False	An RZ format requires twice as much transmission bandwidth as an NRZ format to support the same transmission data rate.
17	False	In a "pulse bipolar" system the light signal is always present, and is varied in intensity above and below a quiescent level, in response to bit logic values.
19	False	Risetime is defined as the time period required for the instantaneous amplitude of a pulse to go from 10% to 90% of peak pulse amplitude.

Chapter 6

Quest.#	Answer	Comment
1	False	Digital multiplexers process and function entirely with electronic or electrical signals, and they have no optical elements at all.
3	True	LEDs typically present a spectral output on the order of 25 to 50 nanometers, while laser sources might be 1 to 6 nanometers.
5	True	Because LED light sources have large active areas and characteristically high capacitance, they are difficult to modulate at higher transmission data rates in digital applications.
7	True	A heat sink merely collects heat from a device and dissipates it over a larger or more effective radiant source. A thermoelectric cooler actually becomes cooler on one surface and warmer on another when an electric current is passed through it. It literally cools the device to which it is thermally attached.
9	True	To control output level, it is necessary to sense the optical output level. Many laser sources control the threshold current from the output sensor, thus stabilizing the transmitter output.
11	False	The basic structure of lasers, plus the necessity for temperature stabilization and bias control, usually produces a more complex and more expensive structure than an LED does.

Quest.#	Answer	Comment
13	True	Due to the more complex structure of APDs, PIN detectors are generally less expensive.
15	False	The decibel (dB) can express a power ratio between two signals, but cannot denote a specific power level. The term for denoting a discrete power level is dBm, the decibel referenced to one milliwatt. Zero dBm equals one milliwatt.
17	False	In the American Hierarchy of digital signals, the basic signal is DS 1, which is a digital transmission rate of 1.544 Mb/s.
19	True	Transmission bandwidth and transmission data rate are directly related and higher data rates require more bandwidth.

Chapter 7

Quest.#	Answer	Comment
1	False	Although theoretically a potential problem, light energy reflected from the ends of unterminated fibers presents few problems in actual practice. In this respect, lightwave technology is different from other transmission systems, where such reflections cannot be casually dismissed.
3	False	Dichroism is the property of a surface reflecting lightwaves of one wavelength while passing lightwaves of a different wavelength. A dichroic mirror will not reflect lightwaves of different wavelengths equally well.
5	True	A graded index (GRIN) filter can be used in this manner. See Figure 7–3 in the text.
7	False	Some optical taps may be passive in nature. These may be employed where a high level of optical power is not required at the tap port output. See Figure 7–7 in the text.
9	False	The "star" type coupler combines different signals without frequency discrimination. In telephone trunking networks signals are generally brought into switching centers individually, and then selectively switched to other legs of the network. In general, the star type coupler is incompatible with telephone trunking philosophy and designs.
11	False	The bandwidth inherent in any transmission facility is equally restrictive or applicable to all lightwave carriers transmitted over that facility. Increasing the transmission capacity by using more than one carrier signal does not require any increase in facility bandwidth.
13	True	When an optical device is designed, a theoretical approach will develop the transmission losses, but in practice, the losses will somewhat exceed these theoretical figures. The extra loss, called excessive losses, are due to the inefficiency of the device design or the intrinsic losses in the device itself. Example: A splitter theoretically divides the input power equally across two outputs. Half of the input power would be 3 dB of loss, but a splitter will actually introduce more than 3 dB of loss, and this amount would be called excess loss.
15	True	It is not possible to eliminate all optical-to-air interfaces, for one example, and although some relief may be provided by using index matching fluids or gels, to some degree, intrinsic losses of this kind must simply be expected and tolerated.
17	True	The spacing and the shape of the etched grating lines is a significant factor in establishing the frequency discriminating characteristics of diffraction gratings.
19	False	The bandwidth provided by a transmission facility is identical for all carriers being transmitted over the facility. Adding a second carrier using WDM techniques does not impose any increase in transmission bandwidth.

Chapter 8

Quest.#	Answer	Comment
1	False	The transmission capacity of a system is determined by the terminal equipment and the facility bandwidth as well. Higher data rate terminal equipment will not function properly over a facility whose bandwidth is inadequate.

Quest.#	Answer	Comment
3	True	Digital Multiplexing consolidates more than one digital signal into a single composite digital signal which has a higher transmission data rate than any of the individual signals.
5	False	A T–1 carrier system generates a DS 1 digital signal, which is the lowest data rate in the North American digital hierarchy. This rate is 1.544 Mb/s.
7	True	A 45 Mb/s equates to a DS 3 signal in the North American digital hierarchy, and a DS 3 signal can support 672 equivalent voice circuits. A DS 3 signal supports 28 DS 1 signals, each of which is a T–1 carrier system with 24 voice channels. 28 × 24 = 672
9	True	Yes. In paired cables, the pairs are closely spaced and twisted together under a single sheath. Signals from one pair may cross-talk into another pair, thus causing interference. By using one pair for East-West transmissions and a second pair for West-East, cross talk can be limited or reduced.
11	True	Yes. An M 12 multiplexer, for example, can accept 4 DS 1 signals as input, and will output 1 DS 2 signal, but it will function quite well if only 2 or 3 DS 1 signals are presented as input, and the output will still be a DS 2 signal.
13	True	When more than one lightwave signal is applied to a single fiber, it is essential that the different lightwave carriers be isolated from each other at terminal ends for individual detection and processing by the terminal equipment. This can be accomplished by using frequency discriminating optical devices such as filters or gratings, etc.
15	False	There is nothing inherent in an optical fiber that restricts transmission to a single direction. Light energy of different wavelengths can be injected into the fiber at different ends, and each such signal will propagate through the fiber and can be retrieved at the other end.
17	False	WDM is particulary well adapted for establishing additional points of service along an optical fiber cable route.
19	True	This is so, and it is necessary because in a bidirectional system, the incoming optical signal is at a relatively low level, having passed through the interconnecting optical fiber. The transmitting signal at this same location is at a relatively high level, and it requires higher isolation in the coupler (filter) to ensure that no interference between the carriers is experienced in the local lightwave receiver.

Chapter 9

1	False	The need for a regenerative repeater in any lightwave system will be determined by two factors. If the fiber attenuation is such that the light power level is too low for positive detection and demodulation of the lightwave signal, then a repeater would be indicated as necessary. If the pulse dispersion at some point in the system is such that error free detection and demodulation is not possible, then a repeater would be necessary. Neither of these conditions is automatically imposed because of any specific transmission data rate, and each system would require individual evaluation regardless of its particular transmission rate.
3	False	Metallic conductor facilities such as copper pairs and coaxial cables introduce more transmission loss at higher frequencies than they do at lower frequencies.
5	True	The term given is correct for multimode fiber and the correct term for a single mode fiber would be ps/nm-km.
7	True	In common practice this is so, although theoretically these items are not identical.
9	True	This is true because of the many propagation modes that will be supported in a multimode fiber. Modal dispersion is more pronounced in optical fibers than is chromatic dispersion.
11	False	The correct formula is: $BW = \frac{440}{T}$

Quest.#	Answer	Comment
13	True	If the fiber length remains the same but the transmission data rate requirement is reduced, a fiber with a lower bandwidth/length characteristic could be employed and still satisfy the system transmission data rate requirement.
15	True	Assuming that the system is not attenuation limited, a fiber introducing less dispersion will permit a longer system while still supporting the same transmission data rate.
17	True	Since a dispersion shifted fiber will reduce the pulse dispersion that is introduced, and since system bandwidth is a function of dispersion, then a dispersion shifted fiber can increase the bandwidth a system presents for use.
19	True	At such a nominal data rate, 90 Mb/s, a single mode fiber will become attenuation limited long before it becomes dispersion limited, assuming the fiber has no excessively large loss characteristics.

Chapter 10

1	True	If the transmission data rate is reduced, the bandwidth requirement is also relaxed; thus a given fiber can be longer in length before it becomes bandwidth limited at a lower transmission data rate.
3	False	The usable length of a fiber is a factor of the transmission data rate but the data rate is not arbitrarily limiting. Certainly the fiber described could handle the 45 Mb/s satisfactorily for some length of fiber, however short that length may have to be.
5	True	This is true for single mode fibers but ignores the length dependent factor that applies to multimode fibers. The process is a bit more complicated in the case of multimode fibers.
7	True	This is so because a connector will incur Fresnel reflection losses from the ends of the fibers, due to the glass-to-air and air-to glass interfaces that exist. These losses may be minimized by the use of a matching index fluid or gel.
9	True	If the optical input signal at the receiver input is too low, the detector may not be able to detect the presence of light energy at all, or it may do so erratically, and produce an unusable quality of electrical output signal.
11	False	The risetime of a device such as a receiver is established by the design and fabrication of the device, and will not be altered by the operating signal level at all.
13	True	Since a lightwave receiver can be over driven with input signal, with a resultant degradation in S/N Ratio or BER, it is sometimes necessary to pad down the receiver input signal. Such a case might be where the system geography presents an unusually short optical fiber length.
15	True	It is not only possible but probable that several combinations of system elements will be technically acceptable in a design. Good engineering will produce not only a technically acceptable solution, but a cost effective design as well.
17	False	There are no compelling reasons that dictate the use of laser light sources above 45 Mb/s. Each application requires individual consideration and design, and there may be applications that do not demand a laser light source even at much higher data rates than 45 Mb/s.
19	True	Since LED light sources are less complex and sophisticated than laser light sources, they are less expensive. When an application can be satisfactorily provided using an LED source, some economy in system cost may be possible.

Chapter 11

1	False	The design engineer bears full responsibility for the technical performance of course, but this in no way relieves him of the additional responsibility to produce cost effective systems.

Quest.#	Answer	Comment
3	True	There are numerous alternative techniques for providing service continuity protection.
5	True	Independent transmission links, as derived through the use of WDM, do provide substantial service protection against service interruptions due to failures of electro-optical or electronic system elements, although they do not protect against fiber failure.
7	True	In many applications, particularly in multiple service point or low density systems, the choice of higher or lower level digital multiplexing is presented, and the final selection in such cases will be heavily influenced by system economics.
9	False	The term ''grade of service'' relates to the number of circuits the system provides for full time use, and redundant equipment or circuits that are provided purely as protective alternatives are not available for full time use. Service continuity protection and grade of service are not synonymous terms.
11	True	In telephone network trunking, trunk circuits are randomly accessed for common usage. Trunks that are incapacitated for service are automatically busied out for selection purposes. Thus a failure or loss of a number of trunks does not entirely preclude switched access to the network, which can be accomplished through those circuits that remain serviceable. The transmission capacity may be reduced for short periods of time, but the service location is not automatically isolated from the network by partial loss of facilities.
13	True	It is technically possible to selectively protect system elements but it may be difficult or impossible with some of the products offered in the market place today.
15	True	To provide independent transmission links for alternative service protection configurations, WDM techniques can be extremely useful.
17	False	The reliability of lightwave transmission systems has been adequately demonstrated to be very acceptable, and in all respects to be the equivalent of other transmission systems such as microwave or cable carrier, etc. Lightwave facilities generally can inherently provide substantially more transmission capacity than other systems.
19	False	The widespread acceptance of protection configurations using four fibers for every case has led to a general acceptance of four fibers as the minimum acceptable service facility. There is no compelling reason for this philosophy however, and as the technology intrudes into lower levels of the network, that is, closer to the individual subscriber loop plant, we may find a more practical approach will prevail.

Chapter 12

1	True	There are a wide range of materials used in fabricating optical fibers, but those fibers employed in telecommunications systems of any length are usually silica glass.
3	True	Optical fibers are more vulnerable to minute defects and flaws than metallic conductors of comparable size.
5	False	The design of optical cables specifically provides for independent movement of the optical fibers within the cable structure.
7	True	To provide for extra fiber length per unit of cable length, tight buffer tube type cables are usually helically wound tighter than in loose buffer tube structures, which accommodate extra fiber length within each buffer tube.
9	True	There is a process for splicing entire ribbons, including all ribbon fibers, as a single work operation in ''ribbon'' type cables.
11	False	In loose tube cable structures, a single fiber or several fibers may be located within each buffer tube. Both techniques are used.
13	True	One technique for providing a moisture barrier in cable structures is to incorporate a metallic membrane in the form of a tape, seamless or seamed tube, etc.

Quest.#	Answer	Comment
15	False	Optical cables can be supplied in a wide range of structures, just as conventional cables can. One type of structure available for multiple paired, coaxial, and optical cables is a self-supporting structure which incorporates both a suspension support member and the transmission cable as a single integral unit.
17	True	Since a lower loss fiber implies fewer impurities in the glass itself, the fabrication process for lower loss fibers is more sophisticated, and generally such fibers are lower in cost also.
19	False	Since the materials and the fabrication techniques used in optical cables are very similar, in some cases identical, with those used in fabricating metallic conductor cables, optical cables can be installed in the same environments. Not only are optical cables suitable for aerial plant, the use of all dielectric optical cables is particularly well suited for such installations.

Chapter 13

Quest.#	Answer	Comment
1	True	Due to the very small dimensions of optical fibers, and the nature of the material itself, glass, optical fiber splicing is a more demanding work operation than splicing metallic conductors is.
3	False	Due to the size, relative inflexibility, and weight of conventional multi-pair cables, it is difficult to handle and place long cable lengths as a single construction unit. Optical cables, on the other hand, are small, light, and quite flexible, and much longer lengths can and are placed as a single work operation.
5	True	Using a split conduit, which has an open slit longitudinally along its entire length, a construction technique has been developed which, in a single work operation, inserts the optical cable within the conduit, pays out the conduit and cable, and plows the composite assembly underground.
7	False	There is no compelling reason to bond or ground the metallic elements of optical cables any differently than similar elements of multi-pair cables would be handled.
9	True	By eliminating any electrical conductive paths through a cable, it is possible to electrically isolate potentially troublesome installations or environments, such as power sub-stations for example, from other transmission networks or plant. Optical cables permit plant of this type to be constructed.
11	True	Disconnect and reconnect capability is the basic justification for optical connectors, and good quality connectors should be capable of re-connection many times without significant changes in transmission loss.
13	True	See Figure 13–3 in the text.
15	True	There are connectors on the market that require the fiber end to be polished as a separate work operation. In some connectors, polishing includes the ferrule or other connector part, as well as the fiber end itself.
17	False	The objective is to minimize the splice losses, thus the precision of light power level measurement is of secondary interest, so long as the measuring technique can adequately demonstrate the minimum loss through the splice being made.
19	False	It is necessary to pass light energy through a splice to measure its loss, but a technique called "local injection and detection" has been developed that permits introducing light into a fiber right at the splice work location, and measuring light out of the fiber past the splice at that same location also.

Chapter 14

Quest.#	Answer	Comment
1	True	An equivalent telephone voice channel requires about 3400 hertz or 3.4 kHz for satisfactory operation. A television signal, unmodulated, requires about 4.5 Mhz for satisfactory operation.
3	False	The distortion introduced into a television signal by an analog transmission system will be increased if more television signals are simultaneously passed through the same system.

Quest.#	Answer	Comment
5	True	To date, analog lightwave transmission systems have been largely limited to CATV hub interconnections. In such applications, the transmission distances are generally distinctly limited in length, and few of these installations require any intermediate repeaters.
7	True	By operating the "feeder" plant at higher transmission levels, tapping efficiency is improved, and a significant cost reduction can be produced when the cost of the entire system is considered.
9	False	The reverse is true. To reduce as much as possible the noise and distortion introduced by the "Super Trunk," it is desirable to reduce the number of amplifiers employed in this plant. Using larger, lower loss cables in Super Trunks does reduce the number of amplifiers required.
11	True	This is the same technique used for multiple channel transmission on coaxial cable systems, and it is equally applicable to lightwave transmission systems.
13	False	One of the advantages of Frequency Modulation in transmission systems of any type is the inherent noise improvement that FM produces as compared to Amplitude Modulation (AM).
15	False	One of the basic advantages of digital transmission systems is that there are no adjustments for carrier modulation. The sampling and digital encoding is established permanently by the device design and fabrication. One reason for the rapid and widespread acceptance of digital transmission technology is the elimination of the critical modulation adjustment.
17	False	No. By far, the bulk of lightwave plant that has been installed in CATV systems to date has been for trunking between service "hubs." There has been no extensive deployment of optical fibers or cables in CATV loop plant or service drops as yet.
19	False	Quite the opposite is true, and the use of optical fibers in CATV has been largely limited to Super Trunk applications also.

Chapter 15

1	False	In most lightwave transmission systems constructed as part of this nation's telecommunications network, the optical cable and fibers represents about 50% of the total system cost.
3	True	Even when cable carrier systems are applied to more efficiently utilize a paired cable, the ultimate traffic-carrying capacity of a cable is a factor of the number of pairs in the cable. Obviously in conventional telephone loop plant, where a single line requires a single pair, the pair count limits the transmission capacity directly.
5	True	This is so, but there is a finite limitation to the number of carrier circuits that a pair, or a multiple pair cable, can support.
7	False	The digital carrier system known as the T–1 system provides 24 equivalent voice circuits for use per system. The digital line rate for a T–1 system is 1.544 Mb/s.
9	True	A coaxial cable can provide hundreds of Megacycles of bandwidth for use, compared to perhaps one or two Megacycles for a pair.
11	False	The transmission losses of optical fibers are such that regenerative repeaters may be spaced many kilometers apart, as much as 40 kilometers or more. In a T–1 carrier system using multiple pair cable, a repeater would be required about every mile.
13	True	Reconfiguring broadband facilities to handle increased traffic requirements is usually much easier, faster, and less expensive than constructing completely new transmission facilities.

Quest.#	Answer	Comment
15	False	Applied in this manner, the optical fiber will most likely be grossly underutilized, and the resulting system would be inefficient, although it may function well enough.
17	True	If the initial installation provided opto-electronic units that could operate at higher rates, then the system growth would be extremely easy to accomplish. Such units should be able to operate at lower data rates without any difficulty at all.
19	True	In a well designed facility, it may not be necessary to reinforce the optical cable at all during its normal service life.

Chapter 16

1	False	It may appear incongruous, but the reverse is true. It is more difficult to design efficient systems that can serve several points, even points that may not have become identified at the time the design must be made.
3	False	The use of subscriber type cable carrier has become very widespread in recent years, and it can effectively extend the usable service life of many paired cables already in place.
5	False	If we consider the alternative of T-1 digital cable carrier, with a repeater required every mile or so, lightwave transmission links make very good sense indeed.
7	True	It will at least stimulate reevaluation of the economic and technical reasons why exchange networks are structured the way they have been in the past.
9	True	By using Wavelength Division Multiplexing techniques, it is certainly technically possible to do so.
11	True	In order to do this, we have to process all traffic signals at all intermediate service points, a technique sometimes referred to as "drop and insert."
13	False	There is no operational history that supports this statement.
15	True	In exchange plant, the use of WDM will permit much more flexible utilization of the facilities. To date, WDM has primarily been considered as a technique to increase transmission capacity. In exchange plant applications, it may well serve a more useful purpose in the flexibility of design that it makes possible.
17	True	From a transmission point of view, it makes little difference what service the circuits provided are employed in.
19	False	If transmission costs can be reduced, then the questions of where or how circuit switching is positioned in the network should be reexamined. We may very well see lightwave transmission having an effect on network structures in future.

Chapter 17

1	False	No it is not practical. Tests of this type, of a new facility, are understandably sophisticated and necessarily so, but less complex tests would be completely adequate to monitor an in-place facility and assure that its operating performance or transmission characteristics had not degraded or changed significantly.
3	True	If the transmission losses and signal levels of a system do not change significantly, it is a strong indication that the system transmission performance has not degraded. There is little reason to expect drastic changes in the performance of terminal equipment that would not be evidenced in changes in transmission levels also. It can happen, of course, but it would be quite unusual if it did.
5	True	Since terminal equipment and even derived circuits are often monitored for unsatisfactory operation, there are sound arguments against performing a lot of sophisticated tests at lightwave frequencies. Tests that have little redeeming merit other than acceptance testing of new facilities might be fiber bandwidth tests, for example.

Quest.#	Answer	Comment
7	True	There is little merit in highly sophisticated testing of in-place, operating systems, and such tests are certainly not essential to the proper operation on a long term basis.
9	True	Rather than make measurements at awkward microwave frequencies, most microwave receivers make the AGC bus voltage available. Measuring this DC voltage provides a level that correlates directly with the RF signal input to the receiver.
11	False	Whenever working lightwave connections are disturbed, the possibility of introducing trouble when they are reestablished is very real. It is poor practice to disturb such connections unnecessarily.
13	True	Optical Time Domain Reflectometers are subject to possible error in interpretation, and they may be limited in operating range also.
15	True	The loop back test technique is easy, fast, and useful in many instances.
17	True	Not only are such tests difficult in the field, they are largely unnecessary, unless the tests are part of a proof of performance or acceptance test program.
19	True	A very practical and useful test technique, and it may often help isolate more subtle or obscure trouble conditions.

INDEX